CELLULAR PATTERNS

CELLULAR PATTERNS

A. Šiber
Institute of Physics, Zagreb, Croatia

P. Ziherl
*Faculty of Mathematics and Physics, University of Ljubljana
and Jožef Stefan Institute, Ljubljana, Slovenia*

CRC Press
Taylor & Francis Group
Boca Raton London New York

CRC Press is an imprint of the
Taylor & Francis Group, an **informa** business

To Mira, Goga, and Katja

Contents

Preface

THE IDEA that living matter can be viewed as a mechanical system is hardly new. In addition to physiological functions characteristic of life alone, animals and plants grow, move, deform, push on their environment, and lift loads, etc., just like a machine, and these processes must involve forces and expended work except that the energy is provided in the form of food rather than fuel. Yet over the past decade or two the mechanistic aspect of biophysics has become increasingly more central both at the macroscopic level revolving around the shape of tissues and at the microscopic level, say in the context of molecular motors and elaborate protein machinery resembling pumps, levers, articulated arms, and other hardware. In part, the recent advances in the field coincide with the advent of extreme mechanics as an umbrella term encompassing large deformations, active materials, instabilities, and nonlinear response of materials.

A hundred years after Thompson's remarkable treatise *On Growth and Form*, the mechanistic view of biological systems is fairly comprehensive, drawing both from the ever more detailed experimental studies and from the theoretical developments in physics, especially in soft condensed matter. In this book, we outline the understanding of tissue structure and form by combining observations and measurements with models, which inevitably differ in the level of description as well as in the premises and objectives. Instead of providing a review of the state of the art, we offer a more pedagogical approach, which amalgamates toy models and sophisticated theories, some of which address the same phenomenon. While we were writing the book our personal opinion of the different approaches matured, and we arrived at the conclusion that few models are completely wrong and none of them is completely correct. We hope that the book will help readers to appreciate the best in each model and encourage them to develop their own theories.

So far no single monograph or textbook has influenced the field as much as *On Growth and Form*, which is still a source of inspiration. Naturally, there exist many fine volumes that discuss similar or related topics, each from a different angle, and our book offers yet another perspective. In terms of length scale, it is one level beyond Boal's *Mechanics of the Cell* which addresses the physical principles behind the polymeric networks within the cell, biological membranes, intermembrane forces, and related phenomena. The topics covered by Davies' *Mechanisms of Morphogenesis* are similar to those in our book but the approach is much more descriptive, going deeper in the biochemical and biological details but hardly addressing the theoretical aspects and physical modeling. Forgacs and Newman's *Biological Physics of the Developing Embryo* is focused specifically on embryogenesis, morphogenesis, and organogenesis with more emphasis on theoretical models than *Mechanisms of Morphogenesis*. Our book is technically somewhat more involved than Forgacs and Newman's but not as detailed as Cowin and Doty's *Tissue Mechanics*, which is concerned primarily with the various tissues as bulk materials and their description using effective theories of elasticity at an advanced engineering level. Bulk tissues are also discussed in Gibson and Ashby's book *Cellular Solids*, which contains an excellent account of the mechanics of foams.

Three more references come to mind. Aste and Weaire's *The Pursuit of Perfect Packing* is dedicated to efficient space-filling arrangements of various objects from disks and spheres to soap bubbles. As this subject matter is best described graphically, *The Pursuit of Perfect Packing* includes many images and our book is similar in this respect. As far as the level of mathematical detail is concerned, our book is reminiscent of Nelson's *Biological Physics*, which in turn deals with biophysics from a more general perspective at a molecular, statistical-mechanical level. Finally, readers eager for a contemporary view of the theory of elasticity may want to look at Audoly and Pomeau's superb *Elasticity and Geometry*.

The material covered here may be of interest to researchers with a background in physics, biology, or engineering as well as to students, possibly as a resource for a graduate course on tissue biophysics or pattern formation. Some prior knowledge in elasticity, hydrodynamics, geometry, and developmental biology is desirable but not required as the book is dotted with boxes summarizing the general concepts in these fields. In many derivations, we work out the main steps but redoing them on a piece of paper will still be a good idea. Also included is a set of 64 homework problems of various degrees of difficulty; these are divided among Chapters 2–6 depending on subject matter and can be found at the end of these chapters. Some of the problems are strongly anchored in the chapter in question and ask for very precise answers, whereas others are thought provoking and meant to extend the discussion a bit beyond the scope of the chapter. All problems are pen-and-paper style but the book also offers many trailheads for a numerical analysis or re-assessment of a given topic.

In the use of mathematical symbols we opted for the conventional choice in a given field, knowing that this implies a non-unique notation so that, e.g., C stands for curvature wherever we talk about elasticity, but in Section 3.3.1 it denotes the contractility of the acto-myosin ring and in the Box on p. 206 dealing with the Euler formula it stands for the number of cells. We made sure that the meaning of a symbol in a given context is still unambiguous. We often use dimensionless quantities, and these too are referred to as reduced [such as the reduced volume defined by Eq. (5.8)], relative [such as the relative density introduced in Eq. (6.17)], or dimensionless [such as the dimensionless surface area in Eq. (6.4)] depending on the practice in a given field. As much of the book is about the mechanics of continuous media where the Einstein summation convention is common, we could have used it but decided not to as not all of the readers may not be completely comfortable with it. Typesetting of the mathematical material follows the convention where scalars, vectors, and tensors are printed using italic S, bold upright **S**, and sans-serif S font, respectively; often it is more convenient to deal with the components of vectors denoted by S_i or tensors denoted by S_{ij}. In a few introductory cases where a single component of a tensor is of interest we make an exception and typeset it as if were a scalar just so as to lighten the discussion. When referring to a certain species, we follow the inconsistent but usual practice where, e.g., *Drosophila melanogaster* is more often referred to by the genus rather than as the fruit fly whereas the sea urchin *Lytechinus variegatus* is more known by the English name. In the index, we compensate for this by also including the other name and referring to that used in the book.

Given that the book is about spatial structures and shapes, it contains a considerable amount of graphical material simply because speaking of a physical form without showing it makes little sense. Much of this material is adapted or reproduced from other sources as acknowledged in figure captions or elsewhere as suitable. Many of the original figures were redrawn, primarily so as to ensure readability in grayscale and a uniform graphical style but also to use the book format as best as possible. In some cases, additional elements were included in the figures in order to emphasize a part that is important for the discussion at hand; where appropriate, figure parts that are inessential or distracting within a given context were removed. Having said this, we stress that the scientific content of all adapted figures is unaltered.

While working on the book we learned to better appreciate the value of freely available public-domain material. We thank all authors whose work included in the book is available under the Creative Commons license, and organizations such as Wellcome Images which facilitate the dissemination of these works. We are very grateful to M. Bačič, W. Drenckhan-Andreatta, R. J. Goodyear, B. Guirao, J. Heuberger, T. Hutton, M. Imai, A. M. Kraynik, M. Leptin, D. R. McClay, J. Nance, J. Nase, M. Rauzi, D. Taimina, and R. C. Wagner for providing their previously unpublished images and letting us use them. We also thank A. Campbell from Above All Images for the aerial photograph of Giant's Causeway and T. Knapič from the Slovenian Museum of Natural History for help with the photographs of sea shells from the Hohenwart collection.

Our understanding of the topic has developed largely through interactions with colleagues, collaborators, and friends, often united in a single person. We are grateful to all of them for sharing their knowledge and ideas, which sometimes inadvertently made us think of a given problem in a different way. In the final stages of writing, we benefited very much from the advice and suggestions of Y. Bellaïche, C. P. Heisenberg, B. Kavčič, M. Kokalj Ladan, M. Krajnc, M. Leptin, P. Mrak, P. C. Nelson, M. Popović, M. Rauzi, J. Rozman, and X. Trepat, and we sincerely thank them all.

We appreciate the continuing support of our institutions—the Institute of Physics in Zagreb and the Faculty of Mathematics and Physics, University of Ljubljana and the Jožef Stefan Institute in Ljubljana—as well as the hospitality of the Erwin Schrödinger International Institute for Mathematics and Physics at the University of Vienna (P. Ziherl) and the Jožef Stefan Institute (A. Šiber) where we spent extended periods of time as visiting researchers. Much of the book ripened both conceptually and physically while we worked next door to each other at the Jožef Stefan Institute, but most of it was written in our homes. We are indebted to our families.

We hope that the readers will like the book and that they will let us know of any inconsistencies and shortcomings.

A. Šiber and P. Ziherl

CHAPTER 1

Introduction

A LL FORMS OF LIFE are marked by some kind of spatial regularity. The evenly spaced arms of the starfish are an example of a high body-scale symmetry (Figure 1.1a). Its many tube feet neatly arranged around the perimeter, all perpendicular to it and equidistant from each other, constitute yet another pattern at a smaller length scale. Equally amazing is the hierarchical order of veins in plant leaves, with the midvein running lengthwise and the secondary veins branching out at similar angles. The edge of some leaves is flat but in others (like in the bay leaf in Figure 1.1b) it is decorated with ripples of a well-defined wavelength. It is easy to find countless examples of patterns in both animals and plants—just think of fish scales, wild goat horns, fern leaves, sunflower seeds, sea shells, etc.

a)   b)

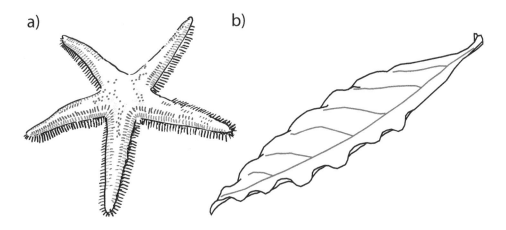

Figure 1.1 *Astropecten aranciacus*, a five-arm starfish found in the Mediterranean and east Atlantic (a). A bay tree leaf with a wavy edge (b).

1.1   EMERGENCE OF PATTERNS

Patterns and other types of symmetric forms and structures are very often seen in inanimate matter too. Take, for example, curtain ripples or waves on the sea, or the von Kármán vortex street in the wake behind an obstacle in flow (Figure 1.2). This sophisticated spatial and temporal pattern consists of pairs of vortices swirling in opposite directions and flushed downstream, with new ones being generated by the backflow behind the obstacle. Even more fascinating than the pattern itself is that it can be explained solely by the equation

Figure 1.2 Von Kármán vortex street in fluid flow past a cylindrical obstacle. The arrow indicates the direction of incoming flow. (Adapted from a photograph by J. Wagner.)

of motion for a viscous fluid known as the Navier–Stokes equation. Having said this, we must admit that in the regime where the vortex street is seen this equation cannot really be solved using pen and paper but this is unimportant. What matters is that we can explain it theoretically.

Moreover, the main principle behind the Navier–Stokes equation is Newton's second law, the basic law of mechanics. In fluids, this simple statement relating force and acceleration— or, in a more abstract sense, cause and effect—assumes a somewhat different form than in motion of rigid or elastic bodies but its essence remains the same. Evidently, neither Newton's law nor the Navier–Stokes equation contain the blueprint for the vortices; they are far too general for this. Yet the vortices do emerge in given circumstances provided that the control parameter of the problem known as the Reynolds number is within a certain range, neither too small nor too large. In a similar fashion, the equilibrium shape of a liquid drop is not determined directly by interactions between the molecules but the drop still assumes the form of a perfect sphere as the minimal-energy state, as we discuss in Section 2.2.

With the discovery of DNA and the spectacular developments of genetic engineering, it has become quite common, at least in the general public, to think that genes encode all the details of an organism. However, we now know that the complex traits such as height and intelligence are neither completely nor simply written in genes; moreover, they do not have to be. To enjoy the beauty of the von Kármán vortex street, one does not need to describe every single feature of each swirl. Just put a cylindrical obstacle in a fast enough flow—and the pattern is there.

Think of clouds (Figure 1.3). The absence of a blueprint, the "cloud code," and even its impossibility are evident. The amount of data needed to encode all structural details of a particular cloud is obviously huge and irrelevant because cloud formation is determined by the thermodynamic state of the atmosphere. Although extremely intricate, the different physical forms of clouds (stratiform, cirriform, stratocumuliform, cumuliform, and cumulonimbiform) result from specific conditions which promote their development and growth. To tune the form of clouds, one thus needs to control parameters that are much more mundane than the form itself—temperature, pressure, humidity, and possibly chemical composition. These physical parameters define a developmental path leading to a certain cloud shape consistent with the underlying physical processes. The form is not encoded; what is encoded is the developmental path that leads to it.

To physicists, relying on general principles rather than on a precise blueprint is just fine. Compress an elastic rod with enough force and it will buckle forming an arch-like shape.

Figure 1.3 Cumuliform cloudscape over Vrh nad Laškim in Slovenia.

Put a drop of mercury on the substrate and it will assume a shape which depends on the nature of the substrate and on the gravitational field of the planet where your experiment is performed. To physicists, shapes are shaped by forces, flows, energies, and instabilities, whereas to biologists, the shapes grow. However these different views are not irreconcilable. When crocheting a potholder, one proceeds in a circumferential fashion, adding one knot at a time. To ensure that the potholder is flat, the number of stitches at a given row compared to the row below must be just right or else the potholder will buckle at the perimeter (Figure 1.4a). An inexperienced crocheter may end up with such a utensil inadvertently but others may well see it as a nice decorative piece.

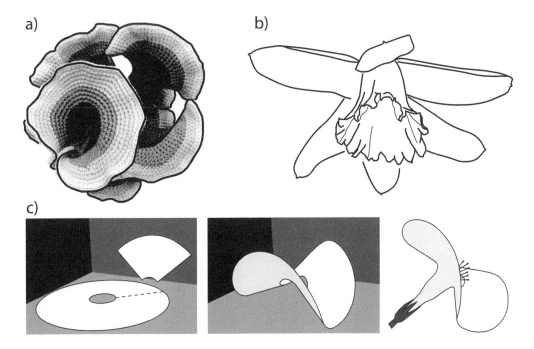

Figure 1.4 Hand-crocheted potholder (a; image courtesy of D. Taimina). Daffodil with a buckled-edge trumpet (b). A circular sector inserted in a paper disk, and a drawing of an imaginary flower of such a buckled shape (c).

A host of biological shapes can be associated with this algorithm, an often-cited example being flower trumpets (Figure 1.4b). The physics of shape formation is relatively simple and can be illustrated by an example requiring only paper, glue, and scissors. If you cut a circular paper disk and insert a paper wedge in the cut, the disk will buckle and form a three-dimensional shape which will minimize its elastic energy (Figure 1.4c). The mathematical background of the process is fairly sophisticated [1] but when making a potholder or some other item, the crocheter only needs to know how many stitches to add in each row. In this case, the form emerges from a simple discrete rule and it is conceivable that such a rule can be materialized by growth processes in a plant tissue, leading to a prescribed number of cells located at a given distance from the edge of the tissue.

How exactly cells do it in a tissue is invariably related to the microscopic, molecular-level processes in each cell and in the plant as a whole. These processes involve consumption of nutrients as a source of energy, chemical synthesis, influx of matter, cell growth and division, etc., all of which is eventually controlled by genes. Yet when considering the tissue at the scale of cells, the operation of this complex machinery can be packed into one or more effective quantities of interest, say cell division rate. In a physical model, the complicated biology at work may, e.g., reduce to the influx of cells into a given tissue as well as friction and stress generated by this influx. All this may eventually cause tissue deformation or even flow as discussed in Section 4.1.3. Many other form-generating physical transformations and deformations can be envisioned, often associated with symmetry breaking. For example, the growth of a thin layer of tissue bound to a substrate is likely to cause formation of wrinkles and folds as shown in Section 4.1.1.

1.2 GROWING A FORM

If we now generalize this scheme to all physical processes that shape cells, tissues, and the organism itself, the final shape of the organism can be viewed as a result of a particular developmental path followed by the fertilized egg and determined by the temporally and spatially resolved switching of the genes. This also removes the need for a detailed map of development of the intricate structure and organization of the different parts of an organism, which is most likely impossible to encode anyhow. Think of both large and fine features of a human body—the network of blood vessels, pulmonary alveolae, kidneys, eyes, auditory ossicles, etc. The amount of information required to describe these structures would be enormous even if one could neglect all details smaller than, say, a millimeter. Putting all of this directly in genes would be very difficult and unnecessary, and relying on physical mechanisms that can mold tissues and organisms is far more economical. The genes and the proteins that they encode appear only as regulators of the physical necessity, switching the flow of matter and energy at the critical points in space and time from one physical possibility to the other. The shape of a tissue is a physical entity, and it can be understood based on the balance of forces molding the tissue. This has been noted by W. His who, in 1874, pointed to similarities in development of the chick embryo and the deformations of a rubber tube (Figure 1.5). At the same time, evolution and embryology as the then-burgeoning fields of biology were infused with the much more abstract ideas of embryonic development known as the recapitulation theory. The two views of life, one emphasizing its mechanical necessity and the other focusing on its evolutionary and teleological aspects, have a long and tense history.

Not everything that one can imagine can be obtained in a developmental process regulated by genes. Much like the thermodynamic and hydrodynamic processes behind cloud formation do not allow the formation of cube-shaped clouds with right angles, straight edges, and perfectly flat faces, genes can only regulate what is physically possible. Not all animals and plants that we can imagine (a horse with a squid head, say?) can be realized

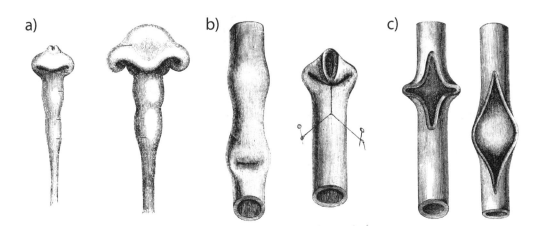

Figure 1.5 Developing brain in a chick embryo two (left) and three (right) days old, with protruding optic lobes (a). A rubber tube with one end is pulled by a thread and fixed as indicated (b). A slit rubber tube with a concave and a convex bend (c). (Reproduced from His' *Unsere Körperform und das Physiologische Problem Ihrer Entstehung* [2].)

by an appropriate genetic code. The precisely scheduled action of genes determines a given developmental path and steers rather than drives the physical flow of mass and energy. If a particular gene is silenced, then the regular developmental path is effectively closed and another one is open instead, potentially leading to a malformed or dysfunctional body part or, more rarely, to an improvement. In some sense, biology and evolution "explore" physics. The forms emerging from physical processes are crafted by the action of genes and utilized as intermediate steps in a developmental path to a fully formed organism. The spectrum of structures that can be generated by physical processes is very broad, and analogies between those seen in inanimate systems and animals or plants have often been sought and considered many times.

A classic reference containing many related examples is the amazing *On Growth and Form* by D. W. Thompson [3]. Among many other things, Thompson's book contains several parallels between fully developed organisms and the shape of fluid bodies, e.g., between shapes of some medusoids and a liquid drop falling through another liquid of a different density and surface tension (Figure 1.6a and b). Both of them have a dome-like body with branching arms and some degree of self-similarity. Inspired by biological form, E. Hatschek coined the name "medusoid bell" for the shapes that he observed in the hanging-drop experiments where liquid gelatin was suspended from a horizontal surface and let drop into a hardening fluid (Figure 1.6c–f). Curiously, drop shape can be altered by slightly changing the density of the surrounding fluid while using the same gelatin. This produces bell patterns with various numbers of radial ribs. Textbooks on physics are typically filled with spheres, cylinders, cubes, and such, but the physical realm also includes the shapes of medusoid bells, falling drops of oil in paraffin, splashes created by pebbles thrown into water, wheel-like colloidal crystals, and shrinking globules of gelatin. All of these shapes have an organic look, but can still be obtained in a carefully steered physical process.

Astounding as they are, the analogies between the shapes of living and non-living matter should not be too surprising on second thought. In many ways, any physical form is subject to restrictions imposed by the properties of the material that it is made of as well as by environmental conditions. Just as no mountain on Earth can be taller than about 10 km,

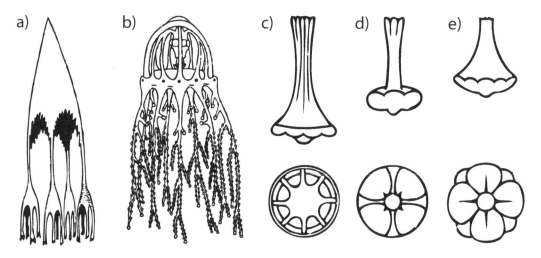

Figure 1.6 Falling drop of fusel oil in paraffin (a) and *Syncoryne* medusoid (b). Hanging medusoid-shape drops of gelatin (c–e). (Reproduced from Thompson's *On Growth and Form* [3].)

no walking animal can be much bigger than an elephant. At the same time, a living tissue does not necessarily always respond to external stimulus or force in the same manner as, say, a piece of paper—living matter is different from the materials that physics typically deals with. For example, accumulation of mechanical stress in a confined growing sheet-like tissue may well change the behavior of cells in the high-stress regions, e.g., by altering the division rate. This in turn can affect the global form of the tissue, which may thus be different from the shape of its inanimate analog. The active nature of living matter is fueled by the energy consumption within the cells that make them and it produces material responses quite different from the ones typically encountered in the physical context. This is discussed throughout this book, most explicitly in Chapter 5 where we focus on morphogenetic movements.

Despite all footnotes and asides, the mechanistic view of the various forms of life is illuminating because it allows one to better appreciate the universality of these forms, which is gratifying. This approach has been pursued for over a century, the classical reference being Thompson's book first published in 1917. With the modern insight into molecular biology, genetics, cell and developmental biology, and, finally, physics, one sees this book through different eyes than when it first appeared on the bookshelves. Today it is easier to treasure the ideas that were elaborated by Thompson and resort to other resources for those that were left out. But it is easy to imagine that a century ago, readers were fascinated by the author's persistent pursuit of a quantitative interpretation of morphology as well as by the sheer volume of the material included as evidence or motivation for his ideas.

Over time, Thompson's all-embracing and often reductionistic style drifted out of fashion, partly because the focus of biology shifted toward genes and partly because of the dawn of modern physics. Nonetheless, the interest in the field never faded and when thinking of the key developments, some of which are listed in Figure 1.7, we see that there exists a continuity in line of thought revolving around the mechanistic and physical view of cells, tissues, and organisms. Some of these developments are described in more detail in this book. Built upon a combination of theoretical concepts and experimental observations obtained using increasingly more sophisticated techniques, this view became quite popular over the past decade or two for multiple reasons. On one hand, the results obtained and the

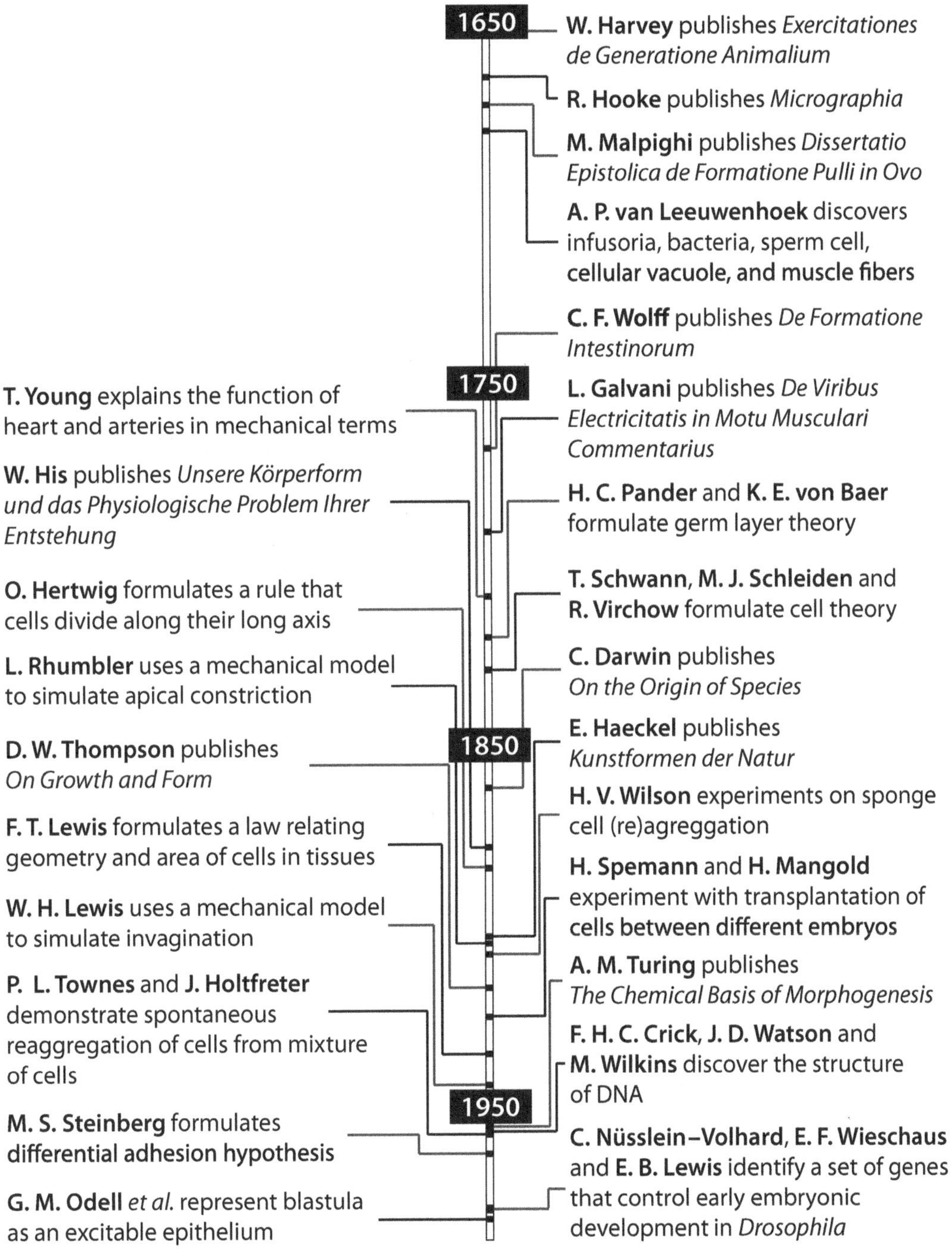

1650 — W. **Harvey** publishes *Exercitationes de Generatione Animalium*

R. **Hooke** publishes *Micrographia*

M. **Malpighi** publishes *Dissertatio Epistolica de Formatione Pulli in Ovo*

A. P. van **Leeuwenhoek** discovers infusoria, bacteria, sperm cell, cellular vacuole, and muscle fibers

C. F. **Wolff** publishes *De Formatione Intestinorum*

1750 L. **Galvani** publishes *De Viribus Electricitatis in Motu Musculari Commentarius*

T. **Young** explains the function of heart and arteries in mechanical terms

H. C. **Pander** and K. E. **von Baer** formulate germ layer theory

W. **His** publishes *Unsere Körperform und das Physiologische Problem Ihrer Entstehung*

T. **Schwann**, M. J. **Schleiden** and R. **Virchow** formulate cell theory

O. **Hertwig** formulates a rule that cells divide along their long axis

C. **Darwin** publishes *On the Origin of Species*

L. **Rhumbler** uses a mechanical model to simulate apical constriction

E. **Haeckel** publishes *Kunstformen der Natur*

D. W. **Thompson** publishes *On Growth and Form*

1850

H. V. **Wilson** experiments on sponge cell (re)agreggation

F. T. **Lewis** formulates a law relating geometry and area of cells in tissues

H. **Spemann** and H. **Mangold** experiment with transplantation of cells between different embryos

W. H. **Lewis** uses a mechanical model to simulate invagination

A. M. **Turing** publishes *The Chemical Basis of Morphogenesis*

P. L. **Townes** and J. **Holtfreter** demonstrate spontaneous reaggregation of cells from mixture of cells

F. H. C. **Crick**, J. D. **Watson** and M. **Wilkins** discover the structure of DNA

M. S. **Steinberg** formulates differential adhesion hypothesis

1950

C. **Nüsslein–Volhard**, E. F. **Wieschaus** and E. B. **Lewis** identify a set of genes that control early embryonic development in *Drosophila*

G. M. **Odell** *et al.* represent blastula as an excitable epithelium

Figure 1.7 Timeline of the key advances related to the development of the mechanistic view of cells, tissues, and embryogenesis (left). The right half of the diagram includes milestones contributing to this framework in a broader context.

predictions made become ever more convincing and detailed, and this is to a large extent due to advanced microscopy techniques which have reached fascinating spatial and temporal resolution at ever smaller invasiveness. The theoretical advances resonate with the modern understanding of the large-deformation elasticity and the advent of novel concepts such as active matter and emergent phenomena, which are inspired in part precisely by biological structures and processes. No less important than the new results and ideas is the change of attitude. While it has been known for a long time that complexity often arises from patently simple models—just think of the logistic map, Mandelbrot set, or nonlinear dynamics in general—its mechanical face was never really as contextualized as it is today.

Active matter

A colony of bacteria in a Petri dish and a school of fish (photograph by H. Sanchez) consist of self-propelled agents which move in space in a locally coordinated fashion such that the velocity of each of them roughly coincides with the average velocity of their not-too-distant neighbors. Despite the diffe-rent length scales, the two systems share many features and display similar spatiotemporal patterns. Such systems are commonly referred to as active matter so as to emphasize that the characteristic types of motion depend on the internally generated propulsion of each agent as well as on the interaction between the agents. Any active system is inherently non-equilibrium and depends on the energy supplied; at the molecular scale, invariably in the form of ATP. Equally interesting are the artificial active particles driven either by a suitable external force (e.g., vibration) or by a suitable chemical reaction.

In biology, understanding the origin of spatial and temporal patterns at various levels and the ensuing structural and functional hierarchy is of paramount importance simply because of the complexity of organisms. The comparison of the *Syncoryne* medusoid to a falling drop of fusel oil in Figure 1.6 may help us rationalize the large-scale body parts but the animal as a whole is much more than that. Irrespective of the species in question, a multicellular organism performs a range of functions that are beneficial to all the cells in it. In animals, some form of a digestive system is needed in order to supply cells with energy. A means of transporting the nutrients to the cells is also required. The organism itself must be both firm and flexible so to withstand forces from the environment, and it is equipped with the locomotion and sensory apparatuses. Finally, the animal must also reproduce. Even in the simplest animals, a comprehensive theoretical understanding of all of these structural and functional features is a very distant objective, and their re-creation in a man-made machine is far beyond any present-day engineering horizon. But there is still much that we can do provided that we choose a suitable functional subunit of an organism, a length scale of interest, and a well-defined phenomenon to study.

In this book, we discuss the mechanical aspects of structure, shape, and shape transformations of simple tissues, primarily single- or multi-layer sheet-like tissues known as epithelia. In doing so, we employ a range of ideas known from theoretical soft condensed matter physics, elasticity, and hydrodynamics including packing considerations, various types of instabilities, minimal-energy arguments, pattern formation theories, elements of topology, etc. Whenever meaningful, we interpret a given phenomenon or feature using more than a single model or theory, often comparing approaches of considerably different nature. This is perhaps most evident in Chapter 4 where we elaborate on models reproducing the shapes of epithelia. This chapter starts with a simple physical picture of a strained elastic band glued to an elastic substrate (Section 4.1), where all biological aspects of the problem are subsumed in the effective energies of these two bodies. As the chapter unfolds, more and more biological aspects of tissues are introduced, including the supply of nutrients needed for cell growth (Section 4.1.3), until we arrive at agent-based models which incorporate rules for cell division and death (Section 4.2.2) combined with a coarse description of cell-level mechanics.

1.3 WHY MODELS?

One may wonder whether there is any use in studying such a diverse range of models, some of which are constructed specifically so as to describe a very particular biological phenomenon. To appreciate the logic pursued in this book, it seems appropriate to pause at the meaning of models in physics as the fundamental natural science. Although essentially an empirical field, physics aspires to provide a succinct formal description of a consistent collection of observations. This is what a physical theory or law is. The applicability of a theory is invariably defined by the spatial, temporal, energy or some other scale of interest, which defines the variables suitable for the problem at hand. For example, the Navier–Stokes equation applies at length scales where the fluid can be treated as a continuous medium so that the quantity of interest is its velocity averaged over all molecules in a volume small enough but still much larger than molecular size. By observing the picture in Figure 1.2, we see that this is not merely sufficient but is, in fact, the only reasonable choice. Naturally, the fluid still consists of molecules but this fact is immaterial as long as we can pack the molecular-level features into effective quantities, say density and viscosity in the case of fluid flow.

The second important trait of physical theories is that they rarely appear in their definitive version from the start. A classical illustration of this point is the development of the understanding of planetary motion. A subject of rather intense studies since antiquity, this problem has occupied scientists and theologians alike. The various views of the motion of celestial bodies eventually led to Copernicus' heliocentric theory. One of the elements of this theory was that the center of the universe is close to the Sun. Now we know that this is incorrect as our Sun itself is on the periphery of the Milky Way, but compared to the geocentric beliefs of the day this was a dramatic paradigm shift. Copernicus postulated that the planetary orbits are circles, but about 70 years later Kepler analyzed the accurate measurements of Brahe and found that the orbits are ellipses, albeit of small eccentricities. Kepler also formulated two other quantitative laws stating that the line between the Sun and a planet sweeps out equal areas in equal times and that the square of the orbital period is proportional to the cube of the major semiaxis of the orbit. Kepler did not provide an explanation for these laws but they are correct.

Analytical mechanics teaches us that Kepler's second law is a simple consequence of conservation of angular momentum for motion in central potential; here we intentionally used a more sophisticated vocabulary than Kepler so as to emphasize how much the view of this particular problem in mechanics has changed since his time. The key development in this story is due to Newton who showed that the shape of orbits is determined by the $1/r^2$ law of gravitation and that Kepler's second and third laws too follow from the solution of the two-body problem with a $1/r^2$ attractive force.

From this perspective, Kepler's laws look more like observations because the ultimate explanation of planetary motion came only with Newton's theory. But Kepler's work was and still is admirable because it provides a concise description of Brahe's measurements, that is an abstract of what one really needs to know about the orbits. There exist many other examples of Keplerian insight, say the Balmer and Rydberg formulae which describe the wavelengths of the hydrogen emission spectrum. These phenomenological formulae offered clues for the development of Bohr's model of atomic structure where classical-mechanical concepts are used in conjunction with the then new elements of quantum mechanics.

Bohr's model is a big step ahead but it is wrong simply because electrons cannot be described as point particles so that it is just not possible to speak of their orbits. Yet one of the key features of the model—the quantization of the angular momentum and the underlying Bohr–Sommerfeld rule—still captures the essence of bound states of quantum-mechanical systems. This is done in a manner that remains appealing from the pedagogical point of view because it does not require the wavefunction-based mathematical apparatus to describe the electron states. By allowing one to appreciate the quantization of phase space and discreteness of energy levels of quantum-mechanical systems within a simplified theoretical framework, Bohr's model is still very insightful, provided that it is used in contexts where it makes sense.

Physics is laden with models seemingly far simpler than Bohr's to the extent that they are referred to as toy models, and yet some of them left an indelible stamp. In the Ising model of ferromagnetism, the magnetic moments of atoms in a solid are reduced to their z components, which are allowed to be either in the up or down state and which interact only with their nearest neighbors. This is evidently a gross simplification, and so are the early studies of the model which focused on one- and two-dimensional versions which too do not seem to be directly related to real magnets.

Despite its stripped-down description of atoms' magnetic moments, the Ising model gave rise to an invaluable series of theoretical results reaching far beyond solid-state physics, eventually transmuting into a paradigm for phase transitions. Curiously, the present role of the model is in stark contrast with Ising's own opinion. After solving the model in one dimension and finding that it does not predict a stable ferromagnetic state, he was unsure of its relevance. Now we know that Ising's intuition was wrong.

Our understanding of the importance of modeling is invariably biased by these and other examples from physics where theoretical models allow one to think about a given phenomenon and analyze it within a transparent mental framework, ideally containing a single mechanism or scenario. This often permits exploring the underlying idea and expanding concepts beyond the realm of the phenomenon in question. At the same time, any predictions obtained must be interpreted with care by paying attention to the premises that the model is based upon as well as to the level of description that we chose. The latter is probably essential for every model, because it is invariably related to its technical side. For example, when analyzing the orbits of planets one can safely treat planets as point particles simply because their size is very small compared to the distance to the Sun. In a similar vein, it is neither practical nor illuminating to think of an animal body or even a cell as an ensemble of molecules. Constructing a description based on the main structural or functional building blocks that are of interest is far more productive—one or more organs in case of the body, and cell membrane, nucleus, organelles, and cytosol in case of cells.

These are the reasons why we view a tissue as a collection of cells and typically disregard subcellular structures. To understand the form and the structure of tissues, we consider cells as the building blocks of interest and operate with the quantities typical for the mechanics of solid bodies or liquids as appropriate. As the immediate cause of motion and deformation of bodies are forces and torques, the quantities of choice in this framework are forces and displacements at the cell-level length scale. The molecular-level workings of cell structures generating these forces, the thermodynamics of conversion of the chemical energy stored in food into various forms of work, and molecular signaling are not included explicitly; instead they enter, e.g., through a suitable variation of model parameters with time or as an effective active force.

The domain covered in this book is illustrated in Figure 1.8 revolving around a cell in an epithelial tissue, which line body surfaces so that the apical side of the cell faces the

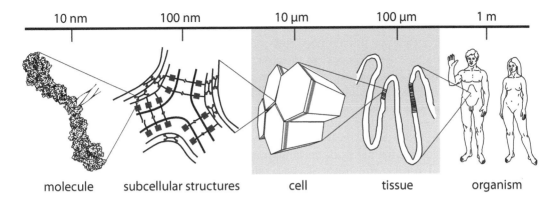

10 nm	100 nm	10 μm	100 μm	1 m

molecule   subcellular structures   cell   tissue   organism

Figure 1.8 Domain explored in this book (shaded area): the basic entity of interest is the cell. Its mechanical behavior originating in molecular-level structures and processes—such as the myosin motor, the acto-myosin ring, and the cadherin cell-cell bonds shown here—is captured by a few effective parameters, which are then used to study the structure and form of tissues.

environment. In many epithelia, cells are equipped with a contractile bundle formed by filaments consisting of a protein called actin and connected by a motor protein known as myosin. Myosins can walk along the actin filaments that they are attached to, resembling a miniature bipedal creature. This pulls the filaments together and generates a contractile force in the acto-myosin ring running around the perimeter of the cell below the apical surface. Naturally, the active motion of myosins requires ATP as the source of energy.

At the level considered in this book, all of this sophisticated molecular machinery is packed into, e.g., the effective ring line tension defined as the ratio of the elastic energy of the ring and its perimeter. The line tension may depend on time so as to reflect the effects of signaling and energy supply during a given process, and is then used as an input that controls the shape of individual cells as well as the shape of the tissue.

This description of tissues is mainly Keplerian but still very rich, employing a range of structure- and pattern-formation mechanisms known from various branches of physics. This transfer of knowledge should not be too surprising. After all, mankind often resorts to the inverse procedure of biomimetics when designing a device or process by copying an animal or plant structure evidently optimized for a given purpose, for example when pondering the possibility of flying. Here the man-made replica of a wing or a fin is devoid of all features of the living model other than the desired function, and could well serve in an experimental study of this function better than the original. The theoretical models of tissue structures and processes described here are like this too, stripped-down formalized reflections of reality.

Needless to say, the interaction between biology and the physics behind such efforts is not unidirectional. While it may seem on first reading that this book tries to explain the observations by browsing through an atlas of whatever physics has to offer, many of the ideas proposed were in fact inspired by the biological phenomenon at hand. The prime example is the theory of active matter, a generalization of elasticity and mechanics which includes effects generated by consumption of nutrients. The development of this field is a nice instance of cross-pollination between sciences, which happens all the time. Just think of Newton's work on calculus as a means of a concise description of physical phenomena, specifically the motion of planets.

1.4   THE MICROSCOPE

The understanding that we pursue here rests on experimental data, and the key tool to obtain them was and still is the optical microscope. The first compound microscopes consisting of an objective lens and an eyepiece lens appeared around 1620, one of them being constructed by Galileo. About 50 years later, the microscope was already employed to observe and examine objects not visible by naked eye. The most famous scientists of the day who used it for investigations of animal and plant samples were Hooke, Grew, Malpighi, and van Leeuwenhoek. Grew was a botanist whose main interest was the structure of plants, Malpighi is best known for his work on human anatomy, but also for the first microscopic investigations and precise drawings of chick embryos, and van Leeuwenhoek discovered microorganisms. But it was really Hooke who, with his *Micrographia* [4], established the microscope as an indispensable tool in studies of what he referred to as "minute bodies."

Published in 1665, Hooke's enthralling book contains a number of exquisite drawings, most of which show microscopic objects from fabrics and crystals to snowflakes and insects. For us, the most important of his drawings is the image of cork cells in Figure 1.9. It is said that Hooke started using the word "cell" in reference to biological cells when seeing the remarkably regular appearance of the cork cross-section partitioned into entities reminiscent of monk's cells. He discussed the porous nature of cork and its "springy nature" in considerable detail, and he went on to calculate the number of cells per unit volume, which was 1259712000 per cubic inch [4]. His estimate of cell size was about 25 μm, which is definitely correct.

Since Hooke, microscopy has come a long way. Differential staining was discovered in the

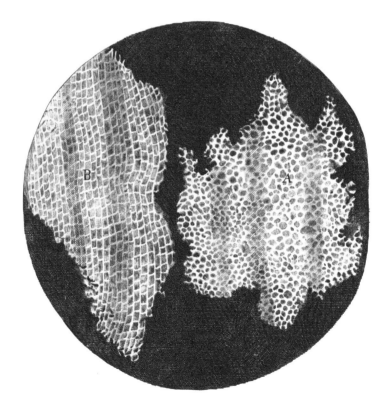

Figure 1.9   Hooke's drawing of a slice of cork seen through the microscope [4].

1850s, allowing a given cell structure such as the nucleus to be distinguished from the rest in a fixed sample. The use of fluorescent dyes, which eliminate the need for a separate source of illumination, took differential staining one step further. The polarization microscope provided a means to image subcellular features and processes in living samples, with the mitotic spindle as the first structure to be identified. Among the next developments, we single out the invention of the confocal microscope as a device that illuminates the sample point by point and produces three-dimensional scans. Equally important as the different ways of sample preparation and image collection are the algorithms used to improve resolution and the overall quality of the micrograph by image processing; the main suite of theoretical and computational techniques used in living specimens is the deconvolution microscopy. In the 1990s fluorescent microscopy in biology was revolutionized by the introduction of green fluorescence protein (GFP) as a tag of very low phototoxicity. At this point, the level of structural and functional details that can be imaged in an almost non-invasive way is indeed amazing as illustrated by the sequence of snapshots of a dividing cell and it neighbors in *Drosophila* embryonic epithelium in Figure 1.10.

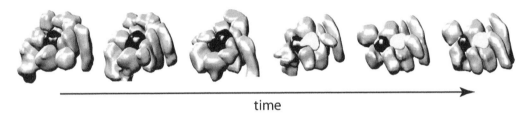

time

Figure 1.10 Sequence of computer-rendered snapshots of a group of cells in the *Drosophila* embryonic epithelium with a dividing cell in the center over a period of about 9 minutes. (Adapted from Ref. [5]).

In some cases, complementary insight can be provided by electron-microscopy images of organisms with a far better spatial resolution than the optical microscope, which is essential in studies of subcellular structures such as organelles. The strikingly convincing appearance of scanning electron micrographs, which stimulate depth perception, is very helpful in visualizing whole embryos and other structures with a complex three-dimensional shape. Some examples of such micrographs are included in this book.

Indispensable as it is, the optical microscope is used not only for observation of cells, tissues, and organisms in their natural, wild-type state but also in conjunction with the various means of physical, chemical, and genetic manipulation of specimens. Cell-level mechanics can be often efficiently studied by micropipette aspiration, diametral compression, and osmotic stress, or by monitoring cell interaction with a suitable natural or artificial substrate. Certain tissues can be excised from the organism so as to study their inherent mechanical or other behavior independently of the surrounding tissues and structures. In some cases, tissues can be dissected or cut in order to examine their recoil, wound-healing behavior, etc.; this is often done using lasers. Also convenient can be cauterization experiments where a part of the tissue is immobilized by laser spot-welds, which can mechanically suppress deformation or displacement.

Equally diverse are the means of chemical manipulation. In the most direct case, we can change the composition of the medium surrounding the tissue or organism which may then affect, e.g., the interaction between cells such as in the seminal Townes and Holtfreter's pH-controlled disassembly and reassembly of the tiger salamander embryo mentioned in Section 2.4. An often very precise way of targeting a specific part of the body, which can be extremely useful in studying morphogenesis, is mutation. Presently, genetic engineering has come a long way and the effects of many mutations have been explored in considerable detail, especially in developmental biology—in fact, it is almost impossible to think of this field without the studies of mutants.

Model organisms constitute yet another pivot of developmental biology. These organisms are easy to cultivate and genetically manipulate, which allows one to explore a given class of phenomena in great detail. The insight obtained can then be used to interpret analogous processes in other organisms including humans. The key animal model organisms in developmental biology are roundworm *Caenorhabditis elegans*, fruit fly *Drosophila melanogaster*, zebrafish *Danio rerio*, African clawed frog *Xenopus laevis*, sea urchin *Lytechinus variegatus*, chick, and mouse, of which we will most often discuss the *Drosophila*. The accumulated body of experimental knowledge of model organisms is indeed comprehensive, and this facilitates modeling because it will, at some point, help one to consider an organism in its entirety rather than as a collection of parts.

This is still not possible. An integrated theoretical view of, e.g., development of even the simplest animals is not yet in sight although there can be little doubt that the successive stages of embryogenesis should be regarded as a continuous transformation rather than as separate events. This book cannot but reflect the current understanding of the mechanics of tissues, mostly focusing on individual structures or processes. But we are certain that more unified theories will emerge. With the impressive arsenal of experimental techniques, ever more precise measurements, and increasing interest in the field itself, this just has to happen.

This book starts with an overview of the basic mechanical features of isolated cells in Chapter 2, which sets the stage by defining the language used later. In Chapter 3 we move to the in-plane structure of epithelial, sheet-like tissues, viewing them as tilings of a flat two-dimensional surface. Chapter 4 discusses the shape of small-scale features of simple epithelia: folds, villi, and crypts. Then we turn to epithelial morphogenesis and embryogenesis and describe theories of the basic morphogenetic movements (Chapter 5). Finally, Chapter 6 is dedicated to the structure of bulk tissues, primarily those consisting of polyhedral cells. The reader will notice that the coverage of the material is uneven. This is partly a reflection of the state of the field and partly our choice, biased by the decision to adopt a pedagogical tone which is more easily done in some topics than in others. Finally, we could not resist including a few historical references where appropriate, largely because of our fascination with the work of pioneers such as F. T. Lewis, A. M. Turing, and, of course, Thompson. *On s'engage et puis. . . on voit.*

CHAPTER 2

Cells as physical objects

A LL CELLS FROM CELLS or *omnis cellula ex cellula* is a statement famously popularized by R. Virchow, emphasizing one of the key properties of cells as the basic units of life— their capability of autonomous replication in the space separated from the environment by their membrane. The cell contains all the information required to replicate itself and to synthesize the molecules necessary for its structure and function. It performs cycles of chemical reactions, which produce energy but also molecules needed for the maintenance of cell structure as well as for cell replication.

Cells come in many different shapes and sizes, but all of them can be classified in only two classes: prokaryotic and eukaryotic. Eukaryotic cells contain internal functional subunits (mitochondria, Golgi apparatus, endoplasmic reticulum, lysosomes, etc.) which are separated from their individual environment by membranes and referred to as organelles. A particularly important compartment within the cell is the nucleus, which contains a neatly packed DNA molecule encoding in its structure all the information that a cell needs in order to synthesize proteins. In eukaryotic cells DNA is compacted by means of specific proteins which bind to it so to reduce its volume. The nucleoprotein structure within the nucleus is called chromatin, and chromatin is further organized in separate structures called chromosomes. The number of chromosomes depends on the organism in question; human cells have 23 pairs of chromosomes.

All plants, animals and fungi are made of eukaryotic cells which group together to form tissues. On the other hand, prokaryotes do not have a cell nucleus and their DNA molecule floats in the cell interior. Bacteria are prokaryotes and so are archaea, which were once thought to be bacteria as they are quite similar in appearance but now they are recognized as a separate domain of life.

Eukaryotic cells of plants and animals have a linear size between about 10 and 100 μm and a volume from a few 10 μm^3 to a few 1000000 μm^3. Their size depends on cell type and on the tissue that they form, but there is also considerable variability within the same type and tissue as shown in detail in Chapter 3. Bacteria are typically smaller than eukaryotic cells. Most of them are between 1 and 2 μm long and enclose a volume of about 1 μm^3. This book deals with eukaryotic cells, mostly animal. Plant cells are different from animal cells primarily in that their membrane is reinforced by the wall made mainly of cellulose. Due to this, their mechanics and rigidity are quite different from that of animal cells.

A cell is usually viewed as a factory or a stage for complicated cycles of biochemical reactions which require specific enzymes, again produced by cells from the information contained in their DNA. The chemical activity of a cell has a pronounced spatio-temporal dependence. Chemicals are produced only in some regions or compartments within the cell, and are either passively or actively transported through the cell into regions where they are needed in other reactions or as building blocks of cell structure. This is why a typical sketch of cell structure emphasizes organelles where specific biochemical tasks are performed, consistent with the view concerned with cell growth, motility, and metabolism. A gene-centered perspective is somewhat different, and emphasizes cell machinery involved in the manipulation and propagation of genetic information, i.e., transcription (production of RNA from DNA), translation (production of proteins from messenger RNA) and post-translation, DNA repair and duplication, cell division, etc.

Both of these views are too complicated for our purposes and goals, as we are interested mostly in the cell shape and the mechanisms that change it, in the response of cells to forcing, and in the interaction between cells which allow them to bind to each other and form tissues but also to get loose and move past their neighbors. All of these mechanisms are molecularly conditioned and depend on the production of proteins which form the cell skeleton and bind the cells together. Specific proteins crosslink the cell skeleton, modify its elastic properties, and anchor it to the membrane and to cell adhesion proteins. The protein filaments of the skeleton can also polymerize and depolymerize, and so over time the cytoskeleton can be eventually completely remodeled under suitable circumstances or restrictions. As a result, the mechanical response of the cell depends on the spatio-temporal features of the protein distribution. While studying the coarse-grained mechanical models of cells and tissues, we must bear in mind that the properties of a cell and its mechanics depend on complicated and dynamic molecular processes.

2.1 CYTOSKELETON AND MEMBRANE

An isolated cell is schematically depicted in Figure 2.1, represented as a round object with a dense interior. Within the cell membrane are the cytosol—a water-based fluid containing ions and proteins and crowded with filamentous protein structures forming the cell skeleton—, the organelles, and the nucleus; the cytosol and the organelles constitute the cytoplasm. The diameter of the nucleus, typically between 2 and 10 μm, defines the secondary length scale. In severely confined cells the nucleus must be deformed too.

The cell membrane, also referred to as the plasma membrane, is made of phospholipid molecules organized in a bilayer. Phospholipid molecules have a characteristic hydrophilic head, which contains a negatively charged phosphate group, and a hydrophobic tail, which consists of two fatty acid chains attached to the head. Embedded in the membrane and attached to it are the membrane proteins, which are involved in the transport of molecules and ions across the membrane as well as in the signal transduction and the cell-cell adhesion. The membrane is liquid-like in the sense that the lipid molecules can freely move in the in-plane directions.

The cell membrane is generally rather rough and laden with wrinkles, folds, protrusions, and pockets which may serve as an area reservoir. If the membrane is stretched, these features can unwrinkle so as to increase the effective exposed membrane area, thus accommodating deformations where this is needed. Such deformations are particularly prominent in neutrophil cells, a type of white blood cells of the immune system of vertebrates. During phagocytosis, i.e., engulfment of foreign material, bacteria, dead cells, etc., neutrophils spread around the particle that they are about to digest. In this process, they can double their effective area [6]. Also indicated in Figure 2.1 are microvilli, easily recognizable finger-like protrusions. They are particularly prominent in some cell types such as the epithelial cells of small intestines where the presence and organization of microvilli in brush-like structures immensely increases the area available for absorption. Microvilli are stabilized by the underlying cytoskeleton, that is, by protruding protein filaments which act as their scaffold.

Not only microvilli, but the whole cell surface is supported by protein filaments whose dynamics and reorganization induce changes of cell shape and surface. These filaments can extend by incorporating more proteins in their structure, or they can shorten by partially

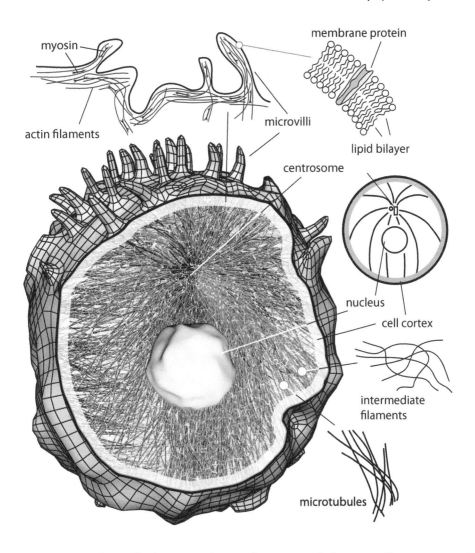

myosin

actin filaments

microvilli

membrane protein

lipid bilayer

centrosome

nucleus

cell cortex

intermediate
filaments

microtubules

Figure 2.1 Sketch of a cell showing the nucleus, cytoskeleton, cell cortex, and the membrane features. Organelles and nuclear pores are not shown for clarity.

disassembling and releasing some of the protein units. The cytoskeleton consists of three types of filaments made of different proteins and characterized by different radii: microtubules, intermediate filaments, and actin filaments. Microtubules are polymers of tubulin dimers and are the thickest of all, with a diameter of ≈ 25 nm. Intermediate and actin filaments have diameters of ≈ 10 and 7 nm, respectively.

These filaments form a tightly interlaced network crosslinked by different proteins [7, 8], each of the three subnetworks being characterized by a distinct topology and spatial distribution. In most animal cells, microtubules radiate from a protein structure called the centrosome, which influences the spatial organization of their network. An imaginary line drawn from the nucleus to the centrosome defines the cell axis. There are several thousands of microtubules in a cell. Intermediate filaments typically form a branched, porous network, stretching from the membrane to the nucleus. Often their density is largest just below the membrane. Not all cells are built of the same intermediate filaments and the expression of the proteins that form them is not the same in all cell types.

Actin filaments are concentrated mostly in a layer below the membrane where they form a shell of thickness from 0.1 to 1 μm [7] also called the acto-myosin or cell cortex. The actin filaments are interlinked by other proteins (fascin, filamin, etc.). Some of the proteins that they interact with, notably spectrin and ERM proteins, bind them to the membrane. Another key protein that attaches to the actin filaments is myosin. This motor protein can rearrange the actin network by using up energy to slide the actin filaments past each other, which makes the cell cortex stretch, contract, loosen, or stiffen. Kinesins perform a somewhat similar role in microtubules, especially during mitosis. In the cell cortex, actin filaments are mostly tangential to the membrane, but in the microvilli they are perpendicular to it so as to form the protein core of the microvilli, much like tent poles (Figure 2.1). The topology of the actin network depends on the forces that the cell is exposed to as well as on the location within the cell.

The elastic properties of the three characteristic filaments of the cytoskeleton are hardly the same, given that the filaments are made of different proteins and that their diameters are different too. Microtubules are the most rigid of all filaments, their persistence length being at least an order of magnitude larger than the typical cell size. Intermediate filaments are the floppiest of all and are more curved and entangled than the other two types [9].

Mitosis and cytokinesis

The contractile force exerted by the acto-myosin network divides one eukaryotic cell into two. A ring-like structure made of actins and myosins known as the contractile ring forms and constricts so to make a membrane bottleneck in the mother cell. The bottleneck eventually closes and separates the two new daughter cells, each containing a complete set of chromosomes. The duplication of chromosomes and formation of two separate cell nuclei is called mitosis, whereas the division of the rest of the cell interior (cytoplasm) is referred to as cytokinesis.

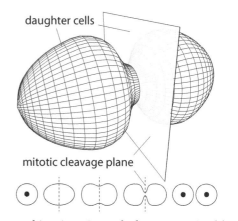

daughter cells

mitotic cleavage plane

While actins and myosins play a decisive part in cytokinesis, microtubules reorganized by motor proteins (kinesins and dyneins) are involved in the separation of the cell chromosomes and in the formation of two daughter nuclei. The plane containing the contractile acto-myosin ring and lying approximately perpendicular to the line connecting the two newly formed nuclei is called the mitotic cleavage plane.

Not all cells in a population divide at the same time, and proliferation of a population is measured in terms of the mitotic index defined as the ratio between the number of cells in mitosis and the number of all cells.

The cytoskeleton supports the membrane and determines cell shape, and it also contributes to the response of the cell to external forcing. It is not, however, the sole factor that controls this response, as the cell also contains the organelles and the cytosol; about 70% of cell volume is water [10]. In the simplest approximation, the cytoplasm can be represented by a volume constraint which restricts the range of shapes that a cell can assume, provided that it does not enter a hypotonic or a hypertonic environment where it absorbs or releases water, respectively, depending on the change of concentration of salt or other solute. Despite the common structural features, there exists a broad range of mechanical behaviors of the different cell types, and each of them is important for a particular function. Some cells are very malleable: erythrocytes, for example, can deform considerably without membrane rupture, which facilitates the passage of blood through small capillaries. Some exert strong traction forces on their neighbors as well as on the matrix around them, and yet others readily sort by preferentially binding to same-type cells. Each of the different mechanical models of cells discussed below revolves around a given aspect of cell structure.

2.2 LIQUID-DROP-LIKE BEHAVIOR

Given that the interior of the cell consists mostly of water, one may try to describe the whole cell as a liquid drop; the cell membrane may be modeled separately or treated simply as a liquid-medium interface. In the latter case, the surface energy E_s is proportional to the drop area A or

$$E_s = \Gamma A. \tag{2.1}$$

Here Γ is the surface tension, which depends on the physico-chemical properties of the liquid, but also on the properties of the medium. The pressure difference across the surface Δp can be obtained from the requirement that the total differential of the free energy $\mathcal{F}$

$$d\mathcal{F} = -SdT - \Delta p dV + \Gamma dA, \tag{2.2}$$

is zero, i.e., that the drop is in thermodynamic equilibrium. Here dV is the volume differential and ΓdA is the mechanical work dW required to increase the surface area of the drop by dA, and so $\Gamma = (\partial \mathcal{F}/\partial A)_{V,T}$. If we assume that the drop is a sphere of radius R and in contact with the environment at constant temperature so that $dT = 0$, the condition $d\mathcal{F} = 0$ gives

$$\Delta p = p_i - p_o = \frac{2\Gamma}{R}, \tag{2.3}$$

where p_i and p_o are the pressures inside and outside the cell, respectively. This result is known as the Young–Laplace law—it relates the pressure difference across the drop surface to the tension of the surface.

To visualize the Young–Laplace law, consider a spherical drop of radius R and a small spherical cap of its surface of radius r_p (Figure 2.2). The cap of radius r_p is under the influence of surface tension forces acting all along its rim and tangentially to the surface of

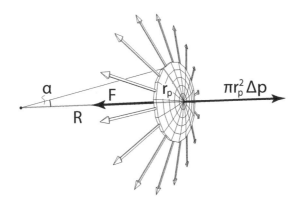

Figure 2.2 Forces at a curved surface: tension forces acting along the rim of a spherical cap of radius r_p (white arrows) amounting to a total force F pointing toward the drop center are balanced by the pressure-difference force $\pi r_p^2 \Delta p$ pointing outward.

the drop. Let the force per unit length of the rim be λ. The net tension force acting on the cap F points toward the center of the drop and is given by

$$F = 2\pi r_p \lambda \sin\alpha = \frac{2\pi\lambda r_p^2}{R}, \tag{2.4}$$

where α is the angle subtended by the cap radius r_p. This force is equilibrated by the pressure difference so that

$$\Delta p = \frac{F}{\pi r_p^2} = \frac{2\lambda}{R}. \tag{2.5}$$

By comparing this result with Eq. (2.3) we see that λ coincides with Γ, i.e., the force per unit length of the surface of the drop acting tangentially is the surface tension. In a surface S of arbitrary shape, the Young–Laplace law can be written as

$$\Delta p = \Gamma\left(\frac{1}{R_1} + \frac{1}{R_2}\right), \tag{2.6}$$

where R_1 and R_2 are the principal radii of curvature at point P. In terms of principal curvatures $C_1 = 1/R_1$ and $C_2 = 1/R_2$, $\Delta p = \Gamma(C_1 + C_2)$.

Principal radii of curvatures

In each point P, the shape of a surface is described by two principal radii of curvatures R_1 and R_2 defined as the radii of the smallest and the largest osculating circles touching the tangent plane τ at P. These osculating circles lie in planes that are perpendicular to each other. In a saddle-shaped surface, one of the principal radii is positive whereas the other one is negative.

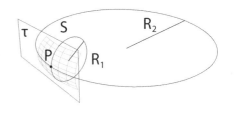

2.2.1 Micropipette aspiration

Equation (2.3) has been used to explain the micropipette aspiration of the neutrophil granulocytes, a type of white blood cells of the immune system (Figure 2.3a). Experiments were performed by lowering the pressure in the micropipette of internal radius R_p between 2 and 7.5 μm, that is, by increasing the suction pressure $\Delta p = p_o - p_p$ [11]; here p_p is the pressure in the micropipette which must be smaller than the pressures in the granulocyte and in the environment denoted by p_i and p_o, respectively. A granulocyte near the tip of the micropipette gradually enters the micropipette. The critical suction pressure p_{cr} corresponds to the stage when a granulocyte forms a hemispherical cap inside the micropipette. Any further increase of the suction pressure draws the granulocyte into the micropipette until it is completely aspirated and starts to flow [11]. The cell volume is conserved during aspiration.

At the critical point, the Young–Laplace law states that pressure differences across the granulocyte membrane inside the micropipette and outside it (Figure 2.3) are given by

$$p_i - p_p = \frac{2\Gamma}{R_p} \qquad \text{and} \qquad p_i - p_o = \frac{2\Gamma}{R}, \tag{2.7}$$

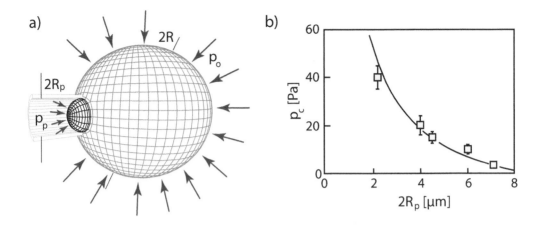

Figure 2.3 Sketch of the micropipette experiment (a). The measured critical suction pressure for granulocytes as a function of the micropipette diameter (b; symbols) and the fit based on Eq. (2.7) (line) (adapted from Ref. [11]).

respectively.[1] This means that the critical suction pressure reads

$$p_c = p_o - p_p = 2\Gamma \left(\frac{1}{R_p} - \frac{1}{R} \right), \tag{2.8}$$

which means that the suction pressure should increase as the micropipette radius is decreased. This is in good agreement with the experimental results in Figure 2.3b, and allows one to determine the surface tension of granulocytes which is $\Gamma = 0.035$ mN/m. Surprisingly, this value is more than 2000 times smaller than the surface tension of water (72 mN/m at 20°C) and about 1000 times smaller than the tension of a phospholipid bilayer (≈ 34 mN/m [12]). This means that the cell surface tension has little direct connection with the rearrangement of molecules like in simple liquids where an increase of surface area means bringing some of the molecules from the bulk to the surface, which breaks a part of the intermolecular bonds. Instead, the surface tension of a cell is an effective quantity describing the response of a complex multiscale entity which involves processes operating on length scales larger than molecular and at much softer energy scales, say unfolding of the wrinkles and folds on the cell surface as well as the deformation and rearrangement of the cell cortex. As we will see, these two effects are difficult to distinguish in experiments.

Experiments show that an increase of the suction pressure beyond the critical pressure induces cell flow [11] where the aspirated end of the cell does not come to rest deeper in the micropipette, but continues to move in until the cell completely enters the micropipette. This is to be expected if Eq. (2.8) is valid. As soon as Δp exceeds p_c, a neck consisting of a hemisphere and a cylindrical section, both of radius R_p, enters the micropipette. As a result, the external radius of the cell is decreased and thus the difference between a given Δp and the pressure within the cell $2\Gamma(1/R_p - 1/R)$ is increased. If the suction pressure Δp is fixed, the length of the neck increases until all of the cell is sucked into the micropipette. However if the pressure is suitably lowered after the neck of a given length has been aspirated, the cell stops moving in but stays aspirated.

This effect corroborates the assumption of the liquid nature of the neutrophil granulocyte

---

[1]Here R is the radius of the spherical portion of the cell outside the micropipette. For $R_p \ll R$, R is similar to the radius of the free, unaspirated cell.

cell [13]. The velocity v_c of the fully aspirated cell when $\Delta p > p_c$ is constant and it increases with Δp. As the cell is under the influence of the suction force, it develops a reaction force which increases with the cell velocity. The reaction force is due to the viscosity of the cell. The velocities of the different parts of a moving and deforming cell vary from point to point but are generally largest in the middle of the cell and close to zero near the membrane. The relative motion of the different layers of the cell gives rise to friction and resistance, which is proportional to the effective viscosity of the cell interior viewed as a homogeneous fluid.

The measurement of the effective viscosity of cells η in micropipette flow relies on the assumption that v_c is inversely proportional to η [11]. In neutrophil granulocytes, $\eta \approx 200$ Pa·s at 23°C, which is about 20 times larger than the viscosity of honey. It should be noted, however, that this is not the viscosity of the cytoplasm neither the viscosity of the intracellular liquid but the effective viscosity of the whole cell, which also includes the deformation and flow of its membrane. Due to the complex architecture of the cell, the effective viscosity depends on the scale of motion. Small particles moving within the cell are expected to experience friction against water, but large particles must collide with the cell skeleton and organelles which too contributes to viscosity. As a result, the macroscopic notion of the viscosity of a cell as a whole is not a particularly well-defined quantity.

The viscosity of cell interior can be determined from the analysis of Brownian motion of fluorescent probes of radius r_f [14] by relying on the Stokes–Einstein relation

$$D = \frac{k_B T}{6\pi\eta r_f},\tag{2.9}$$

where D is the diffusion coefficient, T is temperature, k_B is the Boltzmann constant, and the viscosity η generally depends on particle radius r_f. This analysis indeed shows that the viscosity pertaining to motion of small particles of $r_f \sim 1$ nm is similar to the viscosity of water and is of the order of ~ 1 mPa·s. Large particles experience a larger viscosity. For nanodiamonds of radius $r_f \approx 80$ nm it is about 35 times larger than that of water. The macroscopic viscosity can be extrapolated from these measurements, giving a value of $\eta \approx 4 \times 10^{-2}$ Pa·s [14]. This is three orders of magnitude smaller than that obtained in macroscopic experiments [11,13], but consistent with the analysis of motion of micrometer-size beads in the amoeba *Dictyostelium discoideum* [15].

In micropipette aspiration experiments with a buffer solution for cells that contains calcium ions important for the assembly and disassembly of the cell skeleton, the neutrophil granulocyte is eventually stabilized deeper in the micropipette if the suction pressure is too small to fully aspire it [16]. Chondrocyte cells, which make up cartilage, are also stabilized deeper in the micropipette when $p > p_c$ [13]. This behavior could be interpreted as a failure of the liquid-drop model of the cell and the notion of surface tension itself, suggesting that some cells respond to mechanical deformation as solids. Unlike liquid drops, where the minimal-energy shape corresponds to the minimal surface area under given conditions,[2] solids respond to external force by a deformation associated with a buildup of internal elastic stresses which then restore the undeformed shape after the force is removed.

Within the solid-cell framework, cells are expected to reach a certain finite deformation at a given fixed suction pressure $p > p_c$ because the reaction force increases with the magnitude of deformation, which can be parametrized by the length of the aspirated neck within the micropipette. However, the liquid-drop model can also be adapted to explain a stable aspirated state with a finite neck length. As the area of the cell A increases upon aspiration, the equilibrium reached at a finite neck length may be interpreted as an increase of the cell surface tension in response to the change in the cell area. This view departs from the usual molecular-liquid interpretation of the surface tension as a quantity that does not

---

[2]We tacitly assume that the liquid is incompressible.

depend on the surface area, but this poses no conceptual problems because we have already established that cells should not be regarded as simple liquid drops.

In calcium buffer, the dependence of surface tension of neutrophil granulocytes [16] on the cell area was found to be linear

$$\Gamma = \Gamma_0 + k\frac{A - A_0}{A_0}, \tag{2.10}$$

where Γ_0 and A_0 are the tension and the surface area of an isolated non-aspired spherical cell, respectively, and $\Gamma_0 \approx 0.010$ mN/m [16]. This *ansatz* captures well the experimentally measured surface tension both at small and at large deformations, albeit with a different coefficient k in the two regimes. The data at $(A - A_0)/A_0 < 25\%$ can be described by a quite small $k = 0.16$ mN/m, which implies that Γ is almost constant for such deformations so that $\Gamma \approx \Gamma_0$. At large deformations where $(A - A_0)/A_0 > 30\%$, k increases by more than an order of magnitude ($k \approx 2$ mN/m).

The large difference in the two values of k suggests that there may exist two distinct mechanisms of cell surface tension, one for each regime. Given the structure of the cell, one can indeed think of two such processes. In the small deformation regime, tension can be associated with unwrinkling and flattening of the supported membrane whereas at large deformations it is related to the stretching of the cell cortex (Figure 2.4). The two processes also involve myosins which crosslink the cortex and which may establish bonds at different points in the actin network as the cell is deformed. Finally, we should not forget that the cell cortex is coupled to the cell skeleton, which means that deformations at the surface propagate into the cell interior.

An alternative interpretation of the existence of two regimes postulates two types of membrane wrinkles, the weakly and the strongly crosslinked ones. According to this interpretation (Figure 2.4), the weakly crosslinked wrinkles unfold at a small applied tension whereas the strong wrinkles retain their integrity, unfolding only at large deformations when the tension in the membrane is sufficiently large [16].

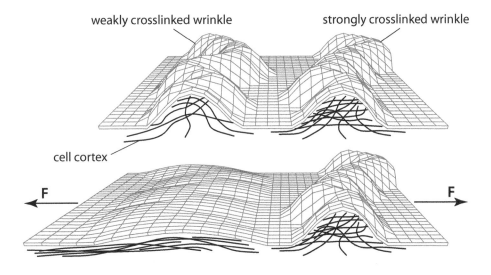

Figure 2.4 Unfolding of the cell membrane under low tension force F and the associated rearrangement of the cell cortex. The weakly crosslinked wrinkle may unfold, whereas the strongly crosslinked one retains its shape and the membrane area that is bound to it.

Experiments show that the area reservoir contained in the wrinkles and folds of a neutrophil cell is at least as large as the projected area of a spherical cell, which is truly remarkable [11,16]. For deformations which require an increase in area larger than about $(A - A_0)/A_0 \approx 120\%$, the cell membrane ruptures. Finally, not all types of cells are characterized by the same effective surface tensions and viscosities. At small deformations, the surface tension of macrophage J774 cells [17] is about $\Gamma = 0.14$ mN/m which is an order of magnitude larger than in neutrophils. These cells also show a significant enhancement of Γ when the area increase exceeds $\sim 10\%$. On the other hand, micropipette aspiration measurements show that the macroscopic viscosity of endothelial cells is about 7500 Pa·s [13], more than 30 times larger than that of neutrophils.

2.2.2 Interpreting experiments

The mechanical response of cells has also been measured in various experimental setups including diametral compression between parallel plates (Problem 2.5) and stretching by optical tweezers [18, 19]. The results of these experiments can often be satisfactorily explained by assuming that the cell is either a solid body or a liquid drop carrying a certain cortical surface tension. The interpretations and physical quantities extracted are, however, different in the two approaches.

Elementary notions in elasticity I

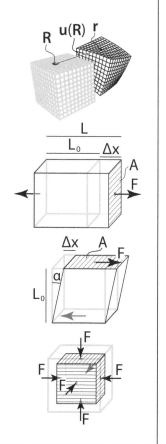

During deformation, a material point initially located at $\mathbf{R}$ moves to $\mathbf{r}$; the displacement vector is given by $\mathbf{u}(\mathbf{R}) = \mathbf{r} - \mathbf{R}$. The strain field measures the nature of deformation in relative terms. The simplest strain pertains to a body uniformly stretched or compressed along a given direction, say x. For two points whose initial and final separations along the x axis are L_0 and L, respectively, strain is given by

$$\epsilon = \frac{\Delta x}{L_0} = \frac{L - L_0}{L_0}. \qquad (2.11)$$

In this case, strain does not vary from point to point. Hooke's law states that

$$\sigma = Y\epsilon, \qquad (2.12)$$

where σ is the uniaxial stress (i.e., the normal force F divided by area A) and Y is Young's modulus of the material.

Bodies can also be deformed by tangential forces giving rise to shear stress τ. Shear strain is described by the shear angle $\alpha = \Delta x/L_0$. The response of the body to shear forces depends on shear modulus μ defined by

$$\tau = \mu\alpha. \qquad (2.13)$$

A uniform hydrostatic pressure compressing the body isotropically reduces its volume V_0 by ΔV. The bulk modulus K relates the corresponding normal stress $\sigma = F/A$ to $\Delta V/V_0$:

$$\sigma = K\frac{\Delta V}{V_0}. \qquad (2.14)$$

Consider, for example, the deformation of a cell by a pair of opposing stretching forces; this is a typical setup in the optical tweezers experiments. Assume for simplicity that the resting shape of the cell is a cube of edge R rather than a sphere and that the stretching forces are uniformly distributed across opposite faces of the cell. For small extensions ΔL, the increase of cell area is quadratic in ΔL so that the force resulting from the surface tension is linear in ΔL. A linear dependence of force on extension is also obtained in a cell represented by a homogeneous elastic cube. However, the physical quantities extracted from the liquid-drop and the elastic-solid model are different. The former is characterized by a surface tension Γ

$$\Gamma \sim \frac{F}{\Delta L} \tag{2.15}$$

whereas the latter is described by Young's modulus Y

$$Y \sim \frac{F}{R\Delta L}. \tag{2.16}$$

From the comparison of Γ and Y we conclude that the effective Young's modulus scales as $Y \sim \Gamma/R$. Similar relations are obtained when the elastic-solid model is applied to micropipette experiments, where one finds that $Y \sim \Gamma/R_p$ [13].

These arguments show that the small-deformation regime cannot discriminate between the liquid-drop and the elastic-solid models. Neither can this be done based on the behavior of cells at large deformations where the theoretical force-deformation relation depends on the details of the model. There exist various types of finite-deformation theories of elasticity, all generally suitable as the basis of the elastic-solid model yet each giving a different force-deformation relation. The same holds for the liquid-drop model where one can think of many different ways of describing the increase of cortex tension, not just Eq. (2.10). In addition, interpretation of experiments is tricky because of the active nature of the cell cortex and its remodeling in response to stress. Nonetheless it is still worthwhile to study the mechanical response of cells and monitor the variation of the parameters of a chosen model, whichever it is, because it may indicate the biochemical or structural changes in the cell [19].

A relevant and molecularly informed insight can be obtained only from precise experiments, with sufficient spatial resolution to modify and perturb the cell in a controlled fashion on a submicrometer or micrometer scale. Figure 2.5 shows a time series of images of

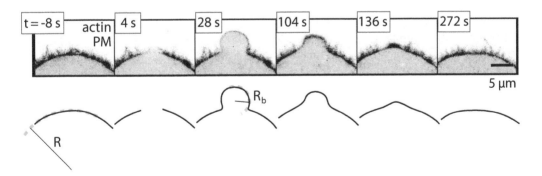

Figure 2.5 Time sequence of micrographs of a part of a HeLa cell near the ablated region below the plasma membrane (PM). The membrane and actin densities are shown in dark and light gray, respectively. The laser pulse was applied at $t = 0$. The contours in bottom row show the reconstructed cell shapes and the fitted radii of curvature of the whole cell (R) and of the bleb (R_b). (Adapted from Ref. [20].)

a part of a HeLa (cervical cancer) cell which was modified by local laser ablation at $t = 0$, so that a spherical piece of cell cortex about 1.5 μm in diameter was partially destroyed. In the before- and after-ablation micrographs of the cell we can monitor the progression of the cell shape and the thickness of the cortex in the ablated part [20].

In this experiment, a bleb starts to grow above the ablated patch of the membrane and reaches its maximum size and radius R_b at about $t = 28$ s after ablation. Afterwards it shrinks and is completely reintegrated in the membrane about 270 s after ablation. By analyzing the images, one can determine the parameters of cell geometry. Fitting the image of an unperturbed cell membrane at $t = -4$ s with a sphere gives a cell radius of $R \approx 13$ μm. The thickness of the cortex of the unperturbed cell was $h = 0.19$ μm. At $t = 28$ s, the radius of the bleb in the image is about 3.2 μm, and the thickness of the cortex below it is about $h_b = 0.05$ μm [20]. The reduction of cortex thickness caused by ablation leads to a localized reduction of cell surface tension and if we assume that the pressure within the cell is the same everywhere, then the reduced surface tension results in the formation of a bleb of a small radius so that Γ_P/R_P is the same in all points P on the cell surface, in accordance with Eq. (2.3). The ratio of radii of curvature in the different points of the surface is thus a measure of the ratio of local surface tensions in these points. This gives $\Gamma_b/\Gamma = R_b/R \approx 0.25$, where Γ_b and Γ are the effective surface tensions of the bleb and of the unperturbed cell surface. The obtained ratio of the tensions is very close to the ratio of the measured cortex thicknesses, $h_b/h = 0.26$, which suggests that the surface tension is directly proportional to the cortex thickness. This can be explained by treating the cortex as an elastic shell and examining the forces in it when it is subjected to a pressure difference.

Elementary notions in elasticity II

A homogeneous rod of radius R is uniformly stretched along the z axis by $\epsilon_{zz} = \Delta_{zz}/L_0 = \sigma_{zz}/Y$, where $\sigma_{zz} = F/A$ and $A = \pi R^2$. As a result, its cross-section contracts isotropically and uniformly in the xy plane so that $\epsilon_{yy} = \Delta_{yy}/R$ and $\epsilon_{xx} = \Delta_{xx}/R$. This is an example of a pure shear deformation.

The transverse strains are related to the lengthwise strain by

$$\epsilon_{yy} = \epsilon_{xx} = -\nu\epsilon_{zz}, \qquad (2.17)$$

where ν is Poisson's ratio of the rod. The relative change in volume of the rod is $\Delta V/V_0 \approx \epsilon_{zz} + \epsilon_{xx} + \epsilon_{yy}$, which vanishes in incompressible materials where $\nu = 1/2$.

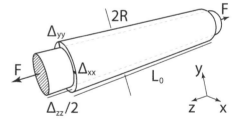

The shear modulus can be expressed in terms of Y and ν and reads $\mu = Y/[2(1+\nu)]$. The bulk modulus is given by $K = Y/[3(1 - 2\nu)]$ and diverges in incompressible materials.

Consider a small, nearly square patch of a cortex shell under isotropic in-plane tension and denote its thickness and area by h and $A_p = x_p y_p$, respectively; here $x_p = y_p$. Were the hatched area $A = x_p h$ in Figure 2.6 under uniaxial tension acting only along the x axis, its relative displacement given by the strain[3] reads $\epsilon_{xx}^0 = \sigma_{xx}/Y = F/(AY)$. But under isotropic tension, this strain is reduced due to the action of the force along the y axis, which makes the patch shrink along the x axis by an amount given by Poisson's ratio ν of the

---

[3]Here the (normal) strains and stresses along the x, y, and z axis are denoted by two subscripts, which may seem redundant. Both subscripts are kept for consistency with the text to follow where the full tensorial nature of stresses and strains are introduced in the Box on p. 28.

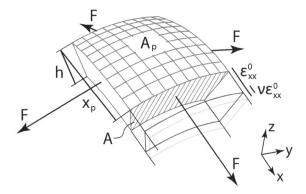

Figure 2.6 A small patch of the elastic cortex shell of thickness h under the influence of isotropic in-plane tension, showing all forces and strain along the x axis; strain along the y axis is not shown for clarity.

cortex.[4] The total increase of the linear dimension of the patch is thus

$$\epsilon_{xx} = \epsilon_{yy} = \epsilon = (1 - \nu)\frac{F}{AY}, \tag{2.18}$$

where ϵ is the magnitude of the in-plane strain. The stretching force per unit length can be written as

$$\lambda = \frac{F}{x_p} = \frac{Yh}{1 - \nu}\epsilon. \tag{2.19}$$

This is the effective surface tension of a spherical elastic shell. Unlike in a liquid surface, it depends on the in-plane strain ϵ. The effective surface tension is proportional to the shell thickness h, which explains the observed dependence of the bleb radius on the thickness of the cortex below it: $R_b/R = \lambda_b/\lambda = h_b/h$.

The energy of the stretched elastic shell can be obtained by calculating the expended work, which shows that the elastic energy U per unit volume of the shell V is given by

$$u_V = \frac{\mathrm{d}U}{\mathrm{d}V} = \frac{\mathrm{d}U}{\mathrm{d}(x_p y_p h)} = 2\int_0^\epsilon \lambda(\xi)\mathrm{d}\xi = \frac{Y}{1 - \nu}\epsilon^2, \tag{2.20}$$

where the factor of 2 comes from stretching along both x and y axes. This result can be expressed in terms of the area strain

$$\frac{\Delta A_p}{A_p} = \epsilon_{xx} + \epsilon_{yy} = 2\epsilon \tag{2.21}$$

so that

$$u_V = \frac{Y}{4(1 - \nu)}\left(\frac{\Delta A}{A_0}\right)^2. \tag{2.22}$$

The elastic energy per unit area of the shell is thus

$$u_A = u_V h = \frac{Yh}{4(1 - \nu)}\left(\frac{A - A_0}{A_0}\right)^2 \equiv \frac{K_A}{2}\left(\frac{A - A_0}{A_0}\right)^2, \tag{2.23}$$

where the last expression defines the area expansion modulus $K_A = Yh/[2(1 - \nu)]$.

---

[4]The derivation rests on the assumption that $h \ll R, R_b$.

Elementary notions in elasticity III

The deformation of an elastic body is described by the strain tensor ϵ_{ij}. During deformation, an infinitesimal material volume element initially at point $\mathbf{R} = \sum_i x_i \mathbf{e}_i$ moves to point $\mathbf{r}$ so that the displacement vector is given by $\mathbf{u}(\mathbf{R}) = \mathbf{r} - \mathbf{R}$. The vector between two nearby points $d\mathbf{R}$ goes over to $d\mathbf{r} = d[\mathbf{R} + \mathbf{u}(\mathbf{R})]$ in the deformed configuration. The change of the squared distance is $(d\mathbf{r})^2 - (d\mathbf{R})^2 = 2 \sum_{ij} \epsilon_{ij}(\mathbf{r}) dx_i dx_j$, where

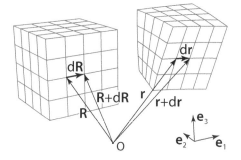

$$\epsilon_{ij}(\mathbf{r}) = \frac{1}{2}\left[\frac{\partial u_i(\mathbf{r})}{\partial x_j} + \frac{\partial u_j(\mathbf{r})}{\partial x_i}\right] + \frac{1}{2}\sum_k \frac{\partial u_k(\mathbf{r})}{\partial x_i}\frac{\partial u_k(\mathbf{r})}{\partial x_j} \qquad (2.24)$$

is the strain tensor. In case of small displacements only the first two terms are retained. The strain tensor encodes both volumetric and shear deformations.

Elastic forces within a deformed body are represented by the stress tensor σ_{ij}. Consider an infinitesimal triangular patch of area dS and normal $\mathbf{n}$, located at point $\mathbf{R}$ within the body and acted upon by a force $\mathbf{f}(\mathbf{R}, \mathbf{n})dS$, where $\mathbf{f}(\mathbf{R}, \mathbf{n})$ is the force density per unit area. A tetrahedron can be constructed so that the patch is one of its sides, and the other three sides of areas $dS\mathbf{n} \cdot \mathbf{e}_i$ lie in perpendicular planes of the coordinate system defined by basis vectors $(\mathbf{e}_1, \mathbf{e}_2, \mathbf{e}_3)$ [21]. The tetrahedron is under the influence of four forces acting on its four sides: $\mathbf{f}(\mathbf{R}, \mathbf{n})dS$ and $\mathbf{f}(\mathbf{R}, -\mathbf{e}_i)dS\mathbf{n} \cdot \mathbf{e}_i$ where $i = 1, 2, 3$. Requiring that the tetrahedron is in equilibrium gives

$$\mathbf{f}(\mathbf{R}, \mathbf{n}) = \sum_i \mathbf{e}_i \sum_j \sigma_{ij}(\mathbf{R})(\mathbf{n} \cdot \mathbf{e}_j), \qquad (2.25)$$

or

$$dF_i = \sum_j \sigma_{ij}(\mathbf{R})(\mathbf{n} \cdot \mathbf{e}_j)dS, \qquad (2.26)$$

where

$$\sigma_{ij}(\mathbf{R}) = \mathbf{e}_i \cdot \mathbf{f}(\mathbf{R}, -\mathbf{e}_j) \qquad (2.27)$$

is the stress tensor.

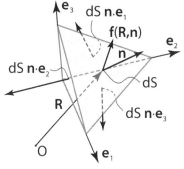

If one insisted on describing the deformation of an elastic shell due to internal pressure as the deformation of a liquid drop, the effective surface tension would have to increase with area, much like in Eq. (2.10). By equating the energies of a drop [11] and a stretched elastic shell at $(A - A_0)/A_0 = 10$ and 2% extension and using $h \sim 0.1\ \mu m$, we can estimate Young's modulus of the cell cortex of neutrophils which is $Y \sim 70$ kPa and 2 MPa, respectively. This is a very small value when compared with other materials—at 10% extension, it is about 1000 times smaller than in rubber whereas at 2% extension, it is about 50 times smaller. One should keep in mind, however, that the elastic moduli of the individual filaments are much larger. The modulus characterizing the elastic deformations of individual microtubules is of the order of ~ 1 GPa. The soft response to forcing is determined by the interconnections and entanglements of the filament network and not exclusively by the elasticity of a single filament [22]. Here we again encounter the problem of describing the response of a cell at many spatial scales by a single number, just like in the case of cell viscosity.

Elementary notions in elasticity IV

Equilibrium of forces for any given volume V within the deformed body requires that $\mathbf{F}^{el} + \mathbf{F}^{ex} = 0$, where $\mathbf{F}^{el}$ and $\mathbf{F}^{ex}$ are the elastic and the external forces, respectively. The elastic forces act across the bounding surface S of the volume in question so that

$$F_i^{el} = \int_S dF_i = \int_S \sigma_{ij}(\mathbf{R})\mathbf{e}_j \cdot \mathbf{n}(\mathbf{R})dS = \int_V \nabla \cdot \left[\sum_j \sigma_{ij}(\mathbf{R})\mathbf{e}_j\right]dV, \qquad (2.28)$$

where the last equality follows from Gauss' theorem. Thus the force equilibrium can be written as

$$\sum_j \frac{\partial \sigma_{ij}(\mathbf{R})}{\partial x_j} + f_i^{ex}(\mathbf{R}) = 0, \qquad (2.29)$$

where $\mathbf{f}^{ex}$ is the volume density of the external force. The requirement that the total torque on each infinitesimal volume be expressible as a surface integral [23] shows that the stress tensor must be symmetric: $\sigma_{ij} = \sigma_{ji}$.

Stress and strain are related by Hooke's law. In isotropic elastic materials,

$$\sigma_{ij} = \frac{Y}{1+\nu}\left(\epsilon_{ij} + \frac{\nu}{1-2\nu}\delta_{ij}\sum_k \epsilon_{kk}\right) \quad \text{and} \quad \epsilon_{ij} = \frac{1}{Y}\left[(1+\nu)\sigma_{ij} - \nu\delta_{ij}\sum_k \sigma_{kk}\right]. \qquad (2.30)$$

The elastic energy of a body of volume V is

$$E_{el} = \frac{1}{2}\sum_{i,j}\int_V \sigma_{ij}(\mathbf{R})\epsilon_{ij}(\mathbf{R})dV. \qquad (2.31)$$

The Hookean elastic energy reads

$$E_{el} = \frac{Y}{2(1+\nu)}\int_V\left[\sum_{i,j}\epsilon_{ij}^2(\mathbf{R}) + \frac{\nu}{1-2\nu}\sum_i \epsilon_{ii}^2(\mathbf{R})\right]dV. \qquad (2.32)$$

2.2.3 Viscoelasticity

A universal feature of the mechanical response of cells is the time dependence. This can be quite clearly seen in Figure 2.5. In less than 300 s, the torn cell skeleton heals itself, the cortex rearranges and reassembles its protein filaments and then returns to the initial state before laser ablation. The characteristic time scale for this process is thus of the order of minutes. In all of the experiments discussed in this chapter, time and relaxation play an important role. A cell released from the micropipette "slowly recovers its initial spherical form" [11]. In diametral-compression experiments (Problem 2.5), cells do not deform instantaneously but approach their final shape on a scale of minutes. Once the force is removed, cells return to their initial state and shape, again within a period of a few minutes.

The observed relaxation behavior is typical for viscoelastic materials of the Kelvin–Voigt type which approach the final state determined by external forces exponentially on both loading and unloading as opposed to ideal solids where the final state is reached instantaneously. The characteristic relaxation time is given by the ratio of the effective viscosity and Young's modulus. Knowing that the cell consists of various elastic elements of different Young's moduli, different degrees of crosslinking, and different overall organization,

---

Kelvin–Voigt and Maxwell model of viscoelasticity

Materials which combine fluid-like and solid-like features in their response to stress are referred to as viscoelastic. This behavior can be described in terms of simple mechanical models consisting of a spring of Young's modulus Y coupled to a dashpot characterized by viscosity η. In the spring, stress is proportional to strain, i.e., $\sigma_S = Y\epsilon_S$, whereas in the dashpot stress is proportional to the strain rate: $\sigma_L = \eta d\epsilon_L/dt$. (Tensor indices are left out for clarity.) In the Kelvin–Voigt model the two elements are coupled in parallel. The strains in both elements are identical ($\epsilon = \epsilon_S = \epsilon_L$) and the total stress is the sum of stresses ($\sigma = \sigma_S + \sigma_L$). This gives

$$\sigma(t) = Y\epsilon(t) + \eta\frac{d\epsilon(t)}{dt} \qquad (2.33)$$

If a constant stress σ_0 is applied at $t = 0$, the strain approaches its final value σ_0/Y with a characteristic time scale $\tau = \eta/Y$:

$$\epsilon(t) = \frac{\sigma_0}{Y}\left[1 - \exp\left(-t/\tau\right)\right]. \qquad (2.34)$$

If the stress is removed at $t = t_R$, the strain relaxes back to zero according to

$$\epsilon(t > t_R) = \epsilon(t_R)\exp\left(-\tau(t - t_R)\right). \qquad (2.35)$$

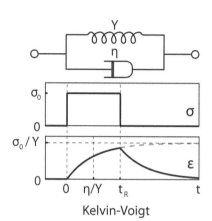

Kelvin-Voigt

The Kelvin–Voigt model can be used to describe viscous solids.

In the Maxwell model the two elements are arranged in series so that the total strain is $\epsilon = \epsilon_S + \epsilon_L$. By differentiating this equation with respect to time and noting that the stresses in both elements are the same, $\sigma = \sigma_S = \sigma_L$, we find that

$$\frac{d\epsilon}{dt} = \frac{1}{Y}\frac{d\sigma}{dt} + \frac{\sigma}{\eta}. \qquad (2.36)$$

The response to stress σ_0 switched on at $t = 0$ is

$$\epsilon(t) = \frac{\sigma_0}{Y} + \frac{\sigma_0}{\eta}t. \qquad (2.37)$$

Once stress is removed at $t = t_R$, the spring stretched by σ_0/Y instantaneously returns to the undeformed state but the displacement of the dashpot $\sigma_0 t_R/\eta$ remains unchanged. This model mimics the behavior of viscoelastic liquids where the short-time solid-like deformation is followed by flow at long times.

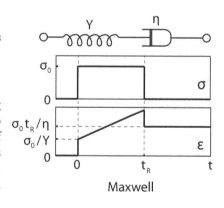

Maxwell

---

one may *a priori* presume that there are several time scales describing the dynamics of cell response, which may, however, be indistinguishable one from another in a time-dependent deformation and relaxation.

In certain experiments one may still be able to study the behavior of a specific cell substructure. For example, actin rearrangement can been monitored in the dynamics of microvilli [24] where the typical relaxation times are again of the order of minutes. The processes involved include bending and rearrangement of actin filaments. Each of these phenomena is characterized by a different relaxation time so that the overall stress relaxation is rather complex.

The cell is a complex structure consisting of elastic and viscous passive matter as well as of active matter. On very short time scales where no cortex remodeling takes place, the cell behaves as an elastic body whereas on long time scales it creeps and flows, slowly approaching the final deformation which may include significant cortex rearrangement. The characteristic relaxation times of the acto-myosin cortex were measured using the laser ablation technique in the one-cell-stage of roundworm embryo and in the gastrulating zebrafish embryo [25]. A laser was used to create precise cuts in the cortex, which were about 20 μm long. The cuts initially relaxed and widened to assume an approximately elliptical shape with the long axis oriented along the cut. The maximum width of the hole was reached about 5 s after cutting so that the relaxation time was of the order of 5 s. After that the wound spontaneously sealed by rearrangement and regrowth of the cell cortex on a scale of ~ 1 minute [25], consistent with the laser ablation experiments shown in Figure 2.5 [20]. Wound healing and cortex remodeling are active processes and cannot be explained using theories of passive viscoelastic materials.

2.3 CELLS IN CONTACT

Cells are never completely isolated. They are at least in contact with the extracellular solution, characterized by osmolarity, pH, temperature, and concentrations of different molecules and ions (e.g., calcium) dissolved in it. The shape and properties of cells depend on their environment. Cells on the surface of the tissue experience two distinct environments—the external medium or the contents of the lumen[5] as well as the cells around and underneath them or, in case of single-cell-thick epithelia, the underlying basement membrane. Adhesive interactions of cells in the tissue are obviously essential for the integrity of the tissue itself. On the other hand, the migratory cells move through their environment, and for those that rely on traction, the establishment as well as dissolution of close contacts with the environment is equally essential.

2.3.1 Adhesion

The elementary energy-based aspects of the cell-environment interaction can be obtained using a generalized version of the liquid-drop model. Consider a cell C at the interface between a flat substrate S and medium M as shown in Figure 2.7. This geometry involves

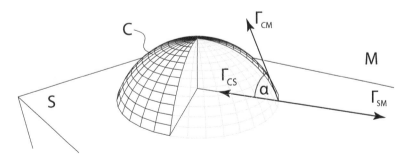

Figure 2.7 Cell C wetting the substrate S and immersed in medium M. The arrows show surface tension forces (per unit length) at the rim and α is the contact angle.

[5]Lumen is the space within a hollow organ or structure, say a blood vessel or intestine.

three surface tensions: cell-substrate tension Γ_{CS}, cell-medium tension Γ_{CM}, and substrate-medium tension Γ_{SM}. In mechanical equilibrium, cell shape is determined by the balance of the three tensions. As the cell spreads across the substrate, it also increases its contact area with the medium but it reduces the contact area between the substrate and the medium. At the rim, equilibrium of forces in the in-plane direction normal to the rim yields

$$\Gamma_{CM} \cos \alpha = \Gamma_{SM} - \Gamma_{CS}, \tag{2.38}$$

where α is the contact or wetting angle. If $\Gamma_{SM} > \Gamma_{CS}$, the contact angle $\alpha < 90°$. In this case, cell energy is decreased by the cell spreading across the substrate because this reduces the substrate-medium contact area and thus the total energy. In the limit where $\Gamma_{SM} = \Gamma_{CS} + \Gamma_{CM}$, the contact angle is $0°$ so that the cell forms an infinitely thin film between the substrate and the medium. If $\Gamma_{SM} < \Gamma_{CS}$, $\alpha > 90°$ and the cell is a truncated sphere sitting on the substrate. With increasing Γ_{CS}, the contact angle increases too and as it approaches $180°$ the drop detaches from the substrate. At any Γ_{CS} larger than $\Gamma_{SM} + \Gamma_{CM}$, Eq. (2.38) cannot be satisfied and the cell does not wet the substrate, assuming a perfectly spherical shape suspended in the medium.

A similar reasoning can be applied to two cells in contact (Figure 2.8). The doublet is characterized by two surface tensions, one for the contact of each cell with the medium, and the interfacial tension at the cell-cell contact zone.[6] The three surfaces separating drops from the medium and from each other are all spherical caps of different radii as shown in the figure. The radii of the interface between cell 1 and medium, cell 2 and medium, and between cells 1 and 2 are denoted by R_1, R_2, and R_3, respectively, and corresponding tensions are Γ_{1M}, Γ_{2M}, and Γ_{12}, respectively.

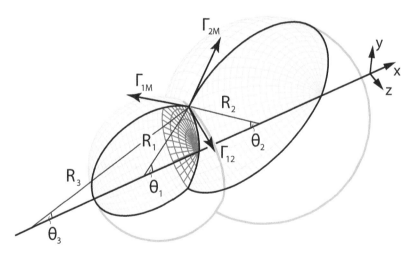

Figure 2.8 Two cells in contact modeled as liquid drops. In the cutaway view containing the axis of rotational symmetry, the contours of the cells are indicated by thick lines.

Let us consider the equilibrium of forces at the rim of the cell-cell contact zone. The balance of forces along a radial direction relative to the axis going through cell centers (say the y axis in Figure 2.8) gives

$$\Gamma_{1M} \cos \theta_1 + \Gamma_{2M} \cos \theta_2 = \Gamma_{12} \cos \theta_3, \tag{2.39}$$

---

[6]Were the two cells of different types, one would need three surface/interfacial tensions to describe the doublet.

whereas the equilibrium of pressures across the cell-cell contact zone reads

$$\frac{\Gamma_{12}}{R_3} = p_1 - p_2 = \frac{\Gamma_{1M}}{R_1} - \frac{\Gamma_{2M}}{R_2}, \tag{2.40}$$

where p_1 and p_2 are the internal pressures in cell 1 and 2, respectively.[7]

Without loss of generality we can assume that $\Gamma_{1M} \geq \Gamma_{2M}$. The critical point then corresponds to $\theta_1 = 180°$, $\theta_2 = 0°$, and $\theta_3 = 180°$, and in this case Eq. (2.39) gives

$$\Gamma_{12}^c = \Gamma_{1M} - \Gamma_{2M}. \tag{2.41}$$

When $\Gamma_{12} < \Gamma_{12}^c$, cell 1 is completely engulfed by cell 2. This process can be interpreted as the spreading of the cell of a smaller cell-medium tension over the cell of a larger cell-medium tension, which remains in a compact spherical form. The process somewhat resembles engulfment or phagocytosis of bacteria and other cells by neutrophil granulocytes, and it may explain why neutrophils need to have an extremely low surface tension [11]. When the low-tension neutrophils are in contact with cells of a larger tension, engulfment takes place spontaneously and does not require any active mechanisms, at least within the simple liquid-drop model.

In case of identical cells, $p_1 = p_2$, $R_1 = R_2$, $\Gamma_{1M} = \Gamma_{2M}$, and $\theta_1 = \theta_2 \equiv \theta$. Now Eq. (2.40) shows that $R_3 \to \infty$ and since θ_3 must be 0 by symmetry, Eq. (2.39) gives $\cos\theta = \Gamma_{12}/2\Gamma_{1M}$. This result leads to a criterion of crucial importance for the formation of tissue: cells adhere to each other and form cell clusters and tissues only when $\cos\theta < 1$, i.e., when

$$\Gamma_{12} < 2\Gamma_{1M}. \tag{2.42}$$

Cell-cell contacts form only if their creation is energetically favorable, or, in other words, if cell-cell adhesion is strong enough to sufficiently lower the cell-cell interfacial tension. In case of cells of different types, cell-cell contact is favorable only if the interfacial tension is smaller from the sum of the two individual surface tensions, i.e., if

$$\Gamma_{12} < \Gamma_{1M} + \Gamma_{2M}. \tag{2.43}$$

Redoing this analysis for clusters of several cells is possible only if the number of cells is small. In clusters of many cells, finding the minimal-energy configuration is an increasingly bigger challenge because of the ever larger number of the different possible topologies of the cluster. For example, a three-cell cluster can be either linear, resembling a necklace, or triangular where each cell is in contact with the other two. In four-cell clusters, the number of possibilities is ever larger (Problem 2.11). For each topology, one should then minimize the total surface energy, which can be done numerically, say using the Surface Evolver package [26]. Figure 2.9 shows a cluster of thirteen cells modeled as liquid drops of fixed volume. One of the cells is in the center of the cluster and the rest are arranged around it in equivalent positions such that the cluster has dodecahedral symmetry. The values of all surface and interfacial tensions in the cluster in the figure are identical but the pressures within the cells are not —the pressure of the central cell is 7% larger than that in the twelve cells around it.

Cell-cell adhesion is the fundamental physical process responsible for the existence of tissues. At the molecular scale, adhesion is caused by transmembrane proteins from the

---

[7]Knowing the structure of the cell, it may seem puzzling that the interface between the cells is treated as a single rather than a double surface, with the corresponding interfacial tension included only once. But there is no inconsistency in this as long as the cell-cell interfacial tension includes the membrane and cortex tensions of the two cells as well as the cell-cell adhesion strength, which decreases the tension of the interface.

a) b)

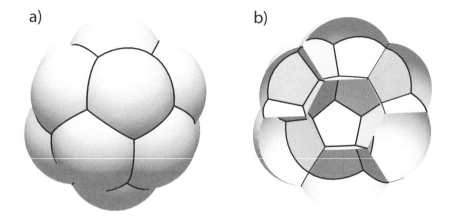

Figure 2.9 Cell cluster with dodecahedral symmetry (a). Cells are approximated by liquid drops and their equilibrium shape is computed by minimizing the total surface energy using Surface Evolver [26]. A cutaway view of the cluster showing one cell in the center and the cells around it (b).

cadherin superfamily. There are hundreds of different cadherin types found in animals [27]. Cadherin molecules are anchored in the membrane (Figure 2.10), with their intracellular component attached to contractile actin-filament bundles of the cell cortex. Their extra-

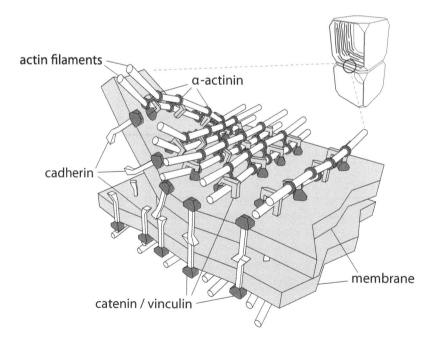

Figure 2.10 Cell-cell adhesion is mediated by membrane proteins called cadherins. Their extracellular parts protrude from the membrane and bind to each other. On the inside, cadherins are attached to the actin filaments by catenins.

cellular part binds to a cadherin on the surface of the neighboring cell [28]. Regions of membrane where cells bind to each other are called adherens junctions. In epithelia, they are arranged in a belt-like formation known as the adhesion belt running around the cell circumference.

2.3.2 Traction

The microscopic workings of cell-cell interaction are rather complex, and so is the interaction of cells with the extracellular matrix. This network of protein fibers such as collagen, elastin, fibronectin, and laminin is a key component of many body structures including connective tissue, and some migratory cells, such as mesenchymal cells discussed in Sections 5.4 and 5.5, move in space by attaching to the matrix. These cells extend protrusions to nearby objects including the extracellular matrix, attach to them, and then actively contract the protrusions so as to move toward the anchor. The large protrusions are known as lamellipodia, thin sheet-like projections of membrane from the leading edge of the cell (Figure 2.11). A lamellipodium consists mainly of a network of actin filaments capable of generating contractile stresses and its leading edge contains many filopodia, smaller finger-like protrusions containing bundles of actin filaments bound together by fascin.

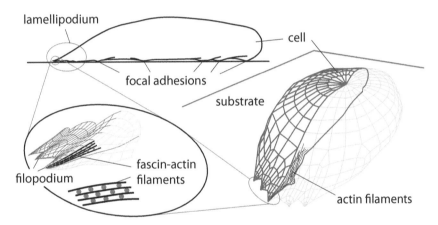

Figure 2.11 Cell traction machinery: lamellipodia and filopodia.

The protrusions connect to the extracellular matrix by engaging specific proteins in the membrane. Transmembrane adhesion receptors known as integrins bind to matrix proteins, forming localized contacts called focal adhesions. The part of the cell membrane adhering to the matrix—or, more generally, to a substrate—is not a contiguous, simply connected region. Instead it consists of many disconnected patches strongly bound to the substrate separated by regions of membrane not involved in adhesion. The binding of the cell to the substrate also remodels the actin network which repolymerizes so as to enable the formation and maintenance of adhesion patches [29]. The patches are dynamic: once an adhesion patch is established at the cell edge below the lamellipodium, patches at the opposite end are dissolved so that the cell can pull itself toward the newly established contact with the substrate. These active mechanisms, fueled by ATP and myosin activation, are involved in the traction force that the cell exerts on the extracellular matrix, neighboring cells, or a substrate.

These processes are important not only for cell migration but also for cell rearrangement within the tissue; these aspects of cell traction will be discussed in Chapter 5. Experimentally, traction forces are conveniently studied using artificial microfabricated substrates consisting of arrays of micropillars 1 or 2 μm in diameter and between 1 and 12 μm tall, with center-to-center separation from 2 to 4 μm [30, 31]. The micropillar array is typically made of poly(dimethylsiloxane) polymer and then covered with fibronectin so as to mimic the extracellular matrix. The regular arrangement of the pillars allows for easy tracking of their displacement as the cell adheres to the substrate (Figure 2.12). Each pillar is an elastic rod and its in-plane displacement u is proportional to the lateral force F exerted by the cell so that $F = ku$, where k is a constant.[8] By measuring pillar displacements one can then map forces generated by the cell, which are typically of the order of tens of nN, whereas pillar deflections are usually no larger than their height L. The forces can be essentially read off the micrographs since pillar deflections and their in-plane projections are proportional to forces. The largest deflections are localized at the edges of cells or cell islands [32]. In Figure 2.12 we can observe that on average, forces at the cell edge are perpendicular to the edge.

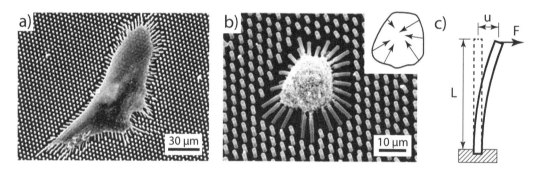

Figure 2.12 Scanning electron micrographs of human mesenchymal stem cells on micropillar arrays. The height of micropillars in panels a and b is 0.97 and 6.10 μm, respectively. The inset in panel b shows the cell border and the forces acting on it, and panel c illustrates pillar deflection. (Adapted from Ref. [31].)

With its dynamical cytoskeleton, the cell behaves as active matter and the stresses that it exerts on the substrate are of the order of 1 nN/μm^2. A simple description of the active stress produced by the cell can be obtained by combining the standard elastic stress σ_{ij} with an active term σ_{ij}^a that is independent of deformation:

$$\sigma_{ij}^t = \sigma_{ij} + \sigma_{ij}^a. \tag{2.44}$$

The effect of the active term can be illustrated by considering a strip of an active material resting on an array of micropillars (Figure 2.13a). For simplicity, we assume that the material contracts only along the x axis and that its dimensions in the other two directions are unaffected by contraction, which is equivalent to setting Poisson's ratio ν to 0. The only component of the total stress tensor of interest is σ_{xx}^t, which reads

$$\sigma_{xx}^t = Y(\epsilon_{xx} - P). \tag{2.45}$$

Here $-YP$ is the active stress independent of deformation and P is the target relative volume change as seen by considering a relaxed, stress-free strip where $\sigma_{xx}^t = 0$ and thus $\epsilon_{xx} = P$. Note that a contractile deformation corresponds to $P < 0$.

An active contractile strip of length $2l_0$ attached to the array of pillars cannot reach the

---

[8]The constant k depends on pillar radius r, height L, and Young's modulus Y of the material in question. For cylindrical rods, $k = 3\pi Y r^4/(4L^3)$. A derivation of this expression can be found in Problem 2.13.

stress-free state. Instead, it exerts a traction force on the pillars which are then displaced along the x axis, the local displacement u being proportional to the force. In equilibrium, the net traction force on a narrow transverse band of the material of length $\mathrm{d}x$ must be equal to the reaction force of the bent pillars:

$$h\frac{\mathrm{d}\sigma_{xx}^{t}}{\mathrm{d}x} = kNu. \tag{2.46}$$

Here h is the thickness of the strip and N is the area density of the pillars. Since $\epsilon_{xx} = \mathrm{d}u/\mathrm{d}x$ this gives

$$\frac{\mathrm{d}^2u}{\mathrm{d}x^2} - \frac{1}{l^2}u = \frac{\mathrm{d}P}{\mathrm{d}x}, \tag{2.47}$$

where

$$l = \sqrt{\frac{Yh}{kN}} \tag{2.48}$$

is the so-called localization length [33]. At either end of the strip, stress must vanish so that the boundary conditions read $\mathrm{d}u/\mathrm{d}x|_{x=\pm l_0} = P$; in addition, the center of the strip is not displaced so that $u(x) = 0$. In the case where P does not depend on position, the equilibrium displacement profile then reads

$$u(x) = Pl\frac{\sinh(x/l)}{\cosh(l_0/l)} \tag{2.49}$$

so that for a contractile material where $P < 0$, the displacement is positive for $x < 0$ and negative for $x > 0$ as it should be [33].

With this solution, the meaning of the localization length is immediately clear. The $l/l_0 \gg 1$ limit corresponds to a soft substrate represented by either pliable pillars or a small pillar density. In this case, $\sinh(x/l) \approx x/l$ and $\cosh(l_0/l) \approx 1$ across all of the strip. This implies that $u(x) \approx Px$ and the actual degree of contraction measured by ϵ_{xx} is equal to P as if the strip were free. On the other hand, if $l/l_0 \ll 1$ then the magnitude of displacement increases exponentially toward either end of the strip. As a result, most of the contraction is localized at the ends, l being the characteristic distance, and the central part of the strip is undeformed. These two regimes are shown schematically in Figure 2.13b and c, respectively.

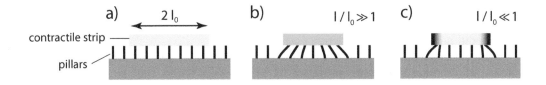

Figure 2.13 Actively deforming strip supported by an array of micropillars in its resting, non-contractile (a) and contractile state in the large and small localization length limit (b and c, respectively). Strain is represented by the shades of gray.

The active-matter model of cell contraction is a patently coarse-grained description of the molecular-level processes at work, simplified but insightful just like the view where we represent cadherin-mediated cell-cell attraction by an effective adhesion strength. Unsophisticated as it may seem, this approach can be used to interpret a range of phenomena as shown below.

2.4 DIFFERENTIAL ADHESION HYPOTHESIS

In the liquid-drop model of cells, we represent the surface and interfacial energies of cells by a term proportional to the area in question. The cell-cell contact zones carry a membrane tension due to the cell cortex as well as an adhesion strength, which effectively lowers the net tension of the zones. The underlying analogy between cells and liquid drops is an obvious simplification because of the dramatic structural differences—yet it is possible that because of the many constraints imposed, this analogy may still well work for moderate deformations of cell shape. One important piece of evidence consistent with this model is that despite their internal structure, many isolated cells are spherical as if they were droplets of simple liquids (Figure 2.14).

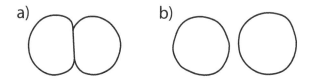

Figure 2.14 Contours of mouse embryonic cells at the two-cell stage in the wild-type embryo (a) and in a mutant embryo (b) where the formation of the adhesion complex containing E-cadherin and β-catenin was disrupted. (Contours reconstructed from fluorescence micrographs in Ref. [34].)

The possibility that cell-cell adhesion and membrane tension can be represented by surface energies has been appreciated and explored for a long time. In his book, Thompson opens the discussion of cell aggregates by saying that when considering the various forces involved, "we may at least begin by assuming, that the agency of surface-tension is especially manifest and important" [3]. The seminal experiments on amphibian neurulae carried out by P. L. Townes and J. Holtfreter [35] demonstrated that these tissues can be disaggregated and reaggregated by changing pH which evidently modified the strength of adhesion between cells (Figure 2.15). This study provided several observations which suggest that the neurulae can indeed be likened to droplets, say the indiscriminate reaggregation of cells of different types and the presence of satellite aggregates formed by peripheral cells after the normal pH was restored. These cells were too distant from the main mass and could not attach to it for kinetic reasons; instead they aggregated into small spherical clumps themselves. Most importantly, the reconstituted neurulae developed much like normal embryos provided that the cells did not dissociate completely.

Many of the phenomena reported by Townes and Holtfreter are explained by the so-called differential adhesion hypothesis proposed by M. S. Steinberg in 1963 so as to explain the possible behaviors of aggregates consisting of two types of cells including mixing, spreading, and isolation of the two populations [36]. When two cell populations are in contact with each other, one cell population typically spreads over a surface of another cell population, which remains in a compact, sphere-like form [35,37]. If the cells which form the tissue are highly mobile as, e.g., in some embryonic tissues, the tissue can be viewed as liquid with a surface tension and the whole process can be viewed as an interaction of two liquids with certain surface tensions and an interfacial tension. The values of the tensions depend on the tissues in question, and the behavior of these tissues is then dictated by the relative magnitudes of the tensions.

The physics of cell-cell interactions studied above within the liquid-drop model can also be used to describe the interaction between liquid-like, fluidized tissues. So far, we have examined the surface energies at the level of an individual cell or cell cluster, but the same

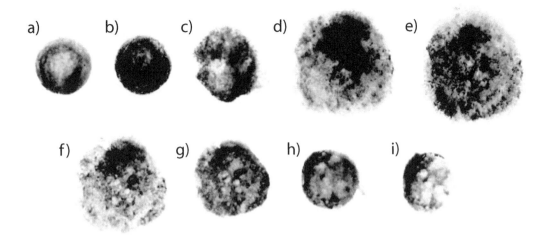

Figure 2.15 Neurula of the tiger salamander *Ambystoma tigrinum* after alkali treatment which increases pH to 9.8 (panels a–e) and after restoration of the normal pH of 8.0 (panels f–i). During the cycle, the initially compact neurula is disintegrated into a mass of disconnected cells and then reaggregated as pH is decreased. (Adapted from Ref. [35].)

concepts can also be applied to liquid tissues, i.e., to tissues in which the cells are mobile enough so that the tissue as a whole can change its shape and area under the influence of forces as if it were a liquid drop.

An isolated drop of a liquid tissue is expected to be spherical so as to minimize the surface energy. This is reasonably well confirmed by experiments [38] involving compression of aggregates of chick embryonic cells [37,39]. Although the surface of a liquid tissue is made of cells and their individual membranes, its tension is not the same as the surface tension of a cell. The reason for this is that the tissue can change its shape not only by changing the membrane area of individual cells but also by relocation of cells from the bulk to the tissue surface, much like molecules in a simple liquid move from the bulk to the surface if surface area is increased. In this process, the exposed surface of the tissue is increased without necessarily increasing the membrane area of individual cells—this issue will be discussed in more detail in Section 2.4.2.

Two different liquid tissues are expected to have distinct surface tensions because both cell-cell adhesion strengths and cell surface tensions are generally not the same. These differences then govern the envelopment behavior of the tissues, that is, the selection of the core and the shell tissue in a core-shell structure suspended in a medium. A neat experiment, which involved five chick embryonic tissues dissociated in a salt solution, pelleted, and incubated in a shaker so that they formed spherical fragments [37], demonstrated that indeed the tissue with the larger surface tension is enveloped by the tissue with the smaller surface tension over and over again. The limb bud mesoderm[9] with $\Gamma^t = 20.1$ mN/m is enveloped by the pigmented epithelium where $\Gamma^t = 12.6$ mN/m whereas the pigmented epithelium is enveloped by the heart tissue with $\Gamma^t = 8.5$ mN/m. In turn, heart tissue is

---

[9]Here the surface tensions pertaining to tissues are denoted with a superscript t to distinguish them from the cell surface tensions. Tissue surface tensions were measured using a variant of the diametral-compression apparatus (Problem 2.5).

enveloped by liver with $\Gamma^t = 4.6$ mN/m which is enveloped by neural retina with $\Gamma^t = 1.6$ mN/m [37]. The measured values of tissue surface tensions are about an order of magnitude larger than the typical surface tension of single cells and about an order of magnitude smaller than the surface tension of water.

2.4.1 Tissue envelopment and cell dispersal

Just like in the case of single cells, the criterion for the complete envelopment of a drop of tissue 1 with surface tension Γ^t_{1M} by tissue 2 with surface tension Γ^t_{2M} ($\Gamma^t_{1M} > \Gamma^t_{2M}$) is

$$\Gamma^t_{12} < \Gamma^t_{1M} - \Gamma^t_{2M}, \tag{2.50}$$

when is Γ^t_{12} is the interfacial tension of tissues 1 and 2.

Tissue-tissue adhesion includes a process that does not have a cell-level analog. Liquid tissues can disintegrate into two or more parts until the smallest parts are individual cells. Consider a drop of tissue 1 consisting of N_1 cells, which is enveloped by tissue 2 consisting of N_2 cells as shown in Figure 2.16. Cells in both tissues are in contact with their neighbors. There are N_{12} cells of tissue 1 in contact with about the same number of cells, N_{12}, of the other tissue.[10] The former are on the surface of the enveloped drop of tissue 1 and are highlighted by dark gray in Figure 2.16, and the latter are in a layer of cells of tissue 2 just above them.

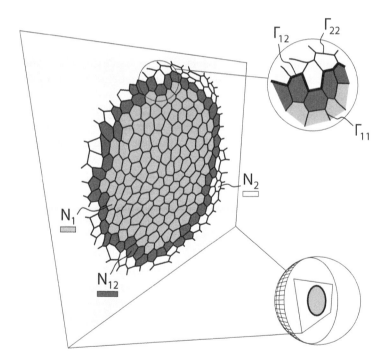

Figure 2.16 Cross-section of a drop of tissue 1 containing N_1 cells enveloped by tissue 2 containing N_2 cells. The cells of tissue 1 that are in contact with tissue 2 are highlighted in dark gray. The inset shows the three surface tensions involved.

---

[10]Strictly speaking, this approximation holds when the radius of the whole drop is much larger than the typical cell diameter.

Let us assume that the average area of each cell in both tissues is the same and equal to A_0. As all cells except those on the surface are surrounded by neighbors, most of the total cell area is buried in cell-cell contacts as shown in the cross-section of the drop in Figure 2.16. The total surface energy of the system reads

$$
\begin{aligned}
E &= (N_1 - N_{12})\Gamma_{11}\frac{A_0}{2} + (N_2 - N_{12})\Gamma_{22}\frac{A_0}{2} \\
&\quad + N_{12}\left(\frac{1}{2}\Gamma_{11}\frac{A_0}{2} + \frac{1}{2}\Gamma_{12}\frac{A_0}{2}\right) + N_{12}\left(\frac{1}{2}\Gamma_{22}\frac{A_0}{2} + \frac{1}{2}\Gamma_{12}\frac{A_0}{2}\right),
\end{aligned}
\tag{2.51}
$$

where Γ_{11} and Γ_{22} are the interfacial tensions between cells in tissues 1 and 2, respectively, and Γ_{12} is interfacial tension at the contact of a cell of tissue 1 and a cell of tissue 2. The first two terms represent the bulk energies of tissues 1 and 2: $N_1 - N_{12}$ and $N_2 - N_{12}$ are the numbers of bulk cells in tissues 1 and 2, respectively, and each piece of cell-cell contact area is equally divided between the two cells.[11] The third term is the energy of cells at the boundary of tissue 1, sharing half of its surface with neighbors from tissue 1 and the other half with neighbors from tissue 2. The fourth term is the tissue 2 analog of the third term. The surface energy of the drop can be rearranged to read

$$
E = N_1\Gamma_{11}\frac{A_0}{2} + N_2\Gamma_{22}\frac{A_0}{2} + N_{12}\left(\Gamma_{12} - \frac{\Gamma_{11} + \Gamma_{22}}{2}\right)\frac{A_0}{2}
\tag{2.52}
$$

so as to combine all terms that depend on the number of boundary cells N_{12}. If

$$
\Gamma_{12} < \frac{\Gamma_{11} + \Gamma_{22}}{2},
\tag{2.53}
$$

the total energy can be reduced by increasing the number of boundary cells N_{12}, which signals an instability of the enveloped drop that leads to dispersal of tissue 1 within tissue 2 until the size of the smallest bit of tissue 1 is comparable to cell size.

Within the liquid-drop model, the formation of a tissue and the nature of its interaction with adjacent tissues depend on the magnitude of cell and tissue surface tensions. In case of two tissues suspended in a medium, there are five different tensions involved and the criteria defining the different regimes are summarized in Figure 2.17, first addressing the stability of tissue 1, then its envelopment in tissue 2, and finally its dispersal within tissue 2. All of these regimes were seen experimentally in mixtures of cells modified so as to express different levels of cadherins, which show that depending on cadherin level these cells either form compact segregated tissues, which may completely envelop one another, or remain well mixed [38].

2.4.2   Cell surface tension vs. tissue surface tension

The surface tension of a tissue and the surface tension of each cell in it are not the same but they are related. The energy of tissue 1 containing N cells, N_s of which are on the surface, is given by

$$
E = \Gamma_{11}(N - N_s)\frac{A_0}{2} + \Gamma_{11}N_s\frac{A_0}{4} + \Gamma_{1M}N_s\frac{A_0}{2},
\tag{2.54}
$$

where the first term is the interfacial energy of cells in the bulk, the second term is the interfacial energy of cells on the surface, and the third term represents the cell-medium interfacial energy of these cells; we assume that half of the area of the surface cells is in

---

[11] We did not include the energy of cells on the surface of tissue 2 which lack some of the neighbors that the cells in the bulk do have. The final result of the analysis, Eq. (2.53), does not depend on this energy.

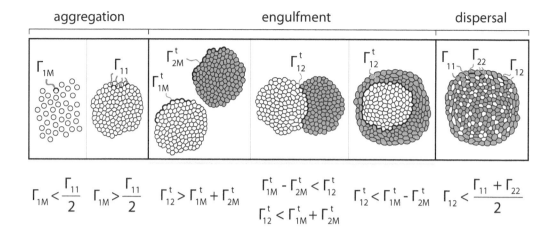

aggregation | engulfment | dispersal

$$\Gamma_{1M} < \frac{\Gamma_{11}}{2} \quad \Gamma_{1M} > \frac{\Gamma_{11}}{2} \quad \Gamma_{12}^t > \Gamma_{1M}^t + \Gamma_{2M}^t \quad \begin{array}{c} \Gamma_{1M}^t - \Gamma_{2M}^t < \Gamma_{12}^t \\ \Gamma_{12}^t < \Gamma_{1M}^t + \Gamma_{2M}^t \end{array} \quad \Gamma_{12}^t < \Gamma_{1M}^t - \Gamma_{2M}^t \quad \Gamma_{12} < \frac{\Gamma_{11} + \Gamma_{22}}{2}$$

Figure 2.17 All possible behaviors of tissue 1 driven by surface interactions from dispersal and stability within the medium (left) to the interaction of a drop of tissue 1 with a drop of tissue 2 (middle) and dispersal of tissue 1 enveloped by tissue 2 (right).

contact with cells in the bulk and half of it is in contact with the medium.[12] When a cell is transferred from the bulk to the surface, the tissue area increases by $\Delta A = A_0/2$ and the tissue energy changes by $\Delta E = -\Gamma_{11}A_0/4 + \Gamma_{1M}A_0/2$. In calculating ΔE, we have neglected all processes related to the relaxation of the cell area once the cell loses some of its neighbors, assuming that the total area of each cell is fixed. Thus the tissue surface tension reads

$$\Gamma_1^t = \frac{\Delta E}{\Delta A} = \Gamma_{1M} - \frac{\Gamma_{11}}{2}. \tag{2.55}$$

If $\Gamma_{11} > 0$ then $\Gamma_1^t < \Gamma_{1M}$, i.e., tissue surface tension is necessarily smaller than cell surface tension, which is, as confirmed by experimental data [37], not true. However, the cell-cell interfacial tension depends on intermembrane attraction and cadherin-cadherin bonds. If these are strong enough, work must be expended so as to decrease the area of cell surfaces in contact, and thus the tension would be indeed negative. If we denote the area density of cadherins and the binding energy of the cadherin pair by ρ_A and Δ, respectively, the interfacial tension can be written as

$$\Gamma_{11} = \Gamma_{1M} + \Delta\rho_A \tag{2.56}$$

and the tissue surface tension reads

$$\Gamma_1^t = \frac{\Gamma_{1M}}{2} - \frac{\Delta}{2}\rho_A; \tag{2.57}$$

note that $\Delta < 0$.

Measurements of surface tension of aggregates of L cells, that is, enteroendocrine cells found, e.g., in large intestine, indeed confirm the linear dependence of Γ_1^t on ρ_A [40]. The cells were modified such that the expression of cadherin levels was varied. The results shown in Figure 2.18 indicate that in these cells the surface tension Γ_{1M} is about 0.64 mN/m, which

---

[12]It may be instructive to calculate this energy relative to the (reference) state when the tissue is completely disintegrated, which would require subtracting a term $N\Gamma_{1M}$ from Eq. (2.54). However, this offset is not important for the derivation of tissue surface tension as the reference energy does not appear in it.

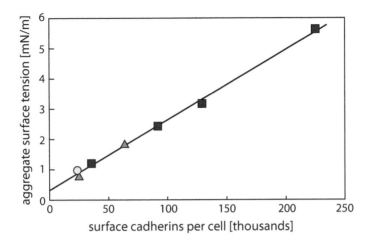

Figure 2.18 Dependence of aggregate surface tension on the number of surface cadherins per cell. Symbols represent the different types of cadherins expressed (N-cadherin: squares, P-cadherin: triangles, E-cadherin: circles). (Adapted from Ref. [40].)

is roughly of the right order of magnitude, and that the binding energy of the cadherin pair Δ is about 3×10^{-17} J. These estimates obtained by assuming the average cell area $A_0 \approx 700$ μm^2 suggest that the energy of a single cadherin bond is about 10000 times larger than the thermal energy. Measurements of cadherin bond energy suggest a much smaller value of about ~ 10 $k_B T$ [41], which is well within the realm of protein-protein interactions. The disagreement of the two values only deepens the problem of relating cell and tissue tensions, which remains an unsolved paradox [42].

However, note that Eq. (2.57) was derived by assuming that the cell-medium tension for cells on the surface of the tissue is the same as that of isolated cells, Γ_{1M}; this is the quantity measured in single-cell micropipette experiments. If there existed a mechanism which would increase the tension of cells on the tissue surface by about an order of magnitude, the paradox could be partly resolved, although the dependence of surface tension on cadherin concentration (Figure 2.18) would still be unexplained. As we have seen in Sections 2.2.1 and 2.2.2, cell surface tension can be modified by the activity of the cell cortex. There are indications that the cortex stiffens when the cell is on the tissue surface [42,43]—this effect is usually referred to as mechanical polarization at the tissue boundary and will be discussed in more detail in Section 3.3.5.

A possible explanation of cortex stiffening relies on the feedback between cadherins and cytoskeletal dynamics. As sketched in Figure 2.10, cadherin is connected to the cytoskeleton, and the engagement of cadherins in cell-cell interaction may act as a molecular signal which eventually results in the stiffening of the cytoskeleton in regions where cadherins in a given cell are either not expressed or not in contact with cadherins from neighboring cells, which happens in cells on the surface of the tissue [42, 44]. This scenario may possibly go far enough to explain the dependence of tissue surface tension on cadherin expression shown in Figure 2.18—the more cadherins in a cell on the tissue surface are engaged in cellular contacts, the more the free part of the cell membrane is stiffened due to the action of the cytoskeleton. In any case, the experimental insights show that simple models of cell mechanics must be used with care, as the molecular mechanisms at work may modify the behavior of the cells to the extent that requires a thorough reassessment of the concepts used, primarily surface tension.

Forty years after Steinberg's differential adhesion hypothesis [36], the role of surface tension in controlling cell shape and in remodeling of tissues is rather widely accepted [45], the tension itself arising from cell cortex tension and cell-cell adhesion. By emphasizing the importance of surface tension, we do not wish to say that this notion is suitable for the description of all microscopic processes involved in cell-cell interaction. Neither do we suggest that the other types of cell energies associated with, say, the cytoskeleton are not important. Yet it should not be too surprising that a model where all interactions associated with the membrane are packed into a simple surface term and all other cell features that determine cell shape except incompressibility are disregarded may still provide a good first-guess description. As shown in the following chapters, most models of tissues include some form of cell surface energy in addition to various other types of energies associated with specific structures such as the acto-myosin ring running around the perimeter of the apical side of epithelial cells.

PROBLEMS

2.1 A neutrophil cell is deformed from an initially spherical shape to a cylindrical pancake of a thickness equal to the diameter of an approximately spherical cell nucleus denoted by $2R_n$. Calculate the relative change in cell area during this deformation, assuming that cell volume is constant. Then use the fact that the maximal change of the area of a deformed neutrophil is about $(A - A_0)/A_0 \sim 100\%$ to estimate R_n.

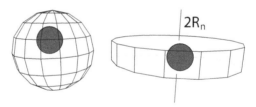

2.2 Derive Eq. (2.6). *Hint:* Evaluate the area and volume differentials, dA and dV, respectively, for a patch of surface of principal radii of curvatures R_1 and R_2, and use Eq. (2.2) to calculate the pressure difference.

2.3 Assume that cell volume is constant during micropipette aspiration and that the shape inside the micropipette can be modeled as a spherical cap of radius R_p attached to the cylinder of length $L_p - R_p$. How does the radius of the cell outside the micropipette change with L_p? Calculate the change in cell area during aspiration.

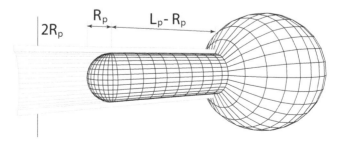

2.4 Show that Eq. (2.10) leads to the following expressions for cell surface energy:

$$
\begin{aligned}
E_s &= C_0 A_0 + C_1 \frac{A - A_0}{A_0} + C_2 \frac{(A - A_0)^2}{A_0} \\
&= D_0 A_0 + D_1 A + D_2 \frac{(A - A_0)^2}{A_0}.
\end{aligned}
\tag{2.58}
$$

Express C_0, C_1, and C_2, as well as D_0, D_1, and D_2 in terms of Γ_0, k, and A_0.

2.5 The response of a cell to external forcing can be measured in the diametral-compression apparatus schematically shown below.

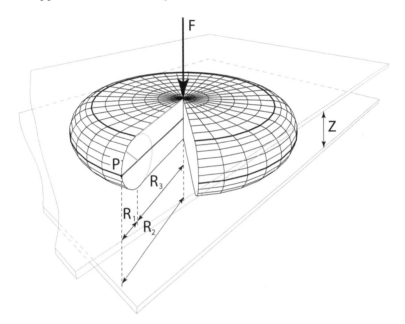

In the experiments reported in Ref. [46], the plates of the apparatus were small pieces of cover glass and the force F was produced by loading the top plate with small pieces of paraffin paper of known weight. When the force exerted by the top plate was $F = 1.6 \times 10^{-8}$ N, the thickness Z of the very nearly spherical sea urchin egg decreased from 95 μm to 60 μm. Use these data to calculate the effective surface tension Γ of the see urchin egg. *Hint*: The equilibrium of pressures on the flat part of the cell at the top plate gives the pressure difference $p_i - p_o = F/R_3^2 \pi$, where R_3 is the radius of the circular contact zone between the plate and the egg. Mechanical equilibrium in point P on the equator requires that

$$
\frac{F}{\pi R_3^2} = \Gamma \left(\frac{1}{R_1} + \frac{1}{R_2} \right)
\tag{2.59}
$$

in agreement with Eq. (2.6). If we approximate the deformed shape by a union of a cylinder and a piece of torus of a major radius R_3 and a minor radius $R_1 = Z/2$ (see below; $R_2 = R_3 + R_1$), we find that its volume reads

$$
V = \pi^2 R_3 R_1^2 + 2 R_1 R_3^2 + \frac{4}{3} R_1^3.
\tag{2.60}
$$

These results can be combined to calculate $\Gamma \approx 0.08$ mN/m. (Ref. [46] reports a value of Γ between 0.07 and 0.15 mN/m for eleven different eggs, obtained by a more accurate description of the shape of the deformed cell.)

2.6 Interpret the surface tension of the sea urchin egg from Problem 2.5 by representing the cell cortex as an incompressible elastic shell of thickness $h \sim 0.3$ μm. Calculate Young's modulus Y of the cortex. Calculate the relation between the pressing force F and plate separation Z, and compare it with the result obtained within the liquid-drop model.

2.7 Thompson proposed that the shape of the cell of *Spirogyra* algae can be approximated as a union of a cylinder and two spherical caps [3], inspired by shapes of oil drops in alcohol mixtures obtained experimentally by Plateau [47]. Thompson argued that the radius of the cap should be twice the radius of the cylinder. This proposition does not seem to be consistent with experiments—panel b of the figure shows the contour of the *Spirogyra* cell wall based on the micrograph in panel a (image courtesy of J. Nance), whereas the construction in panel c corresponds to Thompson's proposition. Explain the rationale behind this proposition in terms of the Young–Laplace law.

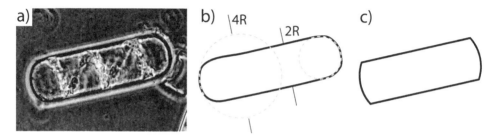

Spirogyra is rarely found as an isolated cell. As a rule, many cells attach so as to bury their caps in contacts, forming thus one-cell-thick filaments. The filaments can be centimeters long and only ~ 50 μm thick (image courtesy of M. Bačič).

2.8 Cells can be deformed by optical manipulation using lasers and gradients of light intensity to produce forces. In a typical experiment, a cell is stretched by a pair of effective forces F acting across the diameter of the equator. Calculate the extension ΔL of a spherical cell by assuming that it is a liquid drop of surface tension Γ. *Hint*: Assume that the shape of a deformed cell is either an ellipsoid or a spherocylinder [48]. Is the spherocylinder a self-consistent model shape?

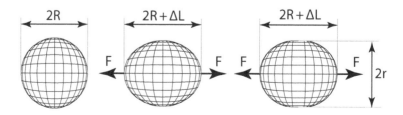

2.9 Examine the radii of curvatures of the bubbles shown in Figure 2.9 and show that the sides of the nearly dodecahedral cell in the center of the cluster are not planar. Prove that their radii are given as $R_c = 14.3R$, where R is the radius of curvature of the bubble surfaces exposed to the environment. *Hint:* Use the fact that the pressure in the central bubble is 7% larger than in other bubbles and assume that the ambient pressure is negligible compared to the pressure within the bubbles.

2.10 Examine the contact zone between the two cells in the two-cell mouse embryo shown in Figure 2.14a and estimate the ratio of cell-cell interfacial tension and cell-medium surface tension.

2.11 Shown below are two possible topologies of a three-cell cluster. In the linear cluster, the cells at the tips only have a single neighbor whereas the middle cell has two neighbors. In the triangular cluster, each cell has two neighbors.

Draw all possible topologies of clusters containing four and five cells.

2.12 A thin, initially straight rod is bent so that its radius of curvature is R. Show that its elastic energy reads

$$E_{el} = \frac{YIL}{2R^2}, \qquad (2.61)$$

where $I = \int_A y^2 \mathrm{d}A$ is the area moment of inertia, A is the area of the rod cross-section, and L is rod length. Show that the cross-section of the rod is deformed when the rod is bent. *Hint:* Bending causes both stretching and compression in different parts of the rod. Imagine a filament running lengthwise through the rod. After deformation, the length of the filament whose distance from the neutral plane[13] is y is $L(y) = (1 + y/R)L$. This gives $\epsilon_{xx} = y/R$ and $\sigma_{xx} = Y\epsilon_{xx}$. Use this in Eq. (2.31) to calculate the bending energy.

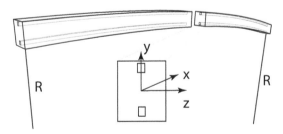

Now employ the above result to show that the elastic energy of a thin rod bent into an arbitrary curve Γ is

$$E_{el} = \frac{D}{2} \int_\Gamma C^2(l)\mathrm{d}l, \qquad (2.62)$$

where l is the natural parameter along the curve Γ, $C(l) = 1/R(l)$ is the (one-dimensional) curvature at l, and $D = YI$ is the bending or flexural rigidity. *Hint:* Divide formally the curve Γ into a sequence of circular arcs and sum their elastic energies to obtain the total energy of the rod.

---

[13]The neutral plane is the plane within the rod which is neither stretched nor compressed upon bending.

2.13 If the deformation of the rod of length L can be described by $y = y(x)$, the curvature is given by

$$C(x) = \frac{d^2y/dx^2}{\left[1 + (dy/dx)^2\right]^{3/2}},\tag{2.63}$$

which can be approximated by $C(x) \approx d^2y/dx^2$ if $dy/dx \ll 1$. The energy of the bent rod is

$$E_{el} = \frac{D}{2} \int_0^L \left(\frac{d^2y}{dx^2}\right)^2 dx.\tag{2.64}$$

Consider the variation of energy and show that the equilibrium condition $\delta E_{el}/\delta y = 0$ leads to [49]

$$\frac{d^4y}{dx^4} = 0.\tag{2.65}$$

Now solve this equation for a rod of length L clamped at $x = 0$ so that $y(0) = 0$ and $dy/dx|_{x=0} = 0$. The boundary conditions at the free end of the rod read $y(L) = y_0$ and $d^2y/dx^2|_{x=L} = 0$. Show that in this case

$$y(x) = \frac{y_0}{2L^3}x^2(3L - x).\tag{2.66}$$

Calculate the energy of the rod bent in a curve described by Eq. (2.66) to find that here

$$E_{el} = \frac{3YIy_0^2}{2L^3}.\tag{2.67}$$

The deflected rod can be thus viewed as a spring with effective displacement y_0. Show that the effective force constant for the circular rod of radius r is

$$k = \frac{3\pi Y r^4}{4L^3}.\tag{2.68}$$

2.14 Calculate the increase of surface area of an incompressible cube-like cell upon uniaxial stretching and show that it is quadratic in extension ΔL for small deformations. Also consider large deformations, examining the deviations from the harmonic behavior.

CHAPTER 3

In-plane structure of single-layer tissues

T HE STANDARD CLASSIFICATION OF TISSUES is based on their structure and function and includes connective, muscle, nervous, epithelial, and mineralized tissues. Connective tissue consists of cells dispersed within an inanimate extracellular matrix, and it sculpts the organs. The extreme examples of connective tissue containing a fluid and a solid matrix are blood and bone, respectively, the latter also illustrating mineralized tissues which incorporate an inorganic component. Muscle tissue generates physical force and drives motion whereas nervous tissue is responsible for the reception and transmission of impulses which control the activity of the different parts of the body. Epithelial tissues are sheet-like and they cover the external and internal surface of the different organs, forming the outermost part of skin and lining the cavities of the digestive tract (gut, esophagus), lungs (alveoli), bronchi, urethra, bladder, blood and lymphatic vessels, etc. The principal function of the epithelia is to protect the underlying tissue but most epithelia are also involved in exchange of matter with the environment, that is, in absorption and secretion.

Single-cell-thick epithelia are known as simple whereas those consisting of two or more layers are referred to as stratified. A further division of epithelial tissues is based on the shape of cells expressed in terms of cell diameter d and height h. Those formed by flattened cells with $h < d$ are squamous whereas in cuboidal epithelia, cells are roughly isometric and their diameter is comparable to height $h \approx d$ (Figure 3.1). Columnar epithelia consist of tall cells with $h > d$. The stratified squamous epithelium contains several layers of flattened cells typically supported by cuboidal cells, and in the pseudostratified epithelium the nuclei of wedge-shaped tall cells are located either at the top or at the bottom of the cell so that by observing the nuclei alone (which are often easily visible using suitable stain) the transverse cross-section looks like a bilayer. The transitional epithelium can readily stretch or expand in response to a change of area or volume of the organ, and thus the shape of cells in its superficial layer depends on the local deformation of the tissue. The superficial cells rest on one or two strata of proliferating cells which ensure tissue renewal.

Many epithelial cells are prismatic in shape and their sides are functionally distinct. The apical side faces the environment or the lumen, that is, the cavity of a structure or an organ such as blood vessel or intestine. The lateral sides of cells are closely attached to each other so as to ensure the integrity of the tissue, and the basal sides are attached to the basement membrane, a thin matrix of protein fibers (collagen, fibrillins, laminins, etc.), underneath the epithelium. Below the basement membrane is a layer of connective tissue called stroma which provides a structural support. Such a configuration allows for a rearrangement of cells within the epithelium so that cells in it may flow when exposed to an in-plane stress;

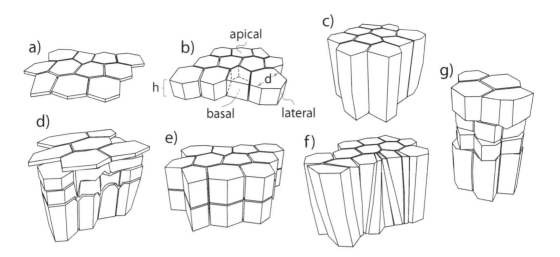

Figure 3.1   Types of epithelial tissues: simple squamous (a), simple cuboidal (b), simple columnar (c), stratified squamous (d), stratified cuboidal (e), pseudostratified columnar (f), and transitional (g). Cell diameter and height are indicated by d and h, respectively.

in some epithelia this behavior is prominent whereas in others it is not. At the same time, epithelial cells are anchored to the basement membrane so that the tissue itself behaves as a solid sheet for out-of-plane deformations such as bending. It is easy to imagine that the mechanical response of the epithelium to both external forces and forces generated within it should depend on the in-plane structure of the tissue, and this is one of the reasons why this particular aspect of organization of the epithelium is interesting. Secondly, the in-plane arrangement of cells provides an insight into the possible equilibrium and dynamic cell-level mechanical processes at work, typically by comparing experimental observations with the results of a theory based on suitable assumptions. Finally, the in-plane structure of simple epithelia is readily visible on the apical side of the tissue which faces the lumen. It is quite natural that the first quantitative studies in the field, say those by Lewis [50,51], were in part motivated by the sheer availability of the data—and by the patterns formed by cells, some of which can be immediately likened to tilings and tessellations known from mathematics, architecture, and art.

3.1   GEOMETRY OF TILINGS

The geometry of a simple epithelial tissue can be approximated by a polygonal tiling of a plane. This is possible because the cells in the tissue are prismatic[1] so that it suffices to examine their arrangement in the apical surface—all other cross-sections of the layer are quite similar. This approximation rests on a few assumptions, the first one being that cell-cell contacts are perfectly straight. As can be seen from the many micrographs of tissues included in this chapter, this is not exactly true but the curvature of the cell-cell contacts is usually small. Replacing the contacts by straight lines seems reasonable and practical, because it allows one to describe cell size and shape using the coordinates of the vertices alone. Here we tacitly assume that the membrane curvature of interest is defined at the cell

---

[1]Except for the cell nuclei in the squamous tissue, which protrude into the lumen and make each cell look like a fried egg.

length scale so that any radius of curvature is of the order of micrometers. The membrane is certainly not straight at a molecular scale where it depends on the shape of the different components—lipids, carbohydrates, and membrane proteins—and on their thermal motion within the membrane. Furthermore, one may expect changes in membrane curvature around any inclusions such as protein channels. In the coarse-grained view pursued here, these effects are not considered.

Bending vs. compression of plates

The elastic energy of a flat plate of thickness h and area A_0 bent so as to form a cylindrical surface of radius R reads

$$E_b = \frac{Yh^3 A_0}{24(1 - \nu^2)} \frac{1}{R^2}, \qquad (3.1)$$

so that

$$D = \frac{Yh^3}{12(1 - \nu^2)} \qquad (3.2)$$

represents the bending or flexural rigidity of the plate. The elastic energy of the plate compressed so that its area decreases from A_0 to A such that it fits, e.g., between the same boundaries as the bent plate is

$$E_s = \frac{YhA_0}{4(1 - \nu)} \left(1 - \frac{A}{A_0}\right)^2. \qquad (3.3)$$

Here $Yh/[2(1 - \nu)]$ is the area expansion modulus of the plate introduced in Eq. (2.23)]. We see that $E_s \propto h$ whereas $E_b \propto h^3$. These scaling laws show that in very thin plates where h is much smaller than the lateral dimension and the radius of curvature, the energy of a bent plate is smaller than the energy of a compressed plate at given boundary conditions. (More precisely, this holds beyond the buckling threshold.) This is the reason why a sheet of paper lying on the table readily buckles when the opposite edges are pushed against each other.

The second assumption involved is that the distance between membranes that adhere to each other is negligible. Indeed, the membranes are typically separated by no more than a few 10 nm which is much smaller than cell size. The third assumption, which allows us to consider the tissue as a tiling, disregards any holes between the membranes of adjacent cells, say those at vertices where three or more cells meet. At a vertex, the vector normal to the polygon edge changes its orientation discontinuously whereas a cell membrane, which is a physical material, cannot be bent over an infinitesimally short distance. To show that the tiling representation of the tissue is still meaningful, we compare the length over which a membrane bends at a vertex to cell size. This length depends on the interplay between cell-cell adhesion, which is characterized by adhesion strength Γ expressed in units of energy per unit area, and membrane bending rigidity D. Cell-cell adhesion is maximized by increasing cell-cell contacts as much as possible, that is, by vanishingly small non-contact zones and thus vanishingly small radii of curvature at the vertices. On the other hand, the bending energy favors large radii. The equilibrium radius of curvature of the membrane at a vertex R_v is determined by minimizing the sum of adhesion and bending energy and reads

$$R_v \sim \sqrt{\frac{D}{\Gamma}}. \qquad (3.4)$$

On inserting the values of D and Γ typical for animal cells ($D \sim 10^{-19}$ J, $\Gamma \sim 10^{-5}$ J/m^2 [52])

we find that $R_v \sim 0.1$ μm. This is a hundred times smaller than the typical cell size of ~ 10 μm, so that the area that is unoccupied by cells is ~ 0.01 % of the total area of the tissue in the en-face view.

After having established that the polygonal-tiling representation of a simple epithelium is reasonable and useful, let us introduce a few basic notions needed to describe the tilings in general. Given that a large majority of cell cross-sections seen in epithelia are convex polygons, we restrict the discussion to convex tilings. The simplest polygonal tilings are periodic tessellations based on a single regular polygon (a triangle, a square as in Figure 3.2a, or a hexagon). In such a tiling, all tiles must have the same size. Size monodispersity is, of course, not required as illustrated by the truncated square tiling with squares of two sizes and by the random rectangle tiling in Figure 3.2b and c, respectively. These two examples also illustrate that a tiling need not be edge-to-edge.

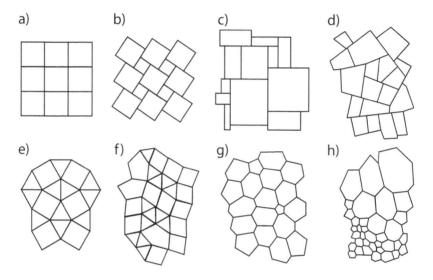

Figure 3.2 Polygonal tilings: simple square tiling (a), non-edge-to-edge truncated square tiling (b), random non-edge-to-edge tiling with rectangles of different shapes and sizes (c), tiling based on fractures on the surface of ceramics (d), snub-square tiling (e), quasi-crystalline polygonal tiling (f), tiling constructed from a photograph of a two-dimensional foam (g), and tiling based on the cell arrangement in flax stem (h).

Figure 3.2e shows a periodic tiling consisting of squares and triangles, which is characterized by translational periodicity, whereas the tiling in Figure 3.2f only possesses bond-orientational order pertaining to the preferred in-plane orientation of the edges. Many tilings typical for animal epithelia are similar to those seen in two-dimensional foams (Figure 3.2g) and contain polygons of different sizes and classes, the latter referring to triangles, quadrilaterals, pentagons, etc. Similar structures are also seen in some plant tissues, e.g., in the flax stem in Figure 3.2h [53]. Some epithelial tissues consist of more than one type of cells. In such a case, the different cell types are often characterized by a particular polygon size and class, which introduces a further dimension in the description of the tissue.

Polygonal tilings consist of faces which share edges, and edges meet at vertices. In the non-edge-to-edge tilings in Figure 3.2b, c, and d, some of the tiles' edges are broken into two by a T vertex, which formally increases the number of vertices in a tile—for example, the large squares in Figure 3.2b are technically octagons. The valence (or the degree) of a vertex is defined by the number of edges radiating from it. The valence of all vertices in Figure 3.2a is four whereas the vertices in Figure 3.2b and e are three- and five-valent, respectively. In most polygonal tilings representing epithelia, vertex valence is almost always three. Four-valent vertices are quite rare and even those that one does see in an experiment are often only apparent, consisting of two unresolved nearby three-valent vertices.[2]

Regular, semiregular, and k-uniform tilings

Polygonal tilings consisting of a single type of polygons are called regular. Tilings which have two or more types of polygons but a single type of vertex are referred to as semiregular, Archimedean, or uniform tilings. k-uniform tilings have k different types of vertices and two or more types of polygonal faces. The vertex figure or vertex configuration specifies the number of sides of faces meeting at a given vertex and the order of the faces. The tiling in Figure 3.2e is Archimedean, featuring only one type of vertex with vertex figure 3.3.4.3.4 or for short 3^2.4.3.4. The tiling on the left is semiregular (3.6.3.6, with two types of polygons) and so is that in the middle (4.6.12, with three types of polygons) whereas the tiling on the right is 2-uniform (3.4.6.4 and 3^2.4.3.4, with three types of polygons). The tiling on the left is known as the trihexagonal tiling.

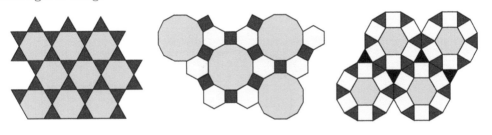

The examples in Figure 3.2 show that tilings must be characterized by the polygon types as well as by polygon sizes. In the tilings that represent epithelia, polygon sizes are generally polydisperse but polydispersity is typically moderate because of physical restrictions. A cell cannot be smaller than, say, the nucleus, nor it can be very big for various reasons. For example, cells in a proliferating tissue grow in size and once their size becomes too large, division is triggered. We note in passing that such restrictions do not exist in a soap froth where the distribution of bubble sizes is much broader (Figure 3.3).

In some tilings, a certain type of regularity such as translational order or bond-orientational order can be mathematically defined and measured. Most tilings seen in epithelia are random, that is, not characterized by an obvious kind of spatial order. Yet they do have a certain degree of regularity. In many cases one may easily conclude, solely on the basis of visual inspection, that a particular tissue is more regular than another—be it that the distribution of polygon classes is more narrow or that the long axes of cells are locally more aligned.

Many aspects of order in random tilings seen in epithelia are universal rather than specific [55]. All polygonal tilings need to tile the plane without gaps, and this imposes a strong topological constraint leading to severe restrictions on the tilings that one could conceive. Try drawing, for example, a tiling consisting only of decagons of about the same size and about the same proportions. You will soon find that the gaps between the decagons

---

[2]At the same time, the impact of the apparent four-valent vertices on the structure of the tissue can be considerable and can lead to a bias in, e.g., the measured distribution of polygon classes [54].

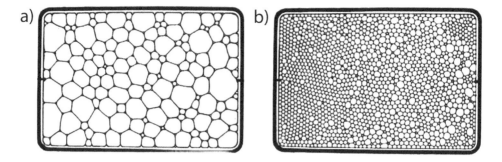

Figure 3.3 Two samples of soap froth trapped between parallel glass plates; in panel a, the largest bubbles have an area about two orders of magnitude larger than that of the smallest bubbles whereas in panel b the variation is smaller. (Images courtesy of W. Drenckhan-Andreatta.)

must be filled by polygons with fewer edges that are smaller than the decagons. The need for polydispersity is even more emphasized in the Apollonian gasket, a fractal structure formed by tangent circles (Figure 3.4).

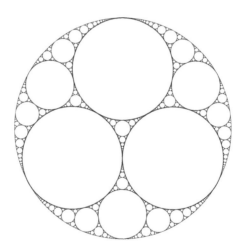

Figure 3.4 The Apollonian gasket consists of tangent circles of ever smaller diameters.

The basic topological constraint imposed on the tilings can be formulated through a relation that the vertices, edges, and faces of the tiling must fulfill. Any closed, non-intersecting path formed of edges of the tiling encloses a certain number of faces (F), vertices (V), and edges (E). These numbers are related by the Euler formula

$$F - E + V = 1. \tag{3.5}$$

The formula holds both for small portions of the tiling including a single polygon where $F = 1$ and $E = V$ as well as for very large portions of the tiling where F, E, and $V \gg 1$. For example, for the piece of tiling shown in Figure 3.2e, $F = 17$, $E = 35$, and $V = 19$, so that $F - E + V = 1$, in agreement with Eq. (3.5).

Consider excising a finite patch of a large tiling with three-valent vertices by choosing a closed path along its edges. Denote the numbers of border edges and vertices on this path by E_p and V_p, respectively. Naturally, $E_p = V_p$. Not all border vertices can be three-valent because some of them lost an edge when we excised the patch and became two-valent (Figure 3.5); let $V_{p,2}$ stand for the number of two-valent vertices. The enclosed patch of the tiling consists of polygons of different number of sides n and if the number of n-gons in the tiling is denoted by F_n, the total number of faces is $F = \sum_n F_n$. In an infinite tiling, two faces meet at each edge but in a finite tiling one needs to pay attention to the border edges. In this case, the relation between the number of faces and edges is

$$F = 2E - E_p, \tag{3.6}$$

which follows from the observation that the border edges belong to a single face. Since each edge connects two vertices, V and E of an infinite three-valent tiling are related by $3V = 2E$. In a finite patch, the left-hand side of this relation is decreased because some vertices are two-valent and

$$3V - V_{p,2} = 2E. \tag{3.7}$$

By combining Eqs. (3.6) and (3.7) with the Euler formula [Eq. (3.5)], one obtains

$$\sum_n F_n(6 - n) = 6 + (E_p - 2V_{p,2}). \tag{3.8}$$

The number of two-valent vertices at the border $V_{p,2}$ and its relation to the number of border edges E_p depends on the construction of the border path itself. To further simplify Eq. (3.8), consider a specific type of border [56] constructed by drawing a circle across the tiling so as to excise the patch (thick gray line in Figure 3.5). The circle will necessarily cut through the edges of the faces and by adjusting its radius, one can ensure that it cuts exactly through two edges of each polygon at the border. The two intersection points (circles in Figure 3.5) now become two new vertices of the patch, and a third new vertex (each of the squares in Figure 3.5) is positioned in between them. All in-between vertices are two-valent, and by connecting all newly obtained vertices we construct the border path denoted by the dashed line in Figure 3.5. In the thus-obtained border, the number of two-valent vertices is equal to the number of three-valent vertices so that

$$E_p = 2V_{p,2}. \tag{3.9}$$

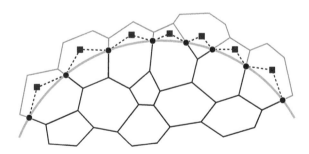

Figure 3.5 Construction of the border path (dashed line), based on the intersection of a circle (thick gray line) and a tiling. New vertices obtained by the construction are denoted by circles (three-valent) and squares (two-valent).

Measures of polygon shape

A common method and intuitive parameter used to quantify polygon shape is the aspect ratio, which may be defined as the ratio of the length L and width W of the rectangle with the smallest area that contains the polygon. Another often-used approach is to fit the polygon by an ellipse and then compute the aspect ratio L/W based on the long and the short semiaxis of the ellipse. For some purposes, elongation defined by $1 - W/L$ can be more suitable than the aspect ratio. In nearly isometric polygons, elongation is close to 0 whereas in very elongated polygons it approaches 1.

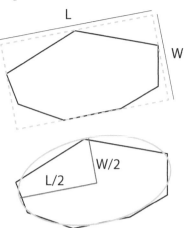

The aspect ratio depends on the linear dimensions of the polygon. In many cases, measures of shape based on integrated quantities such as area A and perimeter P prove to be more robust. Polygon reduced area is defined by $a = 4\pi A/P^2$ and is normalized such that it is equal to 1 in the circle and less than 1 in any other shape. The terminology employed is hardly unique. In some contexts, a is referred to as circularity, shape index, shape factor, or isoperimetric ratio. Moreover, "shape factor" is also used for two related quantities $P^2/(4\pi A) = 1/a$ or $P/\sqrt{4\pi A} - 1 = 1/\sqrt{a} - 1$ and in some cases, "shape index" stands for $\pi A/(2d^2)$, d being cell diameter. Yet another occasionally used measure related to a is elongation $P/\sqrt{4\pi A} = 1/\sqrt{a}$.

After inserting this result in Eq. (3.8), we find that the number of faces contained within the border of the patch now satisfy the relation

$$\sum_n F_n(6 - n) = 6. \tag{3.10}$$

For example, in a tiling made of F_5 pentagons and F_6 hexagons, the number of pentagons must be $F_5 = 6$ whereas the number of hexagons can be arbitrarily large. This means that in a sufficiently large tiling of hexagons and pentagons only, the fraction of pentagons vanishes and the six pentagons that must be there are introduced solely by the construction of the border. If now heptagons are inserted in the pentagon-and-hexagon-only tiling, their number must be equal to

$$F_7 = F_5 - 6. \tag{3.11}$$

In very large tilings where F_5, F_6, and $F_7 \gg 1$, the fraction of heptagons is equal to the fraction of pentagons. For very large tilings, $F_n \gg 1$ for all n and

$$\sum_n (n - 6)p_n = 0, \tag{3.12}$$

where

$$p_n = \frac{F_n}{F} \tag{3.13}$$

is the fraction of polygons with n sides.

This discussion illustrates that topology imposes severe constraints on the structure of the tilings. However, the remaining degrees of freedom are controlled by the physical interactions and processes which give rise to a particular type of in-plane order, average cell shape, or growth pattern of the tissue.

3.2   RANDOM TILINGS

In tilings seen in epithelia, the different types of regularity such as orientational and positional order of cells or their elongation are often associated to each other. For example, in-plane cell polarization may naturally give rise to, or result from, a preferred direction in the tissue. Some of the patterns typical for epithelia are illustrated by four examples seen in *Drosophila*. Panels a and b in Figure 3.6 show two *Drosophila* pupa wing epithelia, one consisting of rather isotropic cells and the other being polarized as witnessed by the easily identifiable long axis of the cells. These examples are from the *pk* mutant and the wild-type animal, respectively, but are not representative of either the mutant or the wild-type animal. On the other hand, the germband of the *khft* mutant consists of pentagonal, hexagonal, and heptagonal cells (Figure 3.6c) and is thus topologically less regular than the examples of the wing epithelium in panels a and b. In terms of cell size, however, it shows a far smaller variability than the wing epithelium of the *shi* mutant (Figure 3.6d).

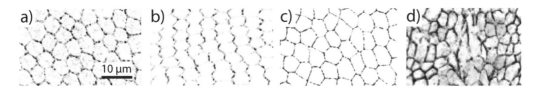

Figure 3.6   Isometric (a) and polarized cells (b) of *Drosophila* wing epithelium during pupation (*pk* mutant and wild-type animal, respectively; adapted from Ref. [57]). The germband of the embryonic epithelium in the *khft* mutant (c) contains quite a few pentagons and heptagons in addition to hexagons (adapted from Ref. [58]), and the *shi* mutant pupal wing (d) is rather disordered both topologically and in terms of cell size (adapted from Ref. [59]). The original micrographs are color inverted.

When quantifying the structure of a tiling, it is natural to first think of its topology and count the number of triangles, quadrilaterals, pentagons, etc. and then divide each of them by the number of all polygons. Thus one obtains the fractions or frequencies of all polygon classes and the vector of these fractions p_n, where n is the number of sides of a given class, is the simplest fingerprint of a tiling. In the common positionally ordered tilings, p_n is non-zero only for a few n simply because they consist of a small number of tile types. For example, in the snub-square tiling in Figure 3.2e $p_3 = 2/3$, $p_4 = 1/3$, and all other p_n are zero. On the other hand, a random tiling admits tiles of all classes. If no particular polygon class is preferred over others, it is reasonable to assume that the distribution of tiles across the classes is smooth such that the fraction of n-gons is in between the fractions of $(n + 1)$- and $(n − 1)$-gons; this is obviously not fulfilled in the most numerous polygon class. Most tilings seen in epithelia are characterized by a monomodal, bell-like distribution of polygon classes. They rarely contain tiles with n greater than 9 and since the smallest possible n is 3, the vector of polygon fractions typically has 7 components. In addition, three-valent vertices are usually far more frequent than four-valent vertices. If the latter can be neglected, the tiling must obey Eq. (3.12).

These features define the distribution of polygon classes rather well even before any specific mechanics of the tissues is included. They are also present in the hexagon-only tiling, which can be viewed as a limiting case of a random tiling where the Euler formula is satisfied trivially. The simplest non-trivial random tiling is obtained by admitting pentagons and heptagons—Eq. (3.12) requires that their fractions must be the same: $p_5 = p_7$. As a result, $p_6 = 1 − 2p_5$. Such a tiling is parametrized by a single variable, say p_6. Any asymmetry of the distribution is a signal that some of the vertices must be four-valent. Random tilings with the fraction of hexagons below about 0.5 usually also contain some quadrilaterals and octagons, which make these tilings more disordered. The distribution typically exhibits a positive skewness which stems from the fact that its left small-n tail terminates at triangles, i.e., at $n = 3$, whereas the right-hand tail is unbounded.

Polygon fractions

Polygon fractions are an intuitive and easily measurable quantity but do not always tell much about the tiling in question. For example, the three tilings shown below all have $p_3 = 2/3$, $p_4 = 1/3$, and $p_{n>4} = 0$ and yet the spatial arrangements of squares and triangles in them are different.

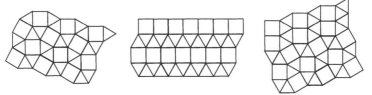

There also exist many different tilings consisting exclusively of pentagons. Here too the positional order is not the same. However, random tilings are devoid of positional order and in practice two random tilings with the same vectors of polygon fractions are rarely as dissimilar as these examples.

In real tissues it is sometimes difficult to count the number of vertices of a cell. This typically happens when one of the sides is very short, i.e., when two barely separated three-valent vertices are mistaken for one four-valent vertex.

The distributions of fractions of polygon classes are nicely exemplified by the *Drosophila* embryonic epithelium during the so-called germband extension when the tissue becomes progressively more disordered. Figure 3.7a shows the images of the germband in the wild-type embryo at stages 6 and 8 as well as the corresponding fractions of polygon classes; also included are the late stage 8 fractions [58]. In stages 6 and 8 the distributions are fairly symmetric about $n = 6$ but in the late stage 8 the skewness is readily visible.

The *Drosophila* embryo germband is an example of a tissue which becomes progressively more and more disordered. The epithelium of the *Drosophila* wing develops in the opposite direction [59]. During the prepupal stage it contains mainly pentagons and hexagons (Figure 3.7b) but in the pupal stages the number of hexagons is gradually increased, reaching about 80% in stage TP5. This case is also illustrative for another reason: it is evident that none of the distributions in Figure 3.7b satisfies Eq. (3.12), which means that this epithelium contains some four-way vertices [59].

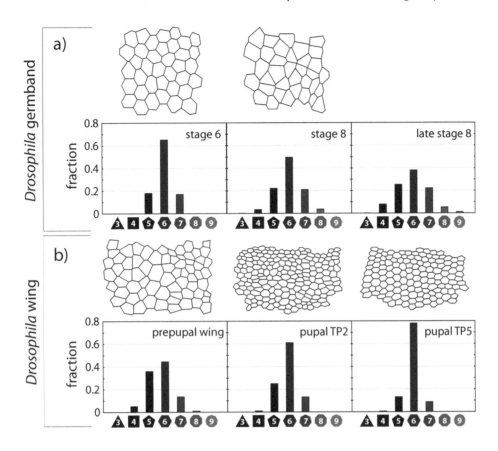

Figure 3.7 Snapshots of stage-6 and stage-8 wild-type *Drosophila* embryonic epithelium and the distribution of polygon fractions in these two stages as well as in the later stage 8 (a; adapted from Ref. [58]). *Drosophila* wing epithelium during prepupal and pupal development in stages TP2 and TP5 (b; adapted from Ref. [59]).

So far the discussion of random tilings has not gone beyond the fractions of polygon classes. Of course, polygon fractions are not the only quantity of interest and they do not uniquely determine the structure of the tiling. But we note that other parameters such as the distribution of polygon areas are correlated with the distribution of polygon fractions. To appreciate this, think of a tiling consisting exclusively of hexagons. In absence of a specific mechanism that would promote hexagons and suppress any other types of polygons, edges of tiles must be of approximately the same length or else topological transformations, typically driven by the tension within the edges, will transform some of the tiles into pentagons and heptagons. Thus it seems that the hexagons must be of approximately the same area simply because otherwise they alone would not tile the plane.

These correlations are related to the two basic constraints in any tiling: the no-gap/no-overlap rule and the edge-to-edge packing rule. From the statistical-mechanical perspective, these rules and the ensuing correlations between the shape and size of individual polygons as well as between the distributions of polygon fractions, their areas, and any other physical parameters pose a considerable challenge because they blur the very identity of the independent degrees of freedom that characterize the tiling. Each of the models which address this issue is a simplification of some kind, and its validity is not easy to prove or disprove.

Topological transformations in tilings

In every network, any topological rearrangement of tiles can be decomposed into two elementary processes referred to as T1 and T2 transformations, which represent neighbor exchange and tile disappearance, respectively. In the first process, an edge is flipped such that the tiles sharing this edge are no longer neighbors; instead, the two tiles tethered to each other by the edge become neighbors. The number of sides of the four tiles is changed—the sides sharing the edge prior to T1 lose 1 vertex each and the other two have 1 vertex more than before the transformation.

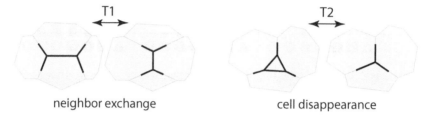

neighbor exchange   cell disappearance

In the T2 transformation, a triangular tile is deleted and replaced by a three-valent vertex. Thus the number of vertices of all neighbors of the triangular tile is decreased by 1. In the inverse transformation a triangular tile is created at a vertex.

Figure 3.8 Giant's Causeway is a coastline rock formation of about 40000 basalt columns in Northern Ireland (panel a, photograph by el ui). Apart from the shadows due to unequal column height, the aerial view (panel b, copyright Above All Images Ltd.) is quite similar to tilings seen in tissues.

At the same time, recognizing the role of the geometrical constraints is important because they apply at all length scales and energy scales, often giving rise to striking similarities between patterns in systems as different as an animal or a plant epithelium and the basalt columns in Giant's Causeway (Figure 3.8). As a result, the physical mechanisms at work operate in a rather restricted phase space, and it is possible that the impact of a given process on the pattern in question may be rather small or that two or more control parameters together determine the structure of the tissue—rather than each of them acting independently of the others.

Before turning to the theoretical understanding of the in-plane patterns observed in epithelia, let us describe the main empirical facts and briefly touch upon the general mathematical considerations applying to all random tilings.

3.2.1 Lewis' law

Historically the first and perhaps the best-known law that appears to apply to a large range of cellular partitions relates the average cell area to the number of vertices. This empirical relationship was discovered in 1928 by F. T. Lewis who studied the cucumber epidermis *Cucumis sp.* [51]. He measured the cross-sectional area of cells in this simple columnar epithelium as well as the number of their vertices n (Figure 3.9), finding that on the average area $\overline{A}_n$ is a linear function of n. Specifically,

$$\overline{A}_n = kA_0(n - 2), \tag{3.14}$$

where k is a constant and A_0 is the overall average cell area. In a more general formulation, $\overline{A}_n = kA_0(n - n_0)$ but many sets of data seem to be consistent with $n_0 = 2$ [60] which

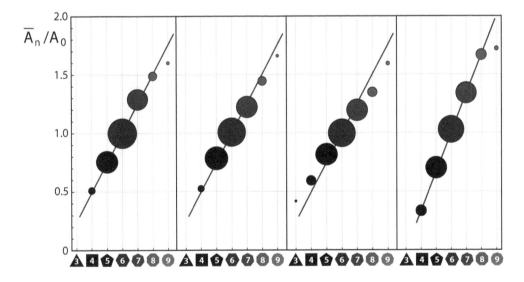

Figure 3.9 Relative average cross-sectional area of n-sided cells, $\overline{A}_n/A_0$, in various tissues vs. polygon class n. The data show two cucumber epidermis and human amnion samples [60] and a tangential section of cork [61] (from left to right). Note that the latter case is better fitted by $\overline{A}_n = 0.33A_0(n - 3)$ rather than by $\overline{A}_n = 0.25A_0(n-2)$ like the other samples. The bubble area is proportional to the fraction of a given polygon class.

is plausible since the smallest number of sides a polygon can have is 3. Experimental data suggest that k is often about 0.25 (Figure 3.9).

Lewis performed his experiments on 1000 cells so that the statistics was rather good. The polygon classes that he observed included quadrilaterals, pentagons, hexagons, heptagons, octagons, and nonagons; he found no triangles or decagons. Like in other epithelial mosaics, the distribution of fractions of polygon classes in his samples was bell-like, peaking at $n = 6$ and falling off rather rapidly toward either end. The numbers of quadrilaterals and nonagons seen were 25 and 4, respectively, both much smaller than the number of hexagons (415). This means that the $n = 4$ and the $n = 9$ datapoints are statistically less reliable. In a more technical language, the confidence intervals of these datapoints are considerably larger than for other n. Yet the linear relation found was quite remarkable.

Since its discovery, Lewis' law has been reported in many other plant and animal tissues including cork [61] (about 8000 cells), *Agave attenuata*, *Aloe arborescens* and *cf. Anthurium* [62] (between 500 and 1500 cells), and larval *Drosophila* wing disc epithelium [63, 64], and it was also seen in non-biological cellular structures such as artificial emulsions. Lewis' law also holds in proliferating tissues [63].

Appealing as it may be for its simplicity and fairly general applicability, Lewis' law is only valid on average rather than for each cell from a given class. If one bothers to look at the raw data prior to averaging (and one should) it is easy to see that the distribution of the cell area in a given polygon class is typically fairly broad. In addition, the raw data also show that the statistics is usually good in pentagons, hexagons, and heptagons but much less impressive in quadrilaterals and octagons; triangles and nonagons are only seen very rarely even in samples containing a few thousand cells. A telling example of a distribution of cells in the (area, polygon class) plane is shown in Figure 3.10, the data representing the fractions of about 8000 cells in the tangential section of cork [61].

These data nicely illustrate all of the above despite the broad 300 μm^2 area bins. It is clear that the correlation between polygon size and number of vertices needs to be interpreted carefully. For example, the distribution of hexagons includes polygons from the

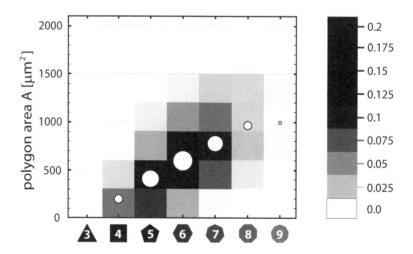

Figure 3.10 Distribution of cells in a tangential section of cork [61] vs. polygon class n and cell cross-section area A. Bubbles represent the average polygon areas in a given class, their areas proportional to the fraction of the class. The grayscale is proportional to the square root of fractions so as to emphasize the smaller fractions.

smallest 150 μm^2 bin as well as from the 1350 μm^2 bin so that the area variations within this polygon class are quite big. In addition, most hexagon areas are in the 450 μm^2 bin indicating that the distribution is skewed rather than symmetric. Naturally, this is expected since the cell area must be positive.[3]

Note that among the six polygon classes seen, quadrilaterals, octagons, and nonagons are rather rare, amounting together to less than 14% of all polygons. The centers of bubbles in Figure 3.10 represent the average polygon area for all n and this relation is indeed linear, the only polygon class not falling on the line being nonagons. Another important piece of information shown is the fraction of each polygon class which is represented by the bubble area—in this sample, the fractions of pentagons, hexagons, and heptagons are 0.249, 0.378, and 0.239, respectively. Finally, notice that these data depart from Eq. (3.14) in that the average polygon area would vanish at $n = 3$ rather than at $n = 2$. In turn, this is consistent with the absence of triangles in this sample.

In this analysis of the size-topology correlation, the number of datapoints is inevitably rather restricted because in any real sample, n typically varies from 3 to 9 so that the number of polygon classes is 7 (or 8, if decagons too are present); in addition, the $n = 5, 6$, and 7 datapoints carry a much bigger weight than the others. An alternative representation is to plot $\bar{n}$ as a function of A [66]. Because cell area is a continuous variable, the range of areas covered can be binned into narrow intervals producing many more datapoints needed to test the linear Lewis' law against alternative hypotheses.

With the provisos mentioned above, Lewis' law is seen rather often in cellular partitions of the plane but there exist several exceptions to it, the most notable and well-studied one probably being the soap froth. Its microscopic mechanics is based on the surface energy and well-understood. This facilitates both the analysis of experimental data and theoretical studies. Of course, a soap froth is three-dimensional and thus by itself not a direct analog of epithelial tissues. Two-dimensional froths can be produced by trapping bubbles between two plates [67] or at the surface of the foam [68]. In both cases, the experimentally observed geometry of the local boundaries is consistent with the Plateau rules suggesting that they are representative of the true two-dimensional foam. The structure of the foam differs from that of typical tissues in several ways. In particular, the distribution of fractions of polygon classes is broader and it includes a non-negligible fraction of triangles and decagons, both rarely seen in tissues. In addition, the average polygon area seems to be described better by an n^2-law rather than by Lewis' law [68].

Plateau rules in two-dimensional froths

In a two-dimensional soap froth, each side of a bubble is a circular arc, and its curvature is determined by the 2D version of the Young–Laplace law [Eq. (2.3)]. The arcs meet at three-valent vertices, and balance of line-tension forces at the vertices requires that the angles between the arcs are all equal to 120°. (The image is adapted from the photograph of a small region of the two-dimensional foam in Ref. [68].)

---

[3]Some experiments [65] suggest that the distributions of polygon areas are consistent with the gamma distribution

$$P(\alpha, \beta, A) = \frac{1}{\Gamma(\alpha)} \beta^\alpha A^{\alpha - 1} \exp(-\beta A), \qquad (3.15)$$

where $\Gamma(\alpha)$ is the gamma function; the mean α/β and the skewness $2/\alpha$ depend on the so-called shape and rate parameters α and β, respectively.

Another empirical size-topology relationship in cellular partitions, known as the Desch law [69] and alternatively as the Feltham law [70], relates the average cell perimeter and the number of sides. Like Lewis' law, the Desch–Feltham law states that average cell perimeter of n-sided cells $\overline{P}_n$ is a linear function of n:

$$\overline{P}_n = P_0(n - n_0), \tag{3.16}$$

where P_0 and n_0 are constants. The Desch–Feltham law applies to fewer real tilings than Lewis' law, and there are only a handful of tissues where it has been reported [62]. On the other hand, it was shown to apply to two-dimensional soap froths where it was explored in detail [68, 71, 72].

3.2.2 Aboav–Weaire law

In addition to size-topology correlations, random tilings are remarkable because of local topological correlations, that is, correlations between the number of sides of neighboring polygons. Lewis himself observed that "there is a marked tendency for the few-sided (polygons) to be in contact with the many-sided, and vice versa" both in periodic tilings of two or more tile types such as the trihexagonal tiling and in tilings seen in his cucumber tissues [60]. This finding is illustrated by an example of *Drosophila* pupal wing epithelium in Figure 3.11, which consists primarily of hexagonal cells with a small admixture of pentagons and heptagons. The neighbors of the left highlighted heptagon in the top part of the image include six hexagons and one pentagon so that on average, they have less than six vertices. The opposite is true for the highlighted pentagon in the bottom of the image, which has one heptagonal and four hexagonal neighbors. In the figure, we have also highlighted four other cells where the correlation pattern is not present just so as to emphasize the statistical nature of the law. In the more disordered tissues such as the *Drosophila* embryonic epithelium during germband extension [73], the local topological correlations are more pronounced.

These topological correlations are generally seen in all random networks including the borders between the columns in Giant's Causeway, grains in polycrystalline solid materials, and biological tissues. Mathematically, they are described by a simple expression giving the average number of vertices in neighbors of an n-sided tile denoted by $m(n)$, which is known as the Aboav–Weaire law [75, 76]:

$$m(n) = 6 - a + \frac{6a + \mu_2}{n}. \tag{3.17}$$

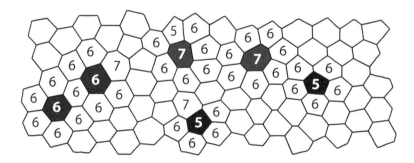

Figure 3.11 Aboav–Weaire law: *En-face* view of the pupal wing epithelium in wild-type *Drosophila* [74]. In a few clusters, cells are labeled by the number of vertices, the central cell carrying a boldface label.

Here $a \approx 1$ is a constant and μ_2 is the second moment of the distribution of the fractions of polygon classes given by $\mu_2 = \sum_n (n-6)^2 p_n$ [77]. In tissues, μ_2 is typically about 1 as one can readily see from the many distributions of polygon classes in this chapter, say in Figure 3.7.

In a perfectly hexagonal tiling where $\mu_2 = 0$, this formula gives $m(6) = 6$ for any value of a. On the other hand, in a tiling with three-valent vertices containing only pentagons, hexagons, and heptagons we have $\mu_2 = 1 - p_6$ where p_6 is the fraction of hexagons. If $p_6 = 0.9$ and $a = 1$, we have $m(5) = 6.220, m(6) \approx 6.017$, and $m(7) \approx 5.871$.

3.2.3 Statistical-mechanical considerations

The functional forms of Lewis' and Aboav–Weaire laws can be understood in terms of arguments unrelated to any specific feature of the physical tiling, which explains the universal validity of these laws. In both cases, the logic pursued is the same—if these laws are obeyed, then the number of constraints restricting the distribution of fractions of the polygon classes is smaller than if they are not obeyed. In this sense, Lewis' and Aboav-Weaire laws are associated with an increase of entropy [78, 79].

For clarity, let us spell out again the requirement that the fractions of polygon classes must be normalized

$$\sum_n p_n = 1 \tag{3.18}$$

and Eq. (3.12)

$$\sum_n (n-6)p_n = 0. \tag{3.19}$$

The average tile area A_0 is defined by

$$\sum_n \overline{A}_n p_n = A_0, \tag{3.20}$$

where $\overline{A}_n$ is the average area of all n-sided polygons, which is some function of n.

For any dependence of $\overline{A}_n$ on n, these three constraints constitute a rank-3 matrix equation for the polygon fractions p_n as the unknowns. However, if $\overline{A}_n$ is a linear function of n as predicted by Lewis' law then Eq. (3.20) is a linear combination of Eqs. (3.18) and (3.19) rather than an additional independent equation, and hence the rank of the matrix equation is reduced to 2 [78]. This in turn implies that the dimension of the hyperplane containing the solution is increased by 1 so that the solution itself is less constrained (or, equivalently, more arbitrary) than it would be if Lewis' law did not hold. Thus if we assume that the structure of a random tiling is determined by the maximal-entropy principle, then Lewis' law simply ensures that the entropy is maximized under fewer constraints than it would be if the law did not hold.

Given that all tiles must have at least three sides, the average area of two-sided tiles vanishes and we can write $\overline{A}_n = kA_0(n-2)$. In this case, Eq. (3.20) can be recast to read

$$kA_0 \sum_n (n-6)p_n + 4kA_0 \sum_n p_n = A_0. \tag{3.21}$$

The first sum on the left-hand side is 0 but the second one is 1 so that $k = 1/4$. As shown in Figure 3.9, this value of k is consistent with many although not all experimental observations. In cork, for example, $k = 1/3$. This can be explained by generalizing the above specific linear form of $\overline{A}_n$ so that it reads $\overline{A}_n = kA_0(n-n_0)$ and reinserting it into Eq. (3.20). This gives

$$k = \frac{1}{6-n_0} \tag{3.22}$$

and

$$\overline{A}_n = A_0 \frac{n - n_0}{6 - n_0} \tag{3.23}$$

By choosing $n_0 = 3$ (i.e., assuming that the tiling is devoid of triangles) we recover the cork version of Lewis' law.

In a similar fashion, we can also rationalize the form of the Aboav–Weaire law [79]. Consider the average number of l-sided neighbors of a tile with n sides denoted by $M_l(n)$. The sum of $M_l(n)$ over all polygon classes must be equal to the number of neighbors of a n-sided tile, which is n. This gives

$$\sum_l M_l(n) = n. \tag{3.24}$$

Another relation that we need is the probability that a given side belongs to an l-sided tile, which can be computed either by counting the l-sided tiles and multiplying their fraction by l or by summing the $M_l(n)$ edges shared by an average n-sided tile and its l-sided neighbor over all polygon classes weighted by their respective fractions. The two results must of course be the same and so we have

$$\sum_n M_l(n)p_n = lp_l. \tag{3.25}$$

If we assume that $M_l(n)$ is a linear function of n, $M_l(n) = a_l + b_l n$, this condition can be rearranged to read

$$(a_l + 6b_l) \sum_n p_n + b_l \sum_n (n - 6)p_n = lp_l, \tag{3.26}$$

which explicitly contains Eqs. (3.18) and (3.19) so that Eq. (3.25) no longer represents an additional constraint to be satisfied. Thus a tiling where $M_l(n) = a_l + b_l n$ must have a larger entropy than any other tiling.

The form of the Aboav–Weaire law follows from Eq. (3.25). The product of the average number of sides of the neighbors of an n-sided cell, $m(n)$, and n is equal to the total number of neighbors can also be written as $\sum_l l M_l(n)$ so that we have

$$nm(n) = \sum_l l M_l(n) = \sum_l l a_l + n \sum_l l b_l. \tag{3.27}$$

In other words,

$$m(n) = \sum_l l b_l + \frac{\sum_l l a_l}{n} \tag{3.28}$$

consistent with the form of the Aboav–Weaire law [Eq. (3.17)].

Developing a more complete statistical-mechanical theory of random tilings is not an easy task. To this end, one should first identify the relevant degrees of freedom of the tiles. The state of a tiling is defined by the vertex coordinates and by the connectivity map specifying the vertices of each tile and their order. Yet this is not very illuminating because any mechanical quantity associated with a tile (e.g., its energy) most likely depends on its area, perimeter, etc., rather than on vertex coordinates themselves. One could try approximating the phase space by the tiles' centroids, areas, perimeters, etc. [80]. This may sound as a considerable simplification, and in some ways it is, but in view of the above general arguments rationalizing Lewis' law and the Aboav–Weaire laws it may still lead to reasonable predictions. In addition, any analytical solution of most if not all statistical-mechanical models is necessarily approximate because the no-gap, no-overlap, and edge-sharing constraints cannot be enforced as in a numerical simulation.

A very interesting theoretical development in the field builds on the analogy between granular materials and cellular structures including tissues [81, 82]. By recognizing that the state of these systems is determined primarily by their volume rather than energy, this approach resolves a fundamental question of applicability of statistical mechanics to materials where the typical energies involved are considerably bigger than the thermal energy—biological cells certainly fall into this category. The key element of this statistical-mechanical theory is the so-called quadron, a quadrilateral formed by the centroid of a cell, a vertex, and the midpoints of the cell sides sharing this vertex. The quadrons are considered as quasiparticles of the network because their number is the same as the number of independent degrees of freedom and because their total area is the area of the tissue.

3.3 PHYSICAL MODELS

After having understood the non-specific features of tilings, we better appreciate the restricted phase space of the physical interactions and processes that determine the in-plane structure of epithelial tissues. Among them, cell division was recognized rather early as a very important component. Lewis' studies keep on returning to the "correlation between cell division and the shapes and sizes" in epithelia [51], and they emphasize that the shape of dividing and resting cells is not the same [50, 51]. This is witnessed by the fractions of polygon classes (Figure 3.12). In resting cells of cucumber epidermis, the distribution is pretty symmetrical around $n = 6$ but in dividing cells it is shifted toward larger n by 1. Lewis also measured the mitotic index in each polygon class and found that the probability of cell division increases very steeply with increasing n, growing from 0.15% in pentagons

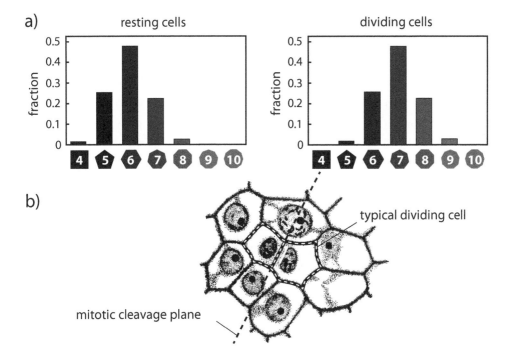

Figure 3.12 Fractions of polygon classes of resting and dividing cells in cucumber epidermis [51] (a). The typical dividing cell is heptagonal, the cleavage plane being perpendicular to the long axis (b; adapted from Ref. [51]).

to 38.5% in nonagons [51]. He also noted that cells divide such that the mitotic cleavage plane cuts along the short axis of the cell cross-section and that the area of dividing cells in any polygon class is larger than that of resting cells in the same class.

Since Lewis' work, this correlation has been studied in some detail, reestablishing, e.g., his observation that the average dividing cell is heptagonal [83–85]. Various rules deciding how cells divide were examined. These rules are then usually combined with a suitable model of cell mechanics, say by first letting a cell grow, then dividing it into two daughters in a plausible way, and finally relaxing the tiling so as to reach equilibrium [63]. Within such a scheme, one should be able to study the relative impact of cell division, rearrangement, and mechanics on the structure of the tissue.

3.3.1 Area- and perimeter-elasticity model

One of the most comprehensive analyses combining all of these effects [63] was used to interpret the patterns seen in the proliferating epithelium of *Drosophila* wing imaginal disc during a stage characterized by dramatic growth resulting in more than a thousand-fold increase in the number of cells. This model relies on several elements of cell structure rather explicitly, and the total energy of a cell in a tissue consists of terms depending on its perimeter P and area A as integral quantities. The first one is the adhesion energy described by a term proportional to the perimeter P

$$-\frac{\Gamma}{2}P, \qquad (3.29)$$

where $\Gamma > 0$ is the adhesion strength; the factor of $1/2$ is needed because the adhesion energy of a cell-cell contact is shared by two cells. The second element to include is the elasticity of the acto-myosin ring, which can be represented by two energy terms, one proportional to P as if the tension of the ring were independent of strain and the other proportional to P^2 like in Hookean springs. The former can be combined with the adhesion energy and their sum reads

$$\frac{T}{2}P. \qquad (3.30)$$

The effective line tension T consists of the positive acto-myosin tension and of a negative adhesion strength, and is negative if adhesion is strong enough. The Hookean ring energy reads

$$\frac{C}{2}P^2, \qquad (3.31)$$

where C is the contractility of the ring.

The other ingredient of the cell mechanical energy is the bulk term related to cell compressibility. If the cytoplasm is treated as an incompressible fluid then its volume should be constant, and if we additionally assume that cell height is fixed then we conclude that the area of cell cross-section should not depart very much from some preferred value A_0. This leads to the so-called area-elasticity energy

$$E_A = \frac{\lambda_A}{2}(A - A_0)^2, \qquad (3.32)$$

where λ_A is the area modulus. The total energy of the tiling is then given by the sum of the two perimeter terms [Eqs. (3.30) and (3.31)] and Eq. (3.32) over all cells. In a tissue containing two or more types of cells, the acto-myosin ring contractility, area modulus, and preferred cell area may vary from cell to cell and the line tensions may be different in each type of edge. We will return to this point in Section 3.3.5 where this model is applied to the patterns in the *Drosophila* retina. However, in an epithelium consisting of a single cell type these generalizations are not needed and the total energy reads

$$E = \sum_i \left[\frac{C}{2} P_i^2 + \frac{T}{2} P_i + \frac{\lambda_A}{2} (A_i - A_0)^2 \right]. \tag{3.33}$$

This form contains four parameters but after expressing perimeters in $\sqrt{A_0}$ as the unit of length and by choosing $\lambda_A A_0^2$ as the characteristic energy, we are left with only two dimensionless parameters: the reduced contractility

$$\mathcal{C} = \frac{C}{\lambda_A A_0} \tag{3.34}$$

and reduced line tension

$$\mathcal{T} = \frac{T}{\lambda_A A_0^{3/2}}. \tag{3.35}$$

Now it is best to recast the dimensionless form of Eq. (3.33) such that the two perimeter terms are combined by completing the square. The result is

$$\mathcal{E} = \frac{E}{\lambda_A A_0^2} = \sum_i \left[\frac{\mathcal{T}^2}{2\mathcal{C}} \left(\frac{P_i}{P_0} - 1 \right)^2 + \frac{1}{2} \left(\frac{A_i}{A_0} - 1 \right)^2 \right] + const., \tag{3.36}$$

where

$$P_0 = -\frac{T}{2C} = -\frac{\sqrt{A_0}\mathcal{T}}{2\mathcal{C}} \tag{3.37}$$

is the preferred tile perimeter. In the following, we dispose of the additive constant.

The numerically obtained phase diagram of minimal-energy tilings is remarkably simple (Figure 3.13), consisting of only two states. At large reduced contractilities where stretching the acto-myosin ring is associated with a considerable energy increase, the preferred tile shape minimizes the perimeter but this is counterbalanced by the area elasticity and the

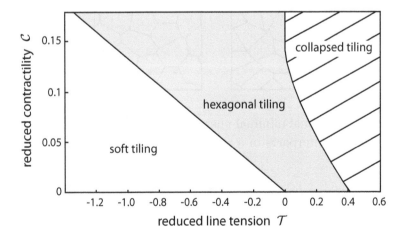

Figure 3.13 Phase diagram of the area- and perimeter-elasticity model consists of two regions—one belongs to soft and the other to the hexagonal minimal-energy states. In the hatched region on the right-hand side of the diagram, the tiling is destabilized by the contractility of the acto-myosin ring. (Adapted from Ref. [63].)

constraint that the plane is tiled edge to edge. Of all tilings containing a single type of polygons, the hexagonal tiling has the smallest perimeter at a given area and thus the large-contractility ground state is hexagonal [63].

At small reduced contractilities, the ground state is random (in the sense that it contains tiles of various shapes) and degenerate (as there exist many realizations of the tiling that have the same energy). A typical snapshot of such a tiling referred to as soft is shown in Figure 3.14a. The tiles do differ in the number of vertices and the relative lengths of edges but their overall shapes expressed using areas and perimeters as integral quantities are very similar. To appreciate this, recall that the preferred tile perimeter P_0 [Eq. (3.37)] is positive at negative line tensions and if $\mathcal{T}^2/\mathcal{C} \gg 1$ the tiles minimize their energy by choosing $P_i = P_0$ and $A_i = A_0$ so that the energy of the tiling is zero. The condition $\mathcal{T}^2/\mathcal{C} \gg 1$ is fulfilled in the whole soft-tiling region. If the preferred area A_0 is small enough compared to P_0^2 so that the polygon preferred reduced area $a_0 = 4\pi A_0/P_0^2$ is smaller than about $\sqrt{3}\pi/6 \approx 0.907$ which corresponds to the regular hexagon, there should exist many equivalent realizations of the tilings, all of them energy-degenerate. In this case, we expect that an in-plane shear stress transforms any given initial configuration to the final one by passing through a sequence of states with identical energy so that their shear modulus is zero. This is why we refer to the tiling where $P_i = P_0$ and $A_i = A_0$ as soft.

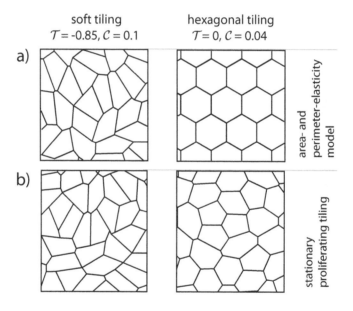

soft tiling
$\mathcal{T} = -0.85, \mathcal{C} = 0.1$

hexagonal tiling
$\mathcal{T} = 0, \mathcal{C} = 0.04$

a)

area- and perimeter-elasticity model

b)

stationary proliferating tiling

Figure 3.14 Soft and hexagonal minimal-energy states of the perimeter-area elasticity model (a) and their counterparts in a stationary proliferating tiling (b). (Adapted from Ref. [63].)

Introducing the preferred tile perimeter [Eq. (3.37)] is quite instructive because it also illustrates why the hexagonal tiling should be the ground state at large contractilities: If $\mathcal{C} \gg \mathcal{T}$ then the preferred area A_0 is necessarily much larger than P_0^2. In this case the minimal-energy shapes have as small a perimeter as possible, which would correspond to circular tiles—but circles do not tile the plane and so the best remaining option is a hexagonal tiling. Obviously, its energy is not zero like in the soft tiling.

Equally interesting is to explore the behavior of the hexagonal tiling at fixed contractility as the line tension is increased. A positive line tension corresponds to a negative preferred perimeter so that cells in fact favor a vanishing perimeter and thus a vanishing area. This case is easily examined by expressing the perimeter of a hexagonal tile in terms of its area, $P = 2^{3/2}3^{1/4}\sqrt{A}$, inserting it into Eq. (3.36), and plotting the energy as a function of A. From this plot one quickly finds that upon increasing $\mathcal{T}$ beyond a certain positive value at a given $\mathcal{C}$, the area of the minimal-energy state vanishes which signals the collapse of the hexagonal tiling. This collapse is caused by the contractility of the acto-myosin ring, which explains why at large $\mathcal{C}$ the tiling is unstable unless the line tension is negative (so that the preferred cell perimeter is positive).

These theoretical predictions are very intriguing but as mentioned above, they were pursued so as to explain the structure of the proliferating *Drosophila* wing disc epithelium. Thus it appears imperative to generalize the model by including cell division. The algorithm implemented starts by gradually increasing the preferred area of a randomly chosen cell until it is doubled (Figure 3.15). During this process, the tiling is relaxed so as to minimize the energy. After the final preferred area is reached, the cell is divided into daughters by a randomly oriented cleavage plane passing through its center. Then each of the daughters is assigned the original preferred area and the tiling is relaxed and rearranged again, the process including T1 and T2 topological transformation as needed [63]. This procedure is repeated for several generations until the tiling reaches a stationary state.

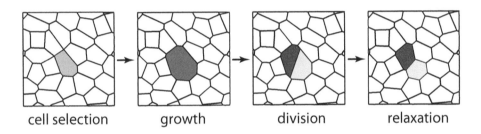

cell selection   growth   division   relaxation

Figure 3.15 Cell division algorithm in the proliferating area- and perimeter-elasticity model: First a cell is randomly chosen, then its preferred area is gradually increased until it is doubled, and in the third step the cell is divided into daughters by a randomly oriented cleavage plane through its center. Finally the tiling is relaxed to reach mechanical equilibrium. (Adapted from Ref. [63].)

As illustrated in Figure 3.14b, cell division completely changes the structure of the hexagonal tiling, which becomes disordered, but the soft tiling appears to be virtually unaffected by division. The proliferating version of the $\mathcal{T} = 0, \mathcal{C} = 0.04$ tiling resembles real tissues rather closely in many ways, which is nicely reflected in fractions as well as in the average areas of the polygon classes (Figure 3.16). The fractions agree quite well with the experimental data and the areas approach the observed Lewis' law. On the other hand, the soft tiling at $\mathcal{T} = -0.85$ and $\mathcal{C} = 0.1$ contains too many quadrilaterals and pentagons and too few hexagons and heptagons, and the areas of all polygons are all the same which is inconsistent with Lewis' law.

These differences can be understood by recalling that the soft tiling does not result from a competition between the perimeter and the area elasticity because the preferred perimeter is large enough compared to the characteristic tile size $\sqrt{A_0}$, and thus its energy is zero in the ground state. This feature of the soft ground state remains unaffected by proliferation. On the other hand, the ground state of the hexagonal tiling is controlled by the opposing forces of area elasticity and perimeter contractility, the latter favoring tiles of area smaller than the preferred value A_0. Thus if some topological disorder is introduced by division,

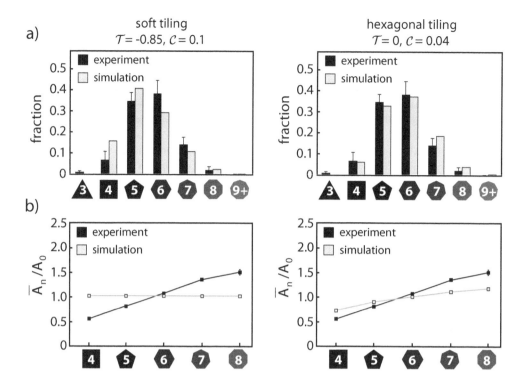

Figure 3.16 Fractions of polygon classes in the stationary proliferating tilings at $\mathcal{T} = -0.85, \mathcal{C} = 0.1$ and $\mathcal{T} = 0, \mathcal{C} = 0.04$ compared to *Drosophila* wing disc epithelium (a). Panel b shows the average cell area in each polygon class for these two theoretical cases and in the tissue itself—notice that the experimental data obey the $k = 1/4$ Lewis' law. (Adapted from Ref. [63].)

tiles with fewer than six neighbors should have an area smaller than the average or else their neighbors should be bigger than the average. This means that quadrilaterals and pentagons should carry a smaller perimeter energy compared to the ground state but their area energy should be bigger as illustrated in Figure 3.17. By the same token, the areas of tiles with more than six neighbors should be larger than the average so that their area and perimeter energies are smaller and larger than in the ground state, respectively. These variations are reflected in tension and pressure differences between tiles, which may well be experimentally observable. In all, introducing proliferation in the hexagonal tiling results in an increase of its energy.

Now we have learned that cell division plays a very important role, and by embellishing the protocol we can hope to even better reproduce the experiments. One of the characteristic features of cell division mentioned above is that the larger cells divide more rapidly than the smaller ones, which can be interpreted as regulation of growth rate by mechanical stress since the larger cells are stretched by their neighbors more than the smaller ones. In one implementation of this idea [84], cells divide as soon as their volume reaches a critical value V_c, and in each time step of the simulation the volume of cell i is increased by

$$\Delta V_i = V_c \left(u + v \frac{A_i - \overline{A}}{A_0} \right), \tag{3.38}$$

where u and v are coefficients that regulate growth rate and its dependence on area whereas

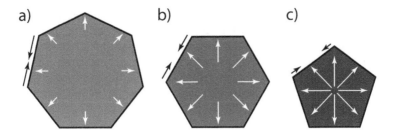

Figure 3.17 In the proliferating area- and perimeter-elasticity model, linear tension around the perimeter shown by black arrows is larger in the larger tiles and smaller in the smaller tiles represented by a heptagon (a) and a pentagon (c), respectively, for qualitative consistency with Lewis' law. Conversely, the in-plane pressure due to area elasticity (white arrows) in the larger tiles is smaller than in the average tile represented by the hexagon (b) and vice versa.

$\overline{A}$ is the average cell area. Much like in the original model [63], this cell division algorithm was then applied to a small cluster of a few dozen cells until the tissue underwent 10000 divisions and reached a stationary state. Division was combined with topological rearrangements and mechanical relaxation, and two-sided cells that appeared as a result of topological transformations were removed.

For a suitable choice of cell growth parameters, the thus obtained fractions of polygon classes in a stationary tiling are very close to the experimental structure of *Drosophila* wing disc epithelium, and the distribution of polygon classes in dividing cells is reproduced even better (Figure 3.18). Much like in Lewis' cucumber epidermis [51], it is essentially the same as in all of the tissue except that it is centered at $n = 7$ rather than at $n = 6$ [84]. Alternative, less complex scenarios of mitosis, where cell division rate was independent of volume or where cell rearrangements were not allowed, fail to account for experimental observations. As both processes can be associated with the effects of mechanical stress, we are led to conclude that this stress is involved in the regulation of cell growth and thus in cell division.

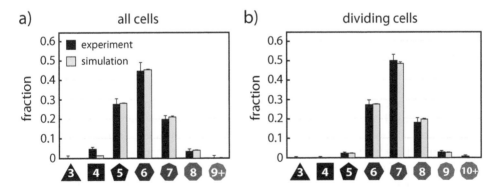

Figure 3.18 Fractions of polygon classes in the stationary state of a proliferating tiling within the area- and perimeter-elasticity model with volume-dependent growth rate compared to the experimental fractions in *Drosophila* wing disc epithelium (a). Plotted separately are the fractions in the dividing cells (b). (Adapted from Ref. [84].)

Yet another aspect of modeling cell division is related to the orientation of the cleavage plane. All of the results discussed so far are obtained by assuming that this orientation is random but in reality, this is rarely the case. Both animal and plant cells divide primarily along the short axis, and this has been recognized for a long time [51, 86]. In turn, the in-plane orientation of the cell is determined by its neighborhood such that the short axis and thus the cleavage plane correlate with the location of quadrilateral and pentagonal neighbors (Figure 3.19a).

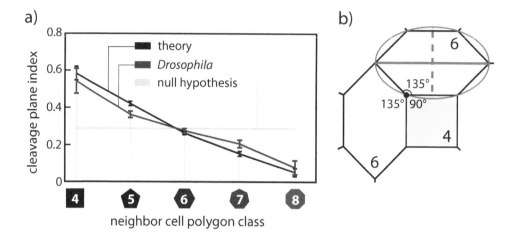

Figure 3.19 The cleavage plane index (defined as the ratio of the cell's n-sided neighbors that are located in the cleavage plane and all n-sided neighbors) vs. neighbor polygon class for the *Drosophila* wing disc epithelium (a; adapted from Ref. [85]). As the cleavage plane index of quadrilaterals and pentagons is larger than the average, the orientation of the cleavage plane is biased rather than random. A simple geometrical argument illustrating why the short axis of a cell points toward the smaller neighbors represented here by a square (b).

Let us illustrate how this happens by first considering a rather idealized situation where the quadrilateral is a square and the neighboring cells are identical hexagons (Figure 3.19b). In this case, the internal angles of a neighbor next to the quadrilateral are larger than the 120° of the regular hexagon and if the lengths of all edges of the neighbor are the same then this cell must be elongated such that its short axis is oriented toward the quadrilateral tile. If the assumptions of equal angles and equal edge lengths used in this packing argument are removed then the correlation is less pronounced but it is still there.

A more detailed analysis of this process carried out by examining small clusters of cells around a heptagon as the most common dividing cell class (Figures 3.12 and 3.18) confirms the existence of correlation. The minimal-energy states of many such clusters were computed within a mechanical model including perimeter elasticity where each edge is represented by a Hookean spring and an area elasticity due to the ideal-gas-type pressure within the cell [85]; although not identical, this model is really not very different from the area- and perimeter-elasticity theory. The equilibrium cluster configurations all unequivocally show that the heptagon is polarized and that to a very good approximation, its short axis determined by approximating the polygon by an ellipse always cuts across it starting from the quadrilateral or a pentagon. On the other hand, in a cluster with no four- and five-sided neighbors the short axis of the dividing heptagon is "repelled" by any seven- or eight-sided neighbors as illustrated in Figure 3.20. These results are consistent with experimental observations [85] so that we are led to conclude that cell division is affected by the structure of the epithelium and vice versa.

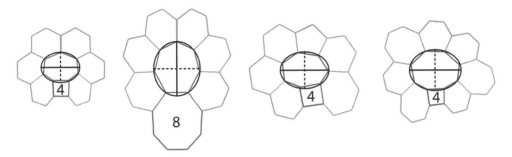

Figure 3.20 A few representative local minimal-energy clusters around the dividing heptagon which demonstrate that the orientation of the heptagon's short axis correlates with four- and five-sided neighbors and anticorrelate with seven- and eight-sided neighbors. (Adapted from Ref. [85].)

We saw how adding increasingly more complex features to the original area- and perimeter-elasticity model altered its predictions and improved the agreement with experimental observations. Yet it seems possible that some of the results are dominated by a single process involved. This can be verified by examining stripped-down variants of the model, which are necessarily limited in scope but still quite instructive because they show that certain aspects of the epithelial structure can be understood much more transparently.

3.3.2 Topological mitosis-only model

The first example of this kind is the topological theory of proliferating epithelia that do not rearrange significantly during development. In these tissues, cell division must be the main process responsible for the structure of the epithelium represented by the fractions of the polygon classes. Knowing how the tissue topology changes during cell division should then be enough to understand the structure of the stationary state reached after several rounds of divisions [83]. This purely topological view does not involve any physical dimension of cells or any form of energy that they carry, so that it obviously cannot reproduce any size-related properties of the tissue such as Lewis' law. Moreover, as the tissue is represented solely by the polygon fractions and not by a concrete tiling, this model is also unable to address topological correlations in the tissue (and thus potentially yield the Aboav–Weaire law). Yet it is remarkable that this quite abstract framework based on fairly general assumptions concerning the development of the tiling still gives very reasonable predictions. A related version of this approach was explored within a model of the basal layer of the epidermis where cells may also leave the basal layer [87].

Let us first examine the evolution of the average number of edges in a tiling due to cell division. On going from generation $t-1$ to generation t, the number of faces is clearly doubled $F_t = 2F_{t-1}$. Figure 3.21 illustrates that the number of edges is $E_t = E_{t-1} + 3F_{t-1}$ and the number of vertices is $V_t = V_{t-1} + 2F_{t-1}$ as each divided cell contributes three new edges and two new vertices to the tiling. Now we calculate the average number of cell sides from F_t and E_t. Each isolated polygon has s edges but in a tiling each edge is shared by

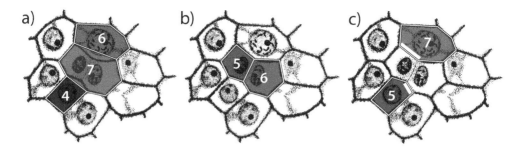

Figure 3.21 During mitosis of a heptagonal cell (a), the number of faces is increased by one as the cell divides into two daughters (b). Three new edges are created: one is the cleavage plane itself and the other two arise from dividing the sides of the highlighted quadrilateral and the hexagonal neighbor of the mother, which turn into a pentagon and a heptagon, respectively (c). The endpoints of the cleavage plane are the two new vertices created. The figure shows cells in cucumber epidermis (adapted from Ref. [51]).

two faces so that we have $s_t = 2E_t/F_t$. Thus

$$s_t = \frac{2\left(E_{t-1} + 3F_{t-1}\right)}{2F_{t-1}} = \frac{1}{2}s_{t-1} + 3. \qquad (3.39)$$

In a stationary state where $s_t = s_{t-1}$, the average number of sides must be 6. The general solution of this dynamical equation $s_t = 6 + (s_0 - 6)\, 2^{-t}$ [83], where s_0 is the average number of sides of the first cell, shows that any initial deviation from this value decays exponentially: given that $s_0 - 6$ is of the order of 1, $s_t \approx 6$ after about eight divisions. Note that this result is a direct consequence of the three-valent topology of the tiling.

A similar line of thinking can be pursued to analyze the evolution of the distribution of polygon classes using a Markov-dynamics model [83]. To this end, first note that the number of vertices of a cell may change either due to its division into daughters or due to the division of a neighboring cell. In both cases the distribution represented by a vector $\mathbf{p}_{t-1} = (p_3, p_4, p_5, \ldots)$ where p_n is the fraction of n-sided cells in generation $t - 1$ will be altered, and this can be formally encoded by multiplying $\mathbf{p}_{t-1}$ by matrices containing the probabilities of transforming an n-sided cell into an m-sided cell in either process.

Determining these probabilities may seem a daunting task, but with a few rather general assumptions it becomes tractable [83, 88]. Consider the vertices of an n-sided cell. As the cell divides, these vertices are distributed between the daughters. Since triangular cells are not seen experimentally, we can assume that each of the daughters must inherit no less than two vertices so that the ways of dividing the mother into daughters depends on how the remaining $n - 4$ vertices can be divided between them. If we regard this as a combinatorial problem and assume that the distribution of these vertices is binomial with probability $p = 50\%$, then the probability that an n-sided mother is transformed into a m-sided daughter is

$$P_{nm} = \left(\begin{array}{c} n-4 \\ m-4 \end{array}\right) \frac{1}{2^{n-4}}. \qquad (3.40)$$

Here we took into account that the total of m vertices of the daughter consist of two vertices received directly from the mother, two vertices formed by the cleavage plane, and the $m - 4$ vertices that are randomly distributed between the daughters. As usual, $\left(\begin{array}{c} n-4 \\ m-4 \end{array}\right) = (n-4)! / \left[(m-4)!(n-m)!\right]$ is the binomial symbol.

The matrix P_{nm} does not cover all of the topological dynamics taking place. The number of vertices in each cell does not change solely because it divides itself but also because its neighbors do so too. On average, this effect can be included by realizing that during mitosis each of the N cells of generation $t-1$ adds two vertices to each of its neighbors connected by the cleavage plane (Figure 3.21). Thus the cells of generation t together have $2N$ more vertices than they would if each of them divided in isolation. Since there are $2N$ cells in generation t, each of them gets one additional vertex on average. This process is conveniently represented by the so-called shift matrix [83]

$$S_{nm} = \delta_{n,m-1} \tag{3.41}$$

and the evolution of the distribution of polygon classes is then described by

$$\mathbf{p}_t = \mathbf{p}_{t-1}\mathsf{PS}. \tag{3.42}$$

This notation reflects the more intuitive visualization of both matrices where we put the mother cells in rows and the daughter cells in columns and analogously for the increase of vertices due to division of neighbors (Figure 3.22).

after cell division

P	4	5	6	7	8	9	10
4	1						
5	1/2	1/2					
6	1/4	2/4	1/4				
7	1/8	3/8	3/8	1/8			
8	1/16	4/16	6/16	4/16	1/16		
9	1/32	5/32	10/32	10/32	5/32	1/32	
10							...

before cell division

after neighbor division

S	4	5	6	7	8	9	10
4	0	1					
5		0	1				
6			0	1			
7				0	1		
8					0	1	
9						0	1
10							...

before neighbor division

Figure 3.22   P and S matrices: each row encodes the transformation of a given polygon class after the division of the mother cell and a neighboring cell, respectively. Note that the numerators of the entries in the P matrix form Pascal's triangle.

Almost independently of the initial state, the stationary distribution of polygon classes is reached fairly quickly and the final state of the above model is rather close to the experimental ones in several proliferating tissues (Figure 3.23). The fractions of all polygon classes are reproduced within a few percent, the only deficiency being the absence of quadrilaterals. In the model, any quadrilaterals generated by mitosis are automatically transformed by the S matrix into pentagons.

To fix this problem, one can think of alternative scenarios of cell division, say by allowing triangles or by equally distributing vertices between the daughters [83]. Each of these models has a different P matrix and thus a different stationary distribution but it seems that the above version is better than the other options explored. Yet another possibility is to abandon the purely combinatorial perspective by also considering the topology of the daughter cells as well as some other features of mitosis such as the preferential orientation of the cleavage plane along the short axis.

The various ways of dividing the cell into daughters are best counted by drawing a figure for each of them. Figure 3.24 shows the division of quadrilaterals, pentagons, and

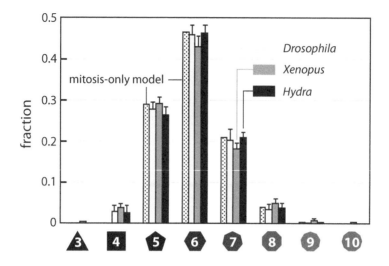

Figure 3.23 Fractions of polygon classes in the mitosis-only model of a proliferating epithelium compared to those seen experimentally in tissues of *Xenopus*, *Drosophila*, and *Hydra*. The overall agreement is rather good, the main discrepancy being the absence of quadrilaterals in the model. (Adapted from Ref. [83].)

hexagons within a model where each daughter has to inherit two or more vertices from the mother—just like in the original model discussed above [83]. In addition, we demand that in each daughter, these two inherited vertices be adjacent to each other and that the two pairs be topologically as far apart as possible, which is consistent with the reported elongated shape of dividing cells. In case of quadrilateral, pentagonal, and hexagonal mothers,

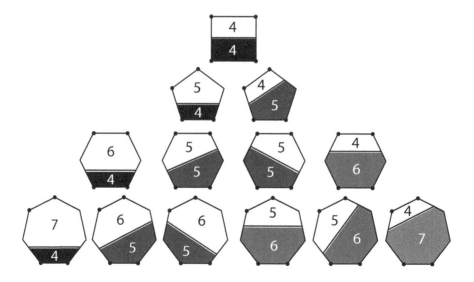

Figure 3.24 Partitions of a 4-, 5-, 6-, and 7-sided cells into daughters within the scheme where vertices inherited from the mother (black circles) must be neighbors and the two pairs of inherited vertices must be as far from each other as possible.

these additional constraints do not show: notice that the pattern leading from an n-sided mother to a shaded m-sided daughter is the same as in the P matrix in Figure 3.22. However, division of heptagonal cells proceeds in a different way. As each of the six possible partitions is assumed equiprobable, the probabilities that a 7-sided mother becomes a 4-, 5-, 6-, and 7-sided daughter are $1/6, 2/6, 2/6$, and $1/6$, respectively. In the original model, these probabilities are $1/8, 3/8, 3/8$, and $1/8$, respectively (Figure 3.22).

As we can expect from the modified P matrix, the stationary distribution of polygon fractions within this scheme should depart from that of the original model, yet the difference turns out to be small. Instead of the original vector of polygon fractions $(p_4, p_5, p_6, p_7, p_8, p_9) = (0, 0.289, 0.464, 0.208, 0.036, 0.003)$ [83] the modified scheme gives $(p_4, p_5, p_6, p_7, p_8, p_9) = (0, 0.302, 0.453, 0.194, 0.044, 0.006)$. Whether this can be considered an improvement when compared to experimental data is a matter of taste.

The mode of cell division depicted in Figure 3.24 is not the only conceivable scenario, and many alternative sets of rules for choosing the cleavage plane can be devised [64]. Some of them lead to patently unnatural, non-bell-like distributions of fractions of polygon classes, which peak at the smallest polygon class. Still there exist several cleavage patterns that describe the observed distributions well. This is reassuring because it is difficult to imagine that the proliferating tissues of all kinds would resort to the same mode of division, and indeed they do not.

3.3.3 Fixed-area/fixed-perimeter model

This discussion has departed quite considerably from the original minimal-energy version of the area- and perimeter-elasticity model, and its mechanical aspects were eventually completely discarded. Let us return to them and reexamine the energy functional Eq. (3.36) from yet another perspective—so as to abandon it again in a rather different manner.

The two terms of the final version of the area- and perimeter-elasticity model [Eq. (3.36)] suggest that while the cells prefer a certain area A_0 and a certain perimeter P_0, an arbitrarily large deviation from the preferred values is possible although energetically costly. In a tiling, this is really not true because the size of neighbors in an edge-to-edge tiling must be comparable. Empirically, this is clearly demonstrated by Lewis' law (Figure 3.9) which shows that the typical area of the largest cells is less than twice the average area whereas in the smallest cells it is between 30 and 50% of the average area. These variations are by no means small but they are not very dramatic either, and considering the limit where they are not allowed at all does seem appealing for two reasons. Firstly, the energy of the soft tiling [63] is zero so that its structure does not really reflect any particular aspect of the underlying energy apart from the fact that it vanishes at $A_i = A_0$ and $P_i = P_0$, and replacing the area and perimeter elasticity energies by hard constraints merely emphasizes this fact. Secondly, such a model has fewer parameters than the original version [Eq. (3.36)] and so the interpretation of its predictions may be more transparent. On the other hand, this stripped-down model obviously cannot reproduce size-dependent features of tissues such as Lewis' law.

So let us consider the model where both C and λ_A in Eq. (3.33) are infinite so that only possible states of the tiling are those where all tiles have the same area and perimeter:

$$A_i = A_0 \qquad \text{and} \qquad P_i = P_0. \tag{3.43}$$

As the structure of the tiling does not depend on the size of tiles, it is instructive to construct a dimensionless parameter called the reduced area and defined by

$$a = \frac{4\pi A}{P^2}. \tag{3.44}$$

The reduced area encodes the overall shape of the tile and it is normalized such that in the circle, that is, in the shape with smallest perimeter at a given area, $a = 1$. In any other shape a is less than 1. For example, a rectangle of sides c and d the reduced area is $a = \pi/[(1+c/d)(1+d/c)]$ so that in a square $a = \pi/4 \approx 0.785$. A nice feature of the reduced area as a shape parameter is that it is not directly sensitive to the number of vertices because it depends on the integral quantities. For example, the rectangle and the elongated hexagon with angles of 120° in Figure 3.25 both have identical areas and perimeters—and thus the same reduced area, which is $2\pi/9 \approx 0.698$.

a) b)

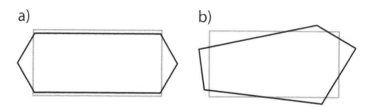

Figure 3.25 Rectangle and a 120°-angle elongated hexagon with identical areas and perimeters showing that shapes of same reduced area a (in this case $a \approx 0.698$) are globally similar although their aspect ratios are evidently different (a). This is further illustrated by the irregular pentagon in panel b which has the same a as the shapes in panel a.

Since in this model the areas and perimeters of all tiles are identical, so is their reduced area which thus remains the sole parameter that controls the structure of the tiling. The equilibrium state can be found by numerical simulations where the individual vertices are randomly displaced at fixed A_i and P_i. T1 topological transformations are performed until the tiling no longer changes on average, which can be judged by monitoring the fractions of polygon classes. The results of such simulations are shown in Figure 3.26 for a few values of the reduced area.

The snapshots in Figure 3.26 illustrate that the fixed-area/fixed-perimeter tilings are more and more ordered as a is increased. The difference between the $a = 0.74$ and the $a = 0.78$ cases does not seem to be very big as both contain polygons with 4 to 9 vertices; triangles and polygons with 10 or more sides are not present. On the other hand, the $a = 0.86$ case shows that at this value of the reduced area, the tiling only includes pentagons,

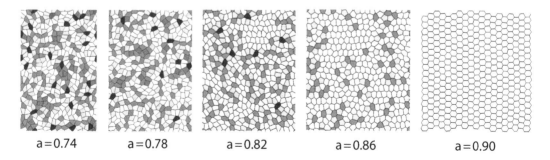

a=0.74   a=0.78   a=0.82   a=0.86   a=0.90

Figure 3.26 Snapshots of representative tilings obtained within the fixed-area/fixed perimeter model of identical tiles. Each class of polygons is colored with the same shade of gray, hexagons being white. (Adapted from Ref. [89].)

hexagons, and heptagons. Pentagonal and heptagonal tiles often occur in pairs, and sizable portions of the tiling consist exclusively of hexagons. Finally, at $a > 0.865$ all tiles are hexagons (albeit not regular) so that in the topological sense the tiling is ordered.

This model exhibits the same phase transition as the minimal-energy version of the area- and perimeter-elasticity model. Tilings with $a < 0.865$ correspond to the soft tiling and those with $a > 0.865$ are hexagonal. Figure 3.27 shows the transition more explicitly in terms of the fractions of the polygon classes, which indicate that one could well distinguish two types of soft tilings. Those at $a \lesssim 0.76$ are rather insensitive to the exact value of a as the fractions do not depend very much on it, and those at $a \gtrsim 0.82$ essentially contain only pentagons, hexagons, and heptagons.

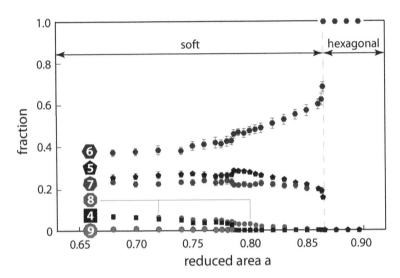

Figure 3.27 Fractions of the polygon classes in fixed-area/fixed-perimeter tilings (adapted from Ref. [89]).

The value of the reduced area at the soft-hexagonal transition can be used to interpret the minimal-energy phase diagram of the area- and perimeter-elasticity model [Eq. (3.33)]. Recall that the preferred perimeter is given by $P_0 = -\sqrt{A_0}\mathcal{T}/2\mathcal{C}$ so that the reduced area of tiles in the soft tiling is

$$a = \frac{16\pi\mathcal{C}^2}{\mathcal{T}^2}. \tag{3.45}$$

In the $(\mathcal{T}, \mathcal{C})$ plane in Figure 3.13, any straight line through the origin corresponds to a fixed value of a. This also applies to the boundary between the soft and the hexagonal tiling where $a \approx 0.89$ which is very close to prediction of the fixed-area/fixed-perimeter model where the transition takes place at $a = 0.865$. Thus the fixed-area/fixed-perimeter model explains the $\mathcal{T} < 0$ part of the phase diagram in Figure 3.13 both qualitatively and quantitatively.

More importantly, this simple model suggests that the structure of epithelial tissues represented by these tilings may be controlled by a single geometric parameter. One can think of many different mechanical theories which involve tile area and perimeter either explicitly as Eq. (3.33) or in a more subtle way. Yet as long as a preferred A_0 and P_0 can be identified and as long as the theory in question permits degenerate solutions with vanishing energy, one should be able to characterize the equilibrium states by a certain preferred tile reduced area.

This hypothesis can be readily tested by analyzing the morphometry of real tilings seen in epithelial tissues, and the results show that there indeed exists an extraordinarily strong correlation between cell area and square of cell perimeter [90]. Figure 3.28 shows the distribution of cells in the *Drosophila* wing epithelium in the (P^2, A) plane. It is obvious that to a very good approximation the area of each particular cell is proportional to its perimeter squared, the coefficient of proportionality being almost the same in all cells irrespective of n. This is reflected in the very high value of the correlation coefficient, which is 0.966. These observations imply that the reduced area of the cells should be well-defined and that its distribution must be very narrow although the tile areas and perimeters themselves are much more scattered. This is indeed the case [90].

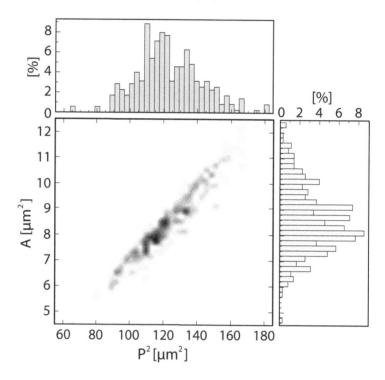

Figure 3.28 Distributions of cells of the *Drosophila* wing epithelium in the (area, perimeter squared) plane. The ratio A/P^2 is virtually the same in all cells and the correlation coefficient of the distribution is 0.966. The top and the right histogram show the distributions of P^2 and A, respectively. (Adapted from Ref. [90].)

The same holds for a very broad range of patterns examined including tilings seen in animal and plant epithelia as well as those in inanimate structures such as the basalt columns in Giant's Causeway [90]. In all cases studied, the correlation coefficient is very close to 1; much too close to ignore despite the finite size of samples. One is led to conjecture that all of these tilings are characterized by a well-defined reduced area, which is quite remarkable given that many of them are of rather different origin and based on very different physical processes. The observed correlation could thus either be a coincidence—or a geometric effect inherent to edge-to-edge tilings, which remains a hypothesis worthwhile exploring.

3.3.4 Dirichlet-tessellation perspective

The logic of shapewise degenerate cells present in all models of tissues containing unpolarized cells can be taken yet another step further by assuming that cell centers behave as point-like particles interacting with each other by a distance-dependent force and that cell-cell boundaries are located halfway between centers of neighboring cells. This framework is admittedly a gross approximation because it includes energies that are only indirectly related to cell size. On the other hand, its main advantage is that by employing the Dirichlet tessellation (also referred to as the Voronoi tessellation or partition) of the plane it automatically takes care of the no-overlap, no-gap constraints, which are among the most challenging issues in any theoretical study of cellular partitions.

Voronoi tessellation

Consider a plane dotted with a set of points arranged in a certain way, and partition it into domains such that the distance of a small region within a domain to the point within it is smaller than the distance to any other point of the set. Such a division of the plane is referred to as the Dirichlet tessellation or partition. Shown here is the Dirichlet tessellation for a set of randomly distributed points [91]. The partition is constructed by drawing a line connecting a given point with its neighbor and then bisecting the line so as to divide the plane 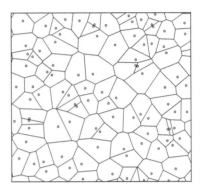 into two half-planes. The Dirichlet cell is the intersection of all half-planes that contain the point. The Voronoi tessellation is an analogous concept in a space of arbitrary dimension.

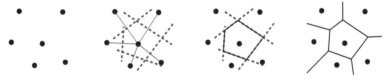

The underlying idea can be explored in two steps: first one should determine whether the observed tissues are consistent with such a partition by inverting the Dirichlet algorithm, and if this is the case then one could compare the partitions predicted for a few reasonable choices of pointwise interaction. In the inverse Dirichlet algorithm, we identify the centers of the domains associated with each tile by subdividing the angle at each vertex of the tile into angles supplementary to the other two angles at the vertex as shown in Figure 3.29a.

In a true Dirichlet tessellation, all of the thus formed rays intersect at a single point but in an arbitrary tiling they do not. Nonetheless, we can approximate the Dirichlet center of a cell by the point that minimizes the distance to all rays [92]. Once the approximate centers of all tiles are found, the Dirichlet tessellation can be constructed. In some tilings the Dirichlet boundaries agree well with the actual ones but in others the disagreement is considerable either in some or in most of the tiles, leading to large deviations of tile areas, perimeters, and shapes (Figure 3.29b).

Despite the disagreements, the logic behind this model is both economical and insightful. In a single step, the Dirichlet construction reduces the number of coordinates needed to describe each tile and it automatically ensures that the plane is neatly divided into edge-

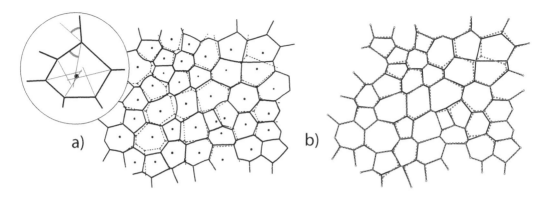

Figure 3.29 Construction of the Dirichlet centers in a polygonal tile (magnified sketch in panel a) and a reconstructed Dirichlet tessellation (dashed lines in panel a) corresponding to cultured retinal pigment cells from chick embryo (solid lines in panel a. The partition after the boundary-shortening procedure (b); this procedure is described in Figure 3.31. (Adapted from Ref. [92].)

to-edge packed tiles. Associating cells to point particles is very close to the physical line of thinking because it implies that one may seek a suitable interparticle interaction which would reproduce the actual positions of the Dirichlet centers. On the other hand, it is unclear what this interaction is like and whether it is isotropic as implied by the point-particle representation of the cells.

One may approach this question by simulating systems of particles characterized by various types of forces. This is doable because any physically relevant force should include a repulsive short-range part so as to ensure that cells do have a certain size and an attractive tail to make them adhere to each other. In zero-order approximation, an ensemble of hard disks should serve this purpose well. As it undergoes a transition from the fluid to hexatic and then to hexagonal phase with increasing density [93,94], the corresponding tessellations should range from disordered to hexagonal [95]. This is indeed the case, and the tessellations obtained seem to be quite similar to patterns seen in tissues (Figure 3.30).

In spite of the similarity of the patterns, the microscopic workings of cells themselves and of cell-cell interaction are too complex to be adequately approximated solely by a potential that depends only on distance between the centers of neighboring cells. A more refined version of the original Dirichlet model involves the so-called boundary shortening procedure [92] whereby the reconstructed Dirichlet tessellation is gradually relaxed locally as if the tile-tile boundaries were under a constant tension. The relaxation is done locally around an arbitrarily selected edge such that the total length of the five edges is minimized while keeping the areas of the two triangles and the two quadrilaterals next to the selected edge unchanged (Figure 3.31).

This procedure leads to symmetric three-valent vertices with all angles equal to 120° and it generally produces patterns that agree with real tissues much better than the original Dirichlet tessellation. In cultured retinal pigment cells shown in Figure 3.29, the correspondence is improved considerably albeit not in all cells to the same extent. In certain cases the boundary-shortening procedure works well even in tissues undergoing dynamical rearrangements involving T1 topological transformations [92].

At a conceptual level, the message of the better agreement obtained by boundary shortening is twofold. Firstly it points to the relevance of surface energy included in the other

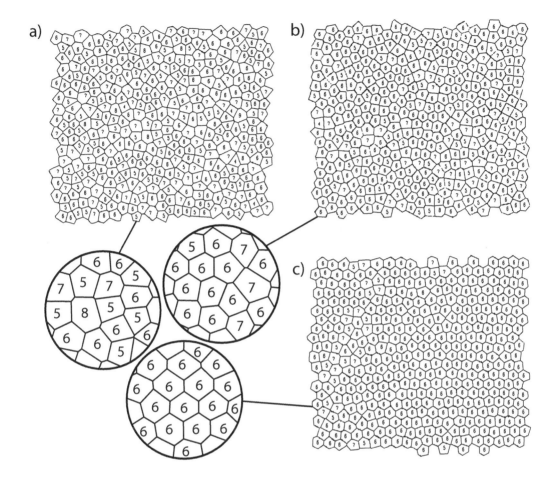

Figure 3.30 Voronoi tessellation of a system of hard disks, the density increasing from panel a to panel c. Polygons are labeled by the number of sides, and insets zoom in on the tilings to better show the differences. (Adapted from Ref. [95].)

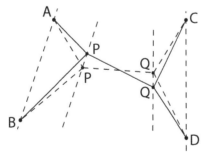

Figure 3.31 Boundary shortening procedure: vertices A, B, C, and D are fixed and vertices P and Q move along lines parallel to AB and CD, respectively [92]. Any displacement of vertex P so as to minimize the total edge length causes a displacement of vertex Q such that the area of the quadrilateral APQC is constant [92].

models discussed here, which is important because it brings the various models closer to each other although it disqualifies the point-particle representation of cells. Secondly, the remaining differences between the observed patterns in tissues and the reconstructed Dirichlet tessellation may seem subtle but this does not necessarily mean that they can be remedied by a minor adjustment—simply because the phase space is so severely restricted. As shown by the analysis of the packing of cone cells in the *Drosophila* retina [96] in Section 3.3.5, seemingly small differences may well require a considerable revision of the model.

3.3.5 Binary and ternary tissues

In tissues consisting of two or more types of cells, their packing geometry is interwoven with the positional order of each cell type. Of particular interest is the so-called cell intermixing [36] where the different cell types alternate by avoiding contacts between same-type cells. A nice example of such order is the epithelium of the basilar papilla in chick embryo shown in Figure 3.32a. The basilar papilla is the hearing organ of lower vertebrates and the epithelium consists of cells of two types: the hair cells, which are more circular, and the supporting cells which form ring-like coronas around the hair cells. Despite some defects, the arrangement of these cells is topologically somewhat reminiscent of the 4.6.12 uniform tiling of rectangular, hexagonal, and dodecagonal tiles (Box on p. 53 and Figure 3.32b).

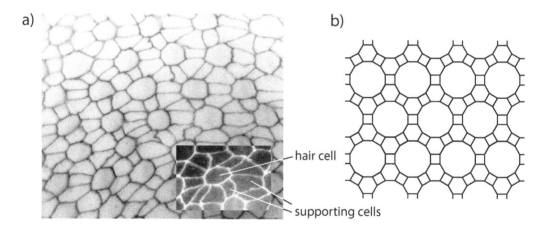

a)

b)

hair cell

supporting cells

Figure 3.32 Basilar papilla epithelium in E12 chick embryo showing a hexagonal array of the predominantly circular hair cells surrounded by the supporting cells (a; image courtesy of R. J. Goodyear is color-inverted for readability except in the bottom-right corner) quite similar to the 4.6.12 tiling in panel b.

A simpler and better known instance of a binary epithelium with an alternating structure is the checkerboard pattern seen in the Japanese quail oviduct epithelium [97]. This pattern can be readily explained using the differential adhesion hypothesis (DAH; Section 2.4), where cell mixing is favored if the interfacial tension of heterotypic cell-cell contacts Γ_{12} is smaller than the mean of the tensions of the two homotypic cell-cell contacts Γ_{11} and Γ_{22}:

$$\Gamma_{12} < \frac{\Gamma_{11} + \Gamma_{22}}{2}. \tag{3.46}$$

This fundamental result is easily derived within the lattice model where we assume that the cells are cubic but it does not depend on the details of implementation of the hypothesis. For example, a finer lattice simulation of DAH where cells occupy several nearby sites and can thus change their shape [98, 99] produces patterns where cells are well-mixed locally, forming a rather ordered checkerboard pattern consistent with the above simple analysis (Figure 3.33). In this more realistic model, there exist quite a few persistent defects leading to inhomogeneities that stretch across many cells.

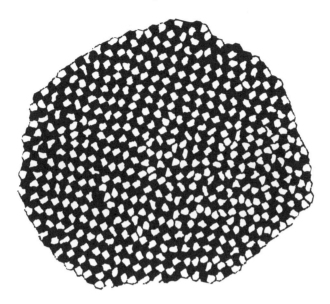

Figure 3.33 Mixed state of the differential-adhesion-hypothesis model in the Potts-model implementation. The local checkerboard structure is readily visible despite the defects that interrupt the pattern here and there. (Reproduced from Ref. [99].)

The DAH criterion [Eq. (3.46)] succinctly describes the main mechanism of cell sorting and it helps us to understand the structure of the *Drosophila* retina as one of the paradigmatic examples of patterning in developmental biology. Like in other arthropods, the compound eye of *Drosophila* consists of clusters of cells called ommatidia. These clusters form a hexagonal lattice like the hair cells of the basilar papilla in Figure 3.32 but their structure is more elaborated: each cluster consists of a few (typically four) photoreceptor cone cells in the center as well as of several types of pigment cells and bristle cells arranged around the cone cells (Figure 3.34a).

The ommatidia of wild-type animals contain four cone cells but in certain mutants their number varies between one and six (Figure 3.34b). The topology of cone cell clusters appears to be the same as in clusters of soap bubbles. For example, ommatidia consisting of three or more cone cells could in principle feature linear or even Y-shaped or branched clusters. Yet this is not the case and as shown in Figure 3.34b. All of the observed topologies are compact much like planar soap-bubble clusters (Figure 3.34c).

The qualitative agreement of cone cell clusters and bubble clusters appears so convincing that one is tempted to conclude that their mechanics are the same. If so, this would indeed be a big step ahead both on the theoretical and on the experimental side, which is why this idea is well worth pursuing further. This can be done by generalizing the surface-tension-type model with an area elasticity term so that the total energy of the cluster reads

$$E = \sum_{ij} T_{ij} P_{ij} + \frac{\lambda_A}{2} \sum_i (A_i - A_{i0})^2. \tag{3.47}$$

Here the first sum is over all cell-cell contacts, and T_{ij} and P_{ij} are the tension and the

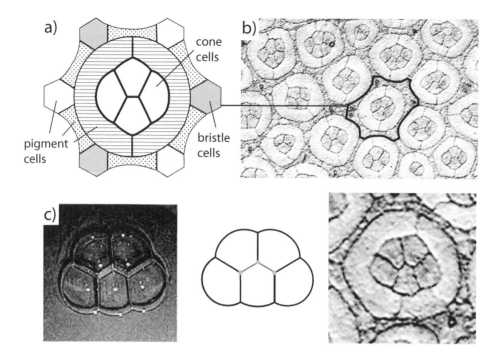

Figure 3.34 In-plane structure of *Drosophila* ommatidium (a) showing four cone cells, several types of pigment cells, and bristle cells. In *Roi* mutants, the number of cone cells varies (b) and their spatial arrangement seems to be consistent with clusters of soap bubbles (c). (Adapted from Ref. [96].)

length of the interface between cells i and j, respectively. The tensions T_{ij} are positive so as to ensure that the shape of cells remains convex. The second term is the area-elasticity term; each cell has a different reference area but their area moduli are the same.

This model has many parameters. To reduce their number, it is plausible to assume that the tensions of all cell-cell contacts of the same type (e.g., cone-pigment) are the same. Within the thus restricted parameter space, one can accurately test the model by setting the preferred area of each cell of the ommatidium to the experimentally measured value. Hard as one may try, the main conclusion of the test is that this variant of the surface-tension model is unable to reproduce the observed geometry of the ommatidia quantified by the angles between the various interfaces [100] which suggests that the model is not suitable.

Fortunately, the area- and perimeter-elasticity model [Eq. (3.33)] can generalized so as to describe heterotypic cell-cell contacts

$$E = -\sum_{ij} \Gamma_{ij} P_{ij} + \frac{\lambda_P}{2} \sum_i (P_i - P_{i0})^2 + \frac{\lambda_A}{2} \sum_i (A_i - A_{i0})^2, \qquad (3.48)$$

which works much better. Here the first term represents the specific cell-cell contacts, Γ_{ij} being the adhesion strength of the interface between cells i and j, and the second one is the perimeter elastic energy cast just like in Eq. (3.36).

Within this framework, the minimal-energy configurations of cone cell clusters agree with the observed geometry very well for a suitable choice of preferred perimeters. This is nicely demonstrated by plotting the predicted and the measured angles that characterize the wild-type four-cell cluster (Figure 3.35), which lie on the diagonal line bisecting the

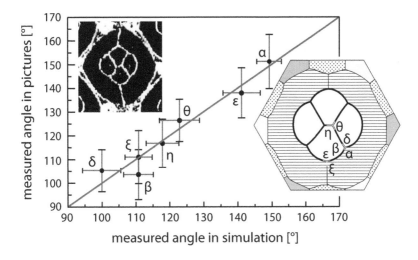

Figure 3.35 Comparison of the observed and reconstructed geometry of the wild-type *Drosophila* ommatidia represented by the seven angles between the various interfaces in the cluster. (Adapted from Ref. [100].)

quadrant. The deviations are all within errorbars,[4] which suggests that the generalized area- and perimeter-elasticity model used in the reconstruction describes the mechanics of these cells rather well. Thus we see that although the surface-tension and the perimeter-elasticity terms may look similar at a qualitative level in that they exert a contractile force around the contour, the differences between them are essential for a proper quantitative interpretation of the shape of ommatidia. Given that it contains an additional term and at least one more parameter per cell, it should not be too surprising that the area- and perimeter-elasticity model is superior.

After having demonstrated how the detailed structure of the ommatidium can be explained, we can now try to identify the key features of the area- and perimeter-elasticity model responsible for the good agreement. Given that in *Drosophila* mutants the cell cross-section area does not seem to change although the packing geometry of the cells does, it seems reasonable to replace the area-elasticity term for each cell by a constraint so that the areas of all cells are fixed. Then one is left with the perimeter elasticity energy and just two adhesion strengths. The parameter space can be further reduced by noting that the cross-section of isolated cone cells is circular so that their reduced area is 1, which connects their preferred perimeter and area. In addition, we can fix P_0 and A_0 in the two primary pigment cells immediately adjacent to the cone cells [101]. As shown in Figure 3.36, the predictions of this simple model compare with observations just as well as those obtained using the more elaborate framework. This is a reassuring message because it re-emphasizes the importance of surface interactions, the common theme of most mechanical theories of tissue structure.

So far the effective properties of cells that enter the physics models were assumed fixed or controlled by an external agent. But cells are not just passive drops and their internal structure may be reorganized because of adhesion, especially by redistributing and reconstituting the acto-myosin network. Imagine two cells soon after they adhere to each

---

[4]The simulation errorbar is estimated by visual comparison of the experimental and the simulated geometries, which shows that a 5–10% variation of the tensions typically does not affect the agreement very much.

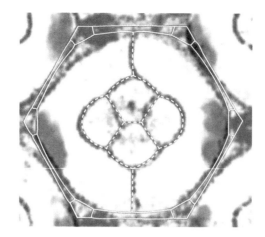

Figure 3.36 Micrograph of a wild-type *Drosophila* ommatidium compared to the theoretical cell cluster geometry obtained within the fixed-area/perimeter-elasticity model (dashed white lines). (Adapted from Ref. [101].)

other (Figure 3.37). At this stage, the main mechanism of adhesion relies on interactions between cadherins in the two membranes. After that, cadherin-cadherin contacts begin to play another role as a part of the signaling cascade. The final outcome of this process is a reduced contractility of the acto-myosin network at the cell-cell contact and an increased contractility in the non-contact parts of the membrane. In the language of coarse-grained models, the tension of the cell faces that are not engaged in cell-cell adhesion is increased relative to those that are in contact with a neighboring cell, which means that cells are mechanically polarized. But if the tensions of the different cell faces are altered then cell shape should change too just as if adhesion strength on the faces in contact were increased [42].

Let us illustrate this idea by a simple two-dimensional example where we consider a rectangular cell of fixed area A with sides carrying tension and adhesion strength. As long as the cell is isolated, tension is the same at all sides and the cell shape is a square. Now

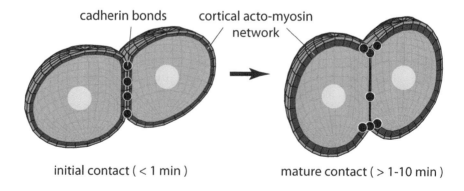

Figure 3.37 Adhesion-induced cell polarization: after the establishment of the initial contact between cells, cadherins redistribute to the non-contact part of the membrane, which leads to an increased tension in these parts and to a reduced tension in the contact zone indicated by a thicker and a thinner black line, respectively [42].

assume that it adheres to two neighbors on opposite sides such that its energy is

$$E = 2T_a L_a + 2T_b L_b - \Gamma L_b, \tag{3.49}$$

where T_a and T_b are the tensions on the non-contact and the contact sides of length L_a and L_b, respectively, and Γ is adhesion strength. In equilibrium,

$$L_b = \sqrt{\frac{2AT_a}{2T_b - \Gamma}} \tag{3.50}$$

which shows explicitly that as far as cell shape is concerned, a simultaneous increase of T_a and decrease of T_b caused by the redistribution of the acto-myosin network is equivalent to an increase of Γ. In fact, it appears that the differential tension due to mechanical polarization can explain the observed values of tissue surface tension, which are orders of magnitude larger than those expected from cadherin-mediated adhesion [42] (Section 2.4.2).

At a more conceptual level, the intra-cellular processes behind the reorganization of the acto-myosin network suggest that cell shape is not the only variable to be considered in cell-cell interaction and thus in tissue organization. Instead, the distribution of the network and, more generally, the distribution of polarity proteins matters too. In addition to its length as the mechanical degree of freedom, each edge is also characterized by the concentration of these proteins which regulate the edge tension.

One implementation of these ideas was designed to model the delicate pattern of the zebrafish retina [102]. The retina is an ordered rectangular lattice of four types of cone cells with photoreceptors for UV, blue, red, and green light, with the red and green cone cells forming doublets. The cone cells are arranged in columns separated by rod cells, and columns consist of repeating UV/red-green/blue/green-red sequences. In addition, the neighboring columns are staggered so that the doublets are arranged side-by-side yet in antiparallel fashion (Figure 3.38). As a result, the unit cell contains a total of twelve cone cells.

The aim of the mechanical model of the zebrafish retina is to explain phenomenologically the predominantly quadrilateral shape of cells and their arrangement in parallel stripes

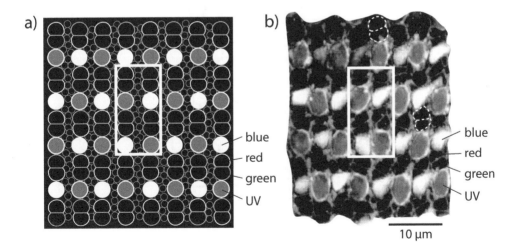

Figure 3.38 Schematic representation (a) and a micrograph of the arrangement of the four types of cone cells in the adult zebrafish retina (b); the rectangles show the unit cell. (Adapted from Ref. [102].)

without resorting to a built-in intrinsic shape or polarization of individual cells; the development of the different spectral types of cone cells is not included [102]. The main assumptions of the model are:

- Cells contain two polarity proteins, A and B, which bind to the membrane and alter the tension of the edge shared by cells i and j such that

$$T_{ij} = T_0 - T_d \left(c_A + c_B \right), \tag{3.51}$$

where T_0 is the bare edge tension in absence of either protein and T_d describes the dependence of tension on the two concentrations. Thus the distribution of polarity proteins within a cell affects its shape.

- The two types of proteins favor opposite sides of a cell, thereby introducing polarity. They preferentially bind to short edges so that cell shape affects their distribution. This closes the cell-shape/protein-distribution feedback loop which also contains external stresses as they too determine the shape of cells. The number of proteins of either type within each cell is constant.

- The mechanical energy of each cell consists of the surface energy and of the area elasticity [Eq. (3.47)]. The displacement of each vertex as well as the evolution of the protein concentrations are governed by overdamped kinetic equations.

This elaborate model contains quite a few additional elements, say progressive growth involving the transformation of precursor cells which lack the cell polarity pathway into cone cells (Figure 3.39) [102]. Instead of describing all of them, let us single out the response of the retina to external stress, which is an essential feature emerging from the coupling between cell shape and the distribution of the polarity proteins. By stretching the cells in a given direction, the proteins are rearranged which alters the tensions of the edges. Most

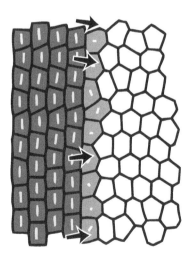

Figure 3.39 During growth, existing cone cells in the zebrafish retina (dark gray) polarize the precursor cells (white) and turn them into presumptive cone cells (light gray), which thus become more ordered. Bars within the cone cells represent the orientation and the degree of cell polarization, and the propagation of the polarization front is indicated by arrows. (Adapted from Ref. [102].)

importantly, since the tension of an edge depends on the concentrations of the polarity proteins on both sides, this mechanism allows the propagation of cell polarity established in a given cell cluster to neighboring cells. In Figure 3.39, the polarized cone cells on the left alter the tension in the neighboring non-polarized cells, which then become presumptive cone cells. In turn, the random tiling of the non-polarized cells is turned into an orientationally ordered tiling of the cone cells.

Numerical simulations of this model of retina development in zebrafish are complicated but the key qualitative conclusions are clear and unambiguous. Figure 3.40 shows snapshots of four representative examples which suggest that both anisotropic stress and gradual column-by-column precursor-cone transformation are needed for a well-ordered lattice of predominantly rectangular cone cells. If all precursor cells turn to cones simultaneously, cells become polarized but their shape is poorly defined. The tissue lacks global orientational order as well as the regular positional order characteristic for the retina. Stretching the tissue uniaxially rather than isotropically improves the orientational order (Figure 3.40) but cell

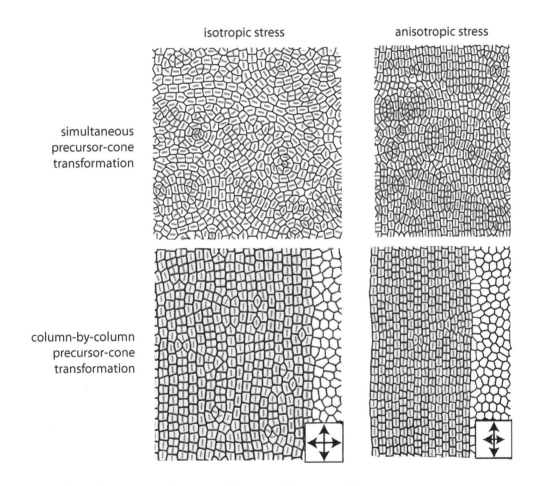

Figure 3.40 Snapshots of the model zebrafish retina. The top row shows its structure in case cell fate is induced in all cells simultaneously whereas in the bottom row this happens in one cell column at a time. The left and right columns show the pattern under isotropic and anisotropic in-plane stress, respectively; the stress pattern is indicated by arrows. (Adapted from Ref. [102].)

shape still exhibits considerable variability and although some short columns can be seen, they are not straight.

The magnitude of orientational order can be increased by a more accurate representation of the dynamics of the precursor-cone transformation. The cones develop during migration to the so-called retinal margin at the edge of the proliferating layer where the polar and the azimuthal directions are nonequivalent. Experiments show that as cells migrate to the retinal margin, they transform into cones in a gradual column-by-column fashion [102]. If this process is included by inducing cell fate in one column at a time, the model indeed reproduces the positional and orientational order as well as the well-defined cell shape seen *in vivo*. At the same time, an isotropically stretched retina displays a reduced degree of positional order of cone cells so that we can conclude that both stress anisotropy and column-by-column dynamics are required so as to ensure the proper pattern.

The tiling patterns typical for epithelial tissues do share many features with those seen in other physical systems characterized by a planar cellular partition, as they are subject to the same hard topological and geometrical constraints. Still the structure of some epithelia depends decisively on phenomena unique to living cells; for example, the key process defining proliferating tissues is cell division. It is reassuring that the theoretical models of these phenomena need not be very complicated, much like the symmetry of crystals of some elemental solids is essentially a matter of packing of spheres. The challenge, of course, is to identify the essential ingredient of the structure or process in question, which invariably relies on accurate and detailed experimental insight both at the cell and at the molecular level. This chapter goes back and forth from the simple to the more complex concepts so as to illustrate how this can be done.

PROBLEMS

3.1 Consider a tiling with a dominant fraction of, say, decagons, such that $10F_{10} \gg nF_n$ for all $n \neq 10$. Such a tiling obviously does not exist, as Eq. (3.10) cannot be satisfied. However, tilings consisting mostly of triangles or quadrilaterals may be constructed as illustrated by Figure 3.2a. The construction of a large border path in such a tiling like in Figure 3.5 will produce a small number of other types of polygons such that $4F_4 \gg nF_n$ for all $n \neq 4$. Thus Eq. (3.10) predicts that $F_4 \approx 0$. What is the source of this problem? What does it tell about tilings made of many triangles and quadrilaterals?

3.2 Prove that the tiling consisting of convex heptagons is impossible. *Hint:* Show that the average valence of vertices in tiling is 2.8.

3.3 In all infinite tilings each edge is shared by two neighboring faces. The mean number of sides of a polygon in an infinite tiling can thus be defined as $\bar{n} = 2E/F$. In tissues containing only three-valent vertices, $3V = 2E$. Use this to show that $\bar{n} = 6$ in these tissues. How does $\bar{n}$ change if four-valent vertices are present?

3.4 Assume that a tiling contains V_3 trivalent and V_4 four-valent vertices, and denote their respective fractions by $v_3 = V_3/V$ and $v_4 = V_4/V$, where $V = V_3 + V_4$. Derive the analog of Eq. (3.12) for this case:

$$\sum_n \left(6\frac{1 + v_4/3}{1 + v_4} - n \right) p_n = 0. \tag{3.52}$$

3.5 A process occurs which transforms all n-sided cells in a tissue into $(n+1)$-sided cells. Show that such process cannot be performed without creating holes in tissues.

3.6 Show that the Desch–Feltham law can be expressed as a linear combination of the topological constraints on the tissue like Lewis' law. In other words, of all possible relations between the average perimeter of cells in a given polygon class $\overline{P_n}$ and the number of sides n which can be used to construct a mathematical model of the tissue, the linear dependence maximizes the entropy of the tissue. Assume that $\overline{P_n} = \nu P_0(n-1)$, and show that $\nu = 1/5$.

3.7 The Aboav–Weaire law can be derived from the planar version of the Gauss–Bonnet theorem which states that

$$\int_{\delta D} C \, dl + \sum_{i=1}^{n} \alpha_i = 2\pi, \tag{3.53}$$

where D is a planar domain bounded by a curve δD containing n discontinuities with exterior angles α_i as illustrated in the image below. Here C is the curvature of the curve in its continuous parts. Apply the Gauss–Bonnet theorem to a domain containing an n-gon G and all its neighbors contained within the boundary δR, and derive the Aboav–Weaire law in the form

$$\frac{1}{n} \sum_{i=1}^{n} m_i = 5 + \frac{6}{n} - \frac{3}{n\pi} \int_{\delta R} C \, dl. \tag{3.54}$$

Here $m_i, i = 1, \ldots, n$ are the numbers of sides of all neighbors of the n-gon G. *Hint:* The continuous parts of the bounding curve can be deformed so that all the exterior angles become $\alpha_i = \pi/3$ for all i. Use this to show that $n = 6 - 3/\pi \int_{\delta G} C \, dl$. Apply this formula to evaluate $n + \sum_{i=1}^{n} m_i$ by integrating over G and all its neighbors; see Ref. [103] for details.

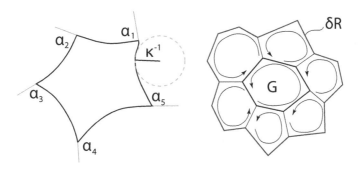

3.8 The elastic response of the acto-myosin ring can be modeled as that of a spring of equilibrium length P_0. The energy of the deformed cell would thus be $E = C_0(P - P_0)^2/2$, where P is the perimeter of the deformed cell and P_0 is the preferred perimeter. When the cell assumes the preferred perimeter, the elastic cellular filaments are unstrained. Show that such a model of the acto-myosin response renormalizes the cell-cell adhesion energy Γ. Discuss the ranges of C_0 and Γ that still allow for the formation of tissue.

3.9 Calculate the equilibrium size of hexagonal tiles in the model given by Eq. (3.33) and express it using reduced contractility $\mathcal{C}$ and reduced line tension $\mathcal{T}$. By considering isotropic deformation of the hexagonal tiling, calculate its bulk and shear modulus.

3.10 Consider a two-step division of cells in the tissue shown below (the cleavage planes are indicated by dashed white lines). How does the division of a neighbor change the number of sides of a given cell? In which case does the number of sides remain unchanged? Calculate the mean number of cell sides in the three phases.

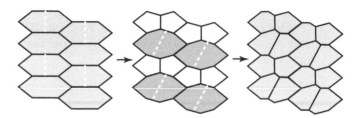

3.11 Construct the P matrix for the division rules illustrated in Figure 3.24.

3.12 Formulate a set of division rules for the cells different from those discussed in Section 3.3.2 and write the corresponding P matrix. (*Hint:* A simple set of rules can be obtained from those encoded in Figure 3.24 by keeping only the most symmetric situations. It is also possible to exclude the division of cells having less than, say, six vertices.) Compute the polygon fractions in the tissue described by the constructed set of cell division rules.

3.13 Show that a single T1 transformation in a perfect hexagonal lattice produces two pentagon-heptagon pairs. Show that a sequence of T1 transformations can be constructed so to separate the thus obtained heptagons and pentagons by an arbitrarily large distance.

3.14 Analyze the Voronoi tessellations pertaining to two-dimensional hard disk systems in Figure 3.30 in terms of polygon fractions. Discuss how the tilings change depending on the density of disks and try to relate them to fixed-area/fixed-perimeter tilings in Figure 3.27.

3.15 Discuss the equilibrium states of the differential-adhesion-hypothesis lattice model of a tissue formed by rectangular cells of aspect ratio b/a.

3.16 In a tissue made of perfectly hexagonal cells, there are 10 cells which are modified so to exhibit a specific adhesion along one of their sides, strongly favoring contact with a surrounding medium. The surface tension along these sides is such that the adhesion strength at the cell-medium contact Γ_m is much larger than the cell-cell adhesion strength. The 10 cells in the tissue are shown in the illustration below and the modified sides are indicated by thick lines. Discuss the formation of cavities in the tissue, assuming that the adhesion strength at the contact of the non-modified cell side with the medium is negligible. Discuss the size and the distribution of cavities in a tissue in which the fraction of specific cells is p.

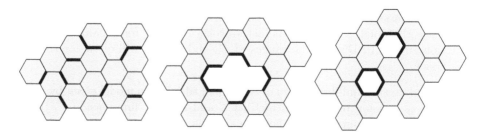

3.17 Assume that the *en-face* cross-section of the cell in a hexagonal tissue can be constructed from straight sections of length L_f and from circular arcs of length L_c and radii R_v as illustrated. The cells adhere to each other along the straight sections and adhesion strength is denoted by Γ. The bending energy of the circular arcs is given by $E_b = D/2 \int_{arcs} \mathrm{d}l/R_v^2$, where $\mathrm{d}l$ is the infinitesimal length along the arc, D is the bending rigidity, and R_v is the radius of curvature. Find R_v that minimizes the energy of the tissue at fixed total perimeter P of the cells (see also Ref. [104]), and relate it to Eq. (3.4). Show that the tissue disintegrates if the bending rigidity is larger than $\Gamma(P/2\pi)^2$.

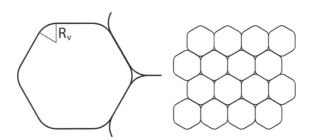

3.18 Estimate the energy barrier for the T1 topological transformation in a regular hexagonal tiling governed by the area- and perimeter-elasticity model [Eq. (3.36)]; specifically, express it as a function of $P_0/\sqrt{A_0}$ for various ratios of $\mathcal{T}^2/\mathcal{C}$. *Hint:* Consider four hexagons around an edge, and calculate the energy difference between the initial state and the state just before neighbor exchange where the edge length is zero. Assume that only the endpoints of the shrinking edge move and that all other vertices of the four hexagons considered are fixed.

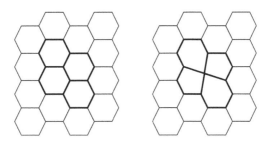

CHAPTER 4

Shape of epithelia

THE MICROSCOPIC TEXTURE OF ORGANS is defined by the form of the epithelium. Some organs are covered by a flat epithelial layer with a smooth surface (like the stratified cuboidal epithelium in an excretory duct of the human salivary gland, or the epithelial lining of the round stingray *Urobatis halleri* testicular duct shown in Figure 4.1a) but in many cases the epithelium is corrugated on a submillimeter length scale larger than the size of the cell yet still smaller than the organ. The corrugation is often strikingly regular, characterized by either one- or two-dimensional periodicity. An example of a one-dimensional modulation of epithelial shape are the longitudinal folds seen, e.g., in the stomach of stalked sea squirt *Styela clava* (Figure 4.1b and c) where the geometry of the waveforms—fold height, crest width, etc.—varies very little from fold to fold. In the gut of a chick embryo, longitudinal folds gradually transform in a zig-zag morphology eventually emerging as an oblique lattice of finger-like protrusions projecting into the lumen [105]. These protrusions known as villi are marked by a well-defined cylindrical shape and form a two-dimensional periodic array. In other cases like in the epithelium of the mammalian small intestine shown in Figure 4.1d, the villi lack a strict positional order. Villi may also be leaf-shaped as illustrated by those found, e.g., in mammalian small intestine (Figure 4.1e).

The corrugated shape of epithelia serves many functions. Their excess area compared to the flat epithelium increases the capacity for absorption or secretion, and this is the reason why corrugated morphologies are typical for gastrointestinal tract, respiratory system, and glands. The corrugated texture may also enhance the retention capacity of the lumen contents at the surface, thereby further facilitating absorption and secretion. In some cases, the excess area accommodates any distension or other deformation of the organ; in others, folding may be caused by contraction of the surrounding tissue. Epithelial folding is important not only in adult animals but also during embryonic development where it is one of the main modes of morphogenesis. The most important developmental phase relying on folding is gastrulation where the initially shell-like blastula is transformed into a more complex invaginated structure with three germ layers eventually giving rise to internal organs, nervous system, and muscles.

To a physicist, the regularity of corrugated epithelial morphologies across large areas is a strong signature of a possible underlying mechanical process simply because these morphologies are very reminiscent of elastic instabilities including folding, wrinkling, creasing, period-doubling, etc., seen in inanimate soft materials at the macroscopic scale [109, 110]. This similarity is the leitmotif of the present chapter and although it will not explain all of the observed phenomenology, it will provide an insight into most of the key features. Our perspective does not exclude the biochemical, genetic, and physiological aspects of the tissues—it merely explores the mechanical scenarios of epithelial morphogenesis as a framework relying primarily on spontaneous symmetry breaking triggered by a uniformly

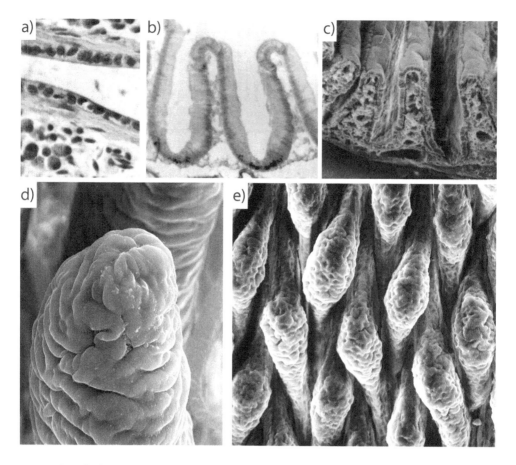

Figure 4.1 Epithelial morphologies: flat epithelium in branch of a longitudinal duct of the testis of *Urobatis halleri* (a; reproduced from Ref. [106]); longitudinal folds in the stomach of *Styela clava* (b and c; adapted from Refs. [107] and [108], respectively); and finger-like (d; image by L. Hirst, Wellcome Images) and leaf-like epithelial villi (e; image by G. Gabella, Wellcome Images) in mammalian small intestine.

distributed control parameter but also on a pre-patterned cell specialization to a differential active deformation and displacement of cells.

This mechanistic perspective is hardly new. Among the earliest studies of this kind, we must single out the experimental model developed and examined by W. H. Lewis just after World War II. He postulated that the various invaginations and cavities in animals including blastopore, neural tube, etc., "be produced by the increased contractive tension of the superficial gel layers of the cell surfaces bordering the concavities" [111]. This led him to construct a tabletop demonstration device consisting of brass bars, rubber tubes, and rubber bands assembled into a series of twelve quadrilateral segments. In each segment, two brass bars were placed opposite one another, the stiff tube keeping them at a fixed center-to-center separation, and their top and bottom ends were connected by rubber bands (Figure 4.2a). Each brass bar except the terminal ones was shared by the adjacent segments and the device was then viewed as a cross-section of the epithelium. The brass bars represented the adhering lateral sides of cells, the stiff tubes mimicked the incompressible cytoplasm by ensuring that the area of each segment is more or less fixed, and the rubber bands played

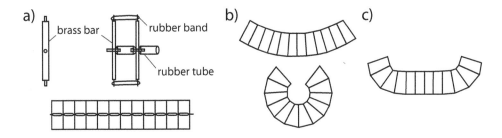

Figure 4.2 Lewis' experimental model of epithelium consisting of brass bars, stiff rubber tubes, and rubber bands assembled into a series of twelve quadrilateral segments each representing the cross-section of a cell (a) [111]. By adjusting the tension of the rubber bands, the series can be deformed into arches of various curvatures (b) or into a bowl-like shape (c).

the role of the contractile apical and basal faces of cells (top and bottom sides in Figure 4.2, respectively).

In each segment, the apical and basal tension can be varied by adding or removing the rubber bands on the respective end. By putting the same number of bands on both sides of all segments, one ensures that the series is straight. On the other hand, if the number of bands on the apical side and the number of bands on the basal side are the same in all segments but different from one another, the series is deformed into a slightly or a very curved arch (Figure 4.2b) depending on the difference. If the apical tension is increased only in the second and in the penultimate segments, a bowl-shaped invagination is created (Figure 4.2c).

Lewis went on and studied the model under various constraints mimicking the neighboring tissues which oppose invagination. He held the end bars fixed and varied the number of rubber bands on the segments, observing that in this case the model does not bend if the tensions on the apical and on the basal side are the same in all segments. He also noted that if the number of bands on the apical and the basal side are different only in the four middle segments, they become somewhat invaginated; also deformed are the remaining segments at either end of the model although they have the same number of bands on the apical and the basal side.

Lewis' mechanical model can be easily re-analyzed numerically. If we assume that the stretched rubber bands can be represented by Hookean springs, the energy of the system reads

$$E = \frac{1}{2} \sum_{i=1}^{N} \left[k_i^a (l_i^a - l_{i,0}^a)^2 + k_i^b (l_i^b - l_{i,0}^b)^2 \right], \tag{4.1}$$

where $l_{i,0}^a$ and $l_{i,0}^b$ are the equilibrium lengths of the rubber bands on the apical and the basal side, respectively, k_i^a and k_i^b are their effective spring constants, and l is the fixed length of the stiff tubes in the middle of the segments. Finally, N denotes the total number of segments. The brass bars and rubber tubes together impose a geometric constraint requiring that the sum of apical and basal sides be fixed: $l_i^a + l_i^b = 2l$. If the equilibrium lengths of the rubber bands are much smaller than l, they can be eliminated from the model by setting $l_{i,0}^a = l_{i,0}^b = 0$ so that the solution depends only on the spring constants.

Solved without any constraints, the model is rather simple as the minimization of the total energy reduces to minimization of energies of each individual segment independently. A few results are shown in Figure 4.3; we do not include the trivial straight shape obtained

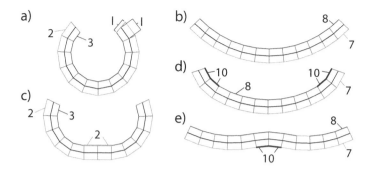

Figure 4.3 Numerically obtained minimal-energy shapes of the unconstrained Lewis' model with $N = 12$ segments. In panels a and b, $k_i^a/k_i^b = 3/2$ and $8/7$, respectively, the labels indicating k_i^a and k_i^b in arbitrary units. The model in panel c is the same as that in panel a except for the two middle segments where the number of bands on the apical side is 2 rather than 3. Panels d and e show variants of the model in panel b with 10 apical bands in the two segments at either end (d) and with 10 basal bands in the two middle segments (e) as indicated. The lengths of the brass bars and the stiff rubber tubes are the same.

in case of identical apical and basal tensions in all segments. In panels a and b, the number of bands on the apical side is the same in all segments (3 and 8, respectively) but different from that on their basal sides (2 and 7, respectively). As a result, the model bends and the degree of bending increases with the ratio k_i^a/k_i^b given by the ratio of the numbers of rubber bands. The other panels show examples of minimal-energy shapes in cases where the number of bands is changed in a few segments as indicated. Each segment acts as an independent hinge, which is best seen by comparing panels b and e which differ only in the curvature of the middle two segments; this, in turn, is enough to considerably alter the overall shape.

The solutions of the model become richer when we constrain it in some way, say by imposing boundary conditions which demand that bars on both ends are vertical and that their midpoints lie on the same horizontal line. Selected periodic solutions with unconstrained wavelength are presented in Figure 4.4. Intriguingly, if the number of rubber bands on the

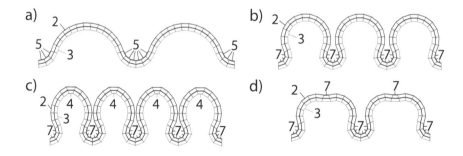

Figure 4.4 A few nontrivial periodic solutions of the $N = 19$ Lewis model with 2 bands on the apical side and 3 bands on the basal side of all segments except where indicated otherwise; the apical or basal sides with a different number of bands are drawn with thick lines.

apical and on the basal side is the same in all segments, the ground state is a straight strip even if these numbers are different from each other. In fact, their ratio may be very large and yet the ground state remains straight (this trivial solution is not shown in the figure). This was also observed by Lewis in his constrained mechanical model. But as soon as some segments are different from others in terms of numbers of bands on either side, the strip buckles and this leads to a folded ground state. Of course, the exact shape of the waveform depends on the distribution of the number of apical and basal bands (Figure 4.4). We will see in Section 4.1.2 that there exist other mechanical models where nontrivial morphologies are possible even if all cells are identical.

Proposed by an established embryologist, this mechanical model is anything but naive. Instead, its succinct form is an expression of Lewis' deep insight into morphogenesis. Nowadays a similar line of thinking may be explored much more easily and in greater detail, and the aim of this chapter is to introduce the theoretical basis for such an endeavor within the context of large epithelial sheets. We first focus on models of epithelia completely devoid of any active, state-dependent behavior of cells and then we gradually turn to the agent-based theories of epithelial deformation. Some of the ideas described will be needed and reused later when discussing morphogenesis in Chapter 5.

4.1 TISSUE AS SUPPORTED LAYER

Many theories of nontrivial epithelial states such as folds are based on buckling. In its simplest variant, buckling or Euler instability is a deflection of a rod due to a compressive lengthwise force. Once this force exceeds a critical value, the initially straight rod bends (Figure 4.5, Problem 4.3) [112]. In the most direct application of the mechanism behind buckling, epithelia are viewed as elastic materials qualitatively somewhat reminiscent of the bimetallic strip used in thermostats (Problem 4.4). The formation of folds and other nontrivial morphologies is triggered by a suitable mechanical control parameter analogous to temperature in the case of the bimetallic strip. Just like heating leads to the misfit strain and bending of the strip, a variation of the control parameter drives the tissue through a sequence of minimal-energy equilibrium states which include the deformed morphologies. The control parameter must reflect a certain property of the tissue or a process within it, and in the two theories discussed below it is related to tissue apico-basal polarity and growth, respectively.

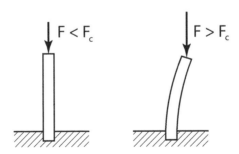

Figure 4.5 The simplest instance of the Euler instability is observed in a column with one end fixed and a compressive force F applied at the free end. The rod remains straight until the force exceeds a critical value F_c, and then it buckles.

4.1.1 Area-mismatch buckling

One of the key features of many epithelia is a rapid cell division needed for tissue renewal, the typical turnover time ranging from a few days to a few weeks. As the epithelium is attached to the basement membrane and stroma underneath it, its in-plane growth can only be accommodated by buckling in a process similar to patterning of a thin metal film deposited on an elastic substrate [113] as a two-dimensional version of the bimetallic strip. The basic idea behind this area-mismatch mechanism can be illustrated using a model where the net effect of cell division is an intra-epithelial pressure [114] which is relaxed by in-plane expansion provided that the homeostatic pressure[1] exceeds a certain threshold (Figure 4.6a). The area increase is stabilized by an elastic deformation of the stroma and the bending rigidity of the epithelium itself (Figure 4.6b), and the interplay of these three mechanisms results in a well-defined fold wavelength.

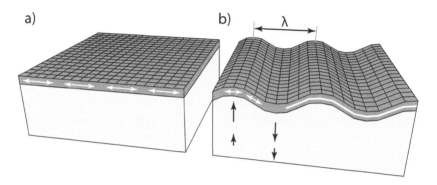

Figure 4.6 Area-mismatch theory of epithelial buckling: homeostatic pressure within the epithelium attached to an elastic stroma drives the in-plane expansion of the epithelium (a), which is counterbalanced by the bending elasticity of the tissue itself indicated by the white contour in panel b and by the deformation of the stroma shown by arrows. The buckled state is characterized by a well-defined wavelength λ.

Monge representation

Here surfaces in three dimensions are represented by their height measured from the horizontal plane: $z = h(x, y)$. Such a representation does not allow for overhangs, i.e., for two or more values of the z coordinate above a point in the (x, y) plane. In the small gradient limit when $|\nabla h(x, y)| \ll 1$, the mean curvature in Monge representation is given by

$$c = \frac{1}{2}\nabla^2 h. \qquad (4.2)$$

---

[1]Homeostatic pressure is the pressure that characterizes the steady state of a proliferating tissue where cell division and cell death or removal are balanced and is exerted on the walls of a semipermeable compartment that allows the passage of water, nutrients, etc., needed for tissue growth.

For simplicity, let us represent the epithelial layer by an elastic plate of fixed thickness h_c, a bending modulus $D_c \propto h_c^3$ [Eq. (3.2)], and a negative surface tension $-\Gamma$ arising from the homeostatic pressure p_h characteristic of the steady state of the tissue where the net growth due to cell division is counterbalanced by programmed cell death. Let us calculate the energy of a longitudinal wavelike deformation of such an epithelium on an elastic half-space. At small deformations we may assume that the shape of the tissue can be described using the Monge representation and a simple sine-wave ansatz

$$z(x) = z_0 \cos(qx), \tag{4.3}$$

where q is the wave vector; here $z(x)$ is used as a shorthand for the vertical component of the displacement vector $\mathbf{u}$ at the surface.

The total energy of a deformed supported epithelium consists of three terms: the bending energy of the tissue itself, its surface energy, and the elastic energy of the stroma. For a simple sinusoidal fold with a height profile modulated only along the x axis, the first term is proportional to the integral of the curvature squared over tissue area and reads $(D_c/2) \int \left(\mathrm{d}^2 z/\mathrm{d}x^2 \right)^2 \mathrm{d}A$, which gives $(D_c/2) z_0^2 q^4 A$. The surface term is proportional to the area of the tissue which is increased due to buckling: $-\Gamma \int \mathrm{d}A = -\Gamma \int \sqrt{1 + (\mathrm{d}z/\mathrm{d}x)^2}\mathrm{d}A$. To the lowest order, the integral can be estimated by $A + (1/2) \int (\mathrm{d}z/\mathrm{d}x)^2 \mathrm{d}A$ so that upon deformation, the area increase compared to the projected area of the (x, y) plane is $z_0^2 q^2 A/2$ and the corresponding change of the surface energy is $-\Gamma z_0^2 q^2 A/2$.

Calculating the elastic energy of the stroma is a little bit more involved but understanding its dependence on the wavevector q is not. In an incompressible medium, the elastic energy density per unit volume reads $(Y_s/3)\epsilon_{ik}^2$ where Y_s is Young's modulus of the stroma and $\epsilon_{ik} \propto z_0 q$ is the strain tensor.[2] In mechanical equilibrium, a modulation of the shape of the surface such as Eq. (4.3) gives rise to a subsurface deformation which decays exponentially with the distance from the surface, the characteristic penetration depth being equal to $1/q$. This means that the effective thickness of the stroma layer that is affected by the bucking deformation is $1/q$, and the integral of the energy density over all half-space reduces to $(Y_s/3)z_0^2 q A$.

Now we can collect and compare the three terms. As all of them are proportional to the projected area of tissue A, it is advantageous to consider the total energy per unit area which reads

$$\frac{E}{A} = \frac{D_c}{2} z_0^2 q^4 - \frac{\Gamma}{2} z_0^2 q^2 + \frac{Y_s}{3} z_0^2 q. \tag{4.4}$$

The terms contain a common factor of z_0^2 expected in the harmonic approximation but are each characterized by a distinct power-law dependence on the wavevector: q^4, q^2, and q. As the energy is measured relative to that of the undeformed state, bucking instability is indicated by a negative value of E/A due to the surface term, i.e., by the intra-epithelial pressure generated by dividing cells.

Simple as it may be, this model clearly emphasizes the roles of three terms and especially the distinction between the bending rigidity of the epithelium and the elasticity of the stroma. The former is dominant at large qs, that is, at small wavelengths, ensuring that the minimal-energy buckled pattern has a finite wavelength and does not lead to arbitrarily fine corrugations. On the other hand, the linear q-dependence of the elastic energy of the stroma stabilizes the epithelium at small qs, i.e., at large wavelengths—without it, the epithelium would necessarily buckle at any value of D_c and Γ. In this sense, the bulk elasticity of the stroma is needed to account for the fact that not all epithelia are folded.

---

[2]This was obtained by assuming that stroma is incompressible so that its Poisson's ratio $\nu = 1/2$ and by using the Hookean elastic energy [Eq. (2.32)].

Subsurface deformations

For an isotropic solid, the equation of mechanical equilibrium [Eq. (2.29)] can be spelled out explicitly in terms of displacement $\mathbf{u}(\mathbf{R})$. To this end, insert the linear part of Eq. (2.24) in the stress tensor σ_{ij} in Eq. (2.30) and then calculate $\sum_j \partial\sigma_{ij}/\partial x_j$. This gives $\sum_j \partial\sigma_{ij}/\partial x_j = Y/[2(1+\nu)]\sum_j \left[\partial^2 u_i/\partial x_j^2 + 1/(1-2\nu)\partial^2 u_j/\partial x_j\partial x_i\right]$. The first and the second term on the right are the i-th components of $\nabla^2\mathbf{u}$ and $\nabla\nabla\cdot\mathbf{u}$, respectively. Thus Eq. (2.29) in vector notation reads

$$\frac{Y}{2(1+\nu)}\left(\nabla^2\mathbf{u} + \frac{1}{1-2\nu}\nabla\nabla\cdot\mathbf{u}\right) + \mathbf{f}^{ex} = 0, \tag{4.5}$$

where $\mathbf{f}^{ex}$ is the body force per unit volume. Consider now an elastic half-space with an imposed deformation of the top surface described by $\mathbf{u}_0(x) = u_0(x)\mathbf{e}_z$. The displacement field $\mathbf{u}(x, z)$ consists of the longitudinal mode with $\nabla\times\mathbf{u}_l = 0$ and the transverse mode with $\nabla\cdot\mathbf{u}_t = 0$. In the absence of body forces, Eq. (4.5) for each of the components of the transverse mode reduces to

$$\frac{\partial^2 u_{ti}}{\partial x^2} + \frac{\partial^2 u_{ti}}{\partial z^2} = 0. \tag{4.6}$$

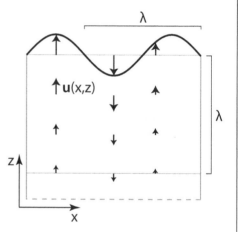

If the surface deformation $\mathbf{u}_0$ is described by $\exp(iqx)$, the dependence of u_{ti} on x must have this form too. Then Eq. (4.5) implies that u_{ti} decays exponentially with distance from the surface so that

$$u_{ti}(x, z) \sim \exp(qz)\exp(iqx). \tag{4.7}$$

The penetration depth of the subsurface deformation thus coincides with its wavelength $\lambda = 2\pi/q$. The same holds for the longitudinal mode.

To calculate the smallest value of the surface tension needed for buckling, the energy of the epithelium [Eq. (4.4)] is recast by introducing the characteristic wavevector $q_c = (Y_s/D_c)^{1/3}$ introduced by equating the two elastic terms. This naturally leads to the energy scale given by $Y_s^{4/3}z_0^2/D_c^{1/3}$ and the final dimensionless energy per unit area reads

$$\mathcal{E} = \frac{1}{2}\left(\frac{q}{q_c}\right)^4 - \frac{\mathcal{G}}{2}\left(\frac{q}{q_c}\right)^2 + \frac{q}{3q_c}, \tag{4.8}$$

where

$$\mathcal{G} = \frac{\Gamma}{D_c^{1/3}Y_s^{2/3}} \tag{4.9}$$

is the reduced surface tension. It is easy to see that buckling takes place when $\mathcal{G} > 3^{1/3} \approx 1.442\ldots$ (Figure 4.7a).

Rewriting the energy in a dimensionless form allows us to readily interpret the predictions of the model. For example, epithelial folds formed by buckling have a well-defined wavelength λ and within the linear analysis presented here it corresponds to the wavevector where the energy is minimal. Figure 4.7a shows that λ must scale as $2\pi/q_c \sim 2\pi(D_c/Y_s)^{1/3}$. Since the bending rigidity of an incompressible epithelium is given by $D_c = Y_c h^3/9$ [Eq. (3.2)

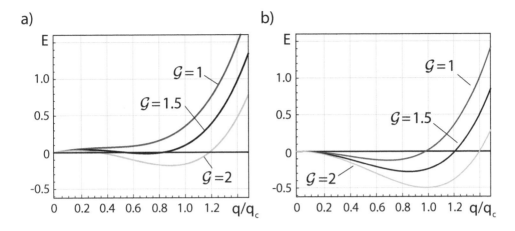

Figure 4.7 Energy of an area-mismatch epithelium vs. wavevector of folds for $\mathcal{G} = 1, 1.5$, and 2 (a): for $\mathcal{G} > 1.442\ldots$, the epithelium buckles with a wavelength which scales as $1/q_c$. A stroma-free version of the area-mismatch buckling theory (b): in this case, the energy is negative at small qs irrespective of the value of $\mathcal{G}$.

with $\nu = 1/2$; here Y_c and h are the effective Young's modulus and the thickness of the epithelial layer, respectively], we find that

$$\lambda \approx \frac{2\pi}{9} h \left(\frac{Y_c}{Y_s} \right)^{1/3}. \tag{4.10}$$

Using this result we can estimate the relative stiffness of the tissue compared to stroma based on the ratio λ/h as a purely geometrical quantity, and it implies that epithelial folds with a wavelength considerably larger than the tissue thickness must be much stiffer than the supporting stroma. For example, the smallest physically relevant wavelength is $2h$ or else the epithelium itself would not fit in the fold. This requires a ratio of Y_c/Y_s of ≈ 24 which seems to be consistent with experimental estimates of the moduli [114]. A somewhat less tightly packed waveform corresponds to a much larger elastic contrast—for example, in a fold with $\lambda = 4h$, $Y_c/Y_s \approx 190$. Of course, the exact value of λ is controlled by the surface tension but its magnitude is dominated by Eq. (4.10). The weak dependence on surface tension and thus on homeostatic pressure may be regarded as a robust feature of the folds: although Γ must exceed a certain threshold for the folds to form, it is of secondary importance for their shape.

Just like any other linear stability analysis, this theory of epithelial folds elaborates a possible mechanism leading to buckling, identifying the conditions when the deformation takes place. The exact shape of the fold is beyond the scope of the theory for several reasons including both mathematical approximations made and physical simplifications such as the assumption that the pressure within the epithelium is constant and does not decrease as the tissue expands. Nonetheless, the scaling results such as Eq. (4.10) do provide an estimate of the length scales and the energies involved. For example, the assumption that stroma is semi-infinite is clearly an idealization but the scaling of the penetration depth $\sim 1/q$ means that any finite-size effects may safely be disregarded if stroma thickness exceeds a couple of fold wavelengths.

The theory also shows why the elasticity of the stroma, soft as it may be, is important. As mentioned above, stroma stabilizes the epithelium against buckling at arbitrarily small magnitudes of the negative surface tension, and this is best appreciated by plotting the dimensionless energy [Eq. (4.4)] without the linear term. Figure 4.7b shows the energy of the stroma-free model for the same values of $\mathcal{G}$ as in panel a. More importantly, the wavelength of the most unstable mode is now given by

$$\lambda_{\text{no stroma}} \approx \frac{2\sqrt{2}\pi}{3} h^{3/2} \left(\frac{Y_c}{\Gamma}\right)^{1/2} \tag{4.11}$$

which depends on the elastic modulus of the epithelium in a different way than Eq. (4.10) and diverges as Γ vanishes. On the other hand, if λ should be of the same order of magnitude at h, then $\Gamma \sim Y_c h$. In turn, Γ can be written as the product of the homeostatic pressure and epithelium height h, which implies that for folds with $\lambda \sim h$ the homeostatic pressure should be of the same order of magnitude as Young's modulus of the epithelium, that is around 10^4 N/m² [114]. This is a reasonable result.

Calculating the dependence of fold wavelength on the elastic moduli and tension may seem to be a dull exercise but it is important because it allows us to understand the limitations of the simple analysis. In particular, finding that the wavelength is larger than but still comparable to epithelium thickness is hardly consistent with our thin-plate treatment of the tissue where we assume that the characteristic lateral dimension (i.e., the wavelength) is much larger than the thickness. Although qualitatively similar, an exact solution of the model where the distribution of stresses within the epithelium and the stroma would be computed numerically may thus well be more representative and closer to the *in vivo* observations.

This was indeed done in cylindrical geometry which represents an epithelium of a tubular organ such as intestine or an airway [105, 115]. The simpler implementation of the model [115] consists of just two structural elements forming a bilayer soft tube: the endoderm faces the lumen and is enclosed within the mesoderm (Figure 4.8). In the reference unstressed state, the tube is characterized by three radii $R_i, R_m,$ and R_0 and thus two ratios

$$H_{en} = \frac{R_m}{R_i} \qquad \text{and} \qquad H_{me} = \frac{R_0}{R_m} \tag{4.12}$$

as the only geometrical parameters of the model encoding the relative thickness of the layers. In a thin-walled tube, H_{en} and H_{me} are both a little larger than 1, respectively, whereas in a tube with a thin endoderm and a thick surrounding mesoderm H_{me} is considerably larger than 1.

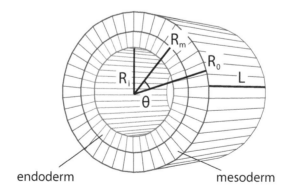

Figure 4.8 Geometry of the bilayer elastic tube in the reference state. The inner endodermal epithelial layer faces the lumen and is surrounded by the mesodermal layer. (Adapted from Ref. [115].)

Buckling is triggered by the differential growth of the two layers. If unconstrained, the endoderm and the mesoderm layer would expand isotropically along the x, y, and z axes by a factor of g_{en} and g_{me}, respectively. If the ratio of the growth factors g_{en}/g_{me} is larger than some critical value, the endodermal layer becomes unstable. The critical growth ratio depends on the geometry of the tube (i.e., on H_{en} and H_{me}) and on the relative stiffness of the two layers. As cell volume is essentially fixed, it is reasonable to treat the endoderm epithelium and the mesoderm as incompressible solid materials each described by a shear modulus alone.[3] The last element of the model is a suitable elastic theory and since deformations are expected to be large, the generic neo-Hookean strain energy function seems a reasonable choice.[4] Within this framework, the critical growth ratio can be calculated by employing the linear stability analysis, and Figure 4.9 shows its dependence on H_{en} at fixed mesoderm thickness corresponding to $H_{me} = 1.8$ and for ratios of shear moduli μ_{en}/μ_{me} ranging from 0.1 to 0.33. Of course, the critical g_{en}/g_{me} must be larger than 1 so that the endoderm expands more than the mesoderm, but the main message learned is that in all cases studied its value is plausibly moderate, ranging from about 1.2 to about 1.6.

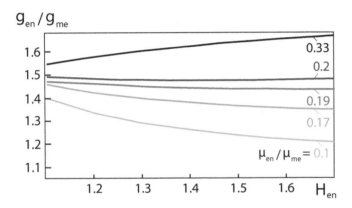

Figure 4.9 Critical growth ratio as a function of H_{en} for $\mu_{en}/\mu_{me} = 0.1, 0.17, 0.19, 0.2$, and 0.33. (Adapted from Ref. [115].)

In the linear analysis, circumferential deformation modes leading to a star-shaped cross-section are treated separately from the longitudinal modes where the lumen diameter varies along the tube. Interestingly, the tube is destabilized by circumferential modes only if μ_{en}/μ_{me} is smaller than a certain critical value. An opposite threshold exists for the longitudinal modes, which lead to buckling only if μ_{en}/μ_{me} exceeds the critical value [115]. These predictions are confirmed by numerical results. Figure 4.10 illustrates the shapes of the tubes corresponding to $H_{en} = 1.3$ and $H_{me} = 1.8$, that is, to a bilayer tube with radii $R_i : R_m : R_o = 1 : 1.3 : 2.34$. At $\mu_{en}/\mu_{me} = 0.1$, the endoderm has undergone a pure circumferential deformation with 15 lengthwise grooves but at $\mu_{en}/\mu_{me} = 0.33$ the unstable mode is longitudinal. For intermediate ratios of the shear moduli both modes are combined, leading to a lattice of bumps and pits that is qualitatively reminiscent of villi and crypts, that is, invaginations of the epithelium around the villi. Naturally, the buckled pattern also

---

[3]Hooke's law [Eq. (2.30)] can also be written as $\sigma_{ij} = \lambda \sum_k \epsilon_{kk} \delta_{ij} + 2\mu\epsilon_{ij}$, where λ and μ are the Lamé constants. In incompressible solids, $\sum_k \epsilon_{kk} = 0$ and the material response is characterized solely by μ. The Lamé constants can be expressed in terms of Young's modulus and Poisson's ratio: $\lambda = Y\nu/[(1+\nu)(1-2\nu)]$ and $\mu = Y/[2(1+\nu)]$.

[4]The Hookean stress is a linear function of strain, which is always the case when strains are small. The neo-Hookean model is one of the large-deformation models with a more complicated stress-strain relation, which reduces to the Hookean model at small strains.

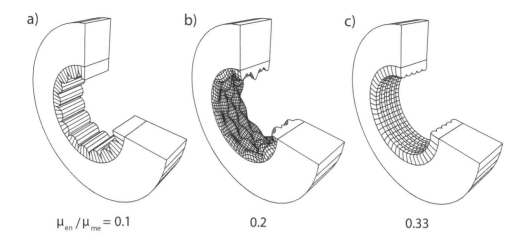

$\mu_{en}/\mu_{me} = 0.1$ 　　　　　　 0.2 　　　　　　 0.33

Figure 4.10 Illustrations of longitudinal, combined, and circumferential patterns of a buckled endoderm in a bilayer tube based on numerical results from Ref. [115]; the ratios of shear moduli are indicated in each panel.

depends on the geometry of the bilayer tube: the circumferential mode is generally characteristic of thick-walled tubes where H_{en} and H_{me} are large whereas the longitudinal mode is seen in thin-walled tubes [115].

These deformation modes are observed in animal tubular epithelia. A convincing example of the longitudinal mode are the ridges seen in eosinophilic esophagitis, a condition caused by accumulation of white blood cells within the epithelium that lines the esophagus (Figure 4.11a). Panels b–d in Figure 4.11 show the cross-section of lengthwise folds characteristic of the circumferential mode in the developing chick gut. Apart from this, an important feature of this model is that the ratios of shear moduli as well as the growth ratios required to reproduce these patterns are realistic, not departing very dramatically from 1. Experimental reports suggest that μ_{en}/μ_{me} may typically range from about 10% to about 70% [115] and that the endoderm/mesoderm growth ratio may reach 1.5 or so [116].

The tubular bilayer tissue is not perfectly analogous to the planar supported epithelium

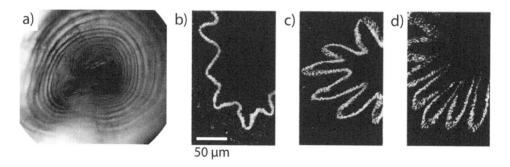

50 µm

Figure 4.11 Longitudinal folds characteristic of eosinophilic esophagitis (a; image by Samir, Wikimedia Commons) and a star-like cross-section of lengthwise folds in the developing chick gut in stage E13, E14, and E15 (b, c, and d, respectively; adapted from Ref. [117]).

because of the cylindrical symmetry, which does affect the detailed features of the circumferential and the longitudinal modes. The most striking difference is the distinction between the two modes, which should be equivalent in a planar epithelium but in a tube they are not. Nonetheless, the essential physical mechanisms involved are the same and the modes should not be too affected by the cylindrical geometry if their wavelength is small compared to the inner tube radius. In turn, choosing to study a tubular rather than a planar epithelium introduces periodic boundary conditions leading to a discrete set of circumferential modes.

The main lesson learned from these results is that by treating tissues as elastic media, one can indeed reproduce the nontrivial folded morphologies. While this agreement is reassuring, it raises the question of whether a tissue can indeed be described using standard concepts from the elasticity of solids. For example, the thin-plate theory behind Eq. (4.4) assumes that the spatial profile of mechanical stress within the tissue is continuous as if cells were homogeneous solids. In simple, single-cell-thick epithelia, this is not true, and one must view cells as discrete entities rather than as structureless materials. Switching from a continuum to a cell-based description is a step toward a more faithful representation which, as we show below, also captures the physics of corrugated epithelia. Moreover, it allows one to naturally introduce the apico-basal polarity as a key feature of these tissues.

4.1.2   Polarity-induced folding

The main elements of such a model are all contained within the differential adhesion hypothesis (DAH) discussed in Section 2.4. To adapt it to epithelia, we merely have to view the lumen and the stroma as two external media so that the tissue is characterized by three tensions encoding forces along the lateral cell-cell contacts, the basal cell-basement membrane interface, and the apical cell-lumen interface. Each of these three tensions is a coarse-grained idealization of the forces acting along a certain cell face. The lateral tension subsumes the cortex tension and the cell-cell adhesion, which can be regarded as a negative contribution to the tension favoring a cell-cell contact area as large as possible. The basal and the apical tensions also depend on the cortex tension yet are generally dissimilar simply because the respective interfaces are functionally distinct.

To show how this model leads to epithelial elasticity, we approximate the cell by a prismatic cuboid of height h and a square base of edge a to reflect the predominantly isometric en-face cross-section of cells. The DAH free energy then consists of three surface terms

$$E = \Gamma_a a^2 + 2\Gamma_l ah + \Gamma_b a^2, \tag{4.13}$$

where Γ_a, Γ_l, and Γ_b are the apical, lateral, and basal tension, respectively. Note that the second term includes a factor of $1/2$ as the lateral surface energy is shared by two cells. Assume that $\Gamma_a = \Gamma_b$ and that cells are incompressible so that their volume $V = a^2 h$ is constant. Before spelling out the energy, it is convenient to introduce $a_0 = V^{1/3}$ as the characteristic length scale and $\Gamma_l a_0^2$ as the characteristic energy scale. Then the dimensionless version of Eq. (4.13) reads

$$\mathcal{E} = 2\alpha \frac{a^2}{a_0^2} + 2\frac{a_0}{a}, \tag{4.14}$$

where

$$\alpha = \frac{\Gamma_a}{\Gamma_l} \tag{4.15}$$

is the so-called reduced apical tension. This form shows that $\mathcal{E}$ has a minimum representing the equilibrium cell shape (Figure 4.12). The energy of the cell is minimized at base edge length of $a_{eq} = a_0/(2\alpha)^{1/3}$, which decreases with α. The corresponding cell height h_{eq} is

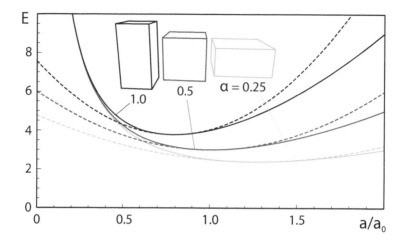

Figure 4.12 Energy of a cuboidal cell within the tension-based model as a function of the edge of the base for reduced apical tensions $\alpha = 0.25, 0.5$, and 1 (solid lines). Each dashed line represents a harmonic expansion of the energy around the minimum, and the inset shows the equilibrium shapes of cells for each value of α.

given by $a_0(2\alpha_{eq})^{2/3}$ so that in equilibrium the cell aspect ratio reads

$$\frac{h_{eq}}{a_{eq}} = 2\alpha. \tag{4.16}$$

As sketched in Figure 4.12, this model predicts that at $\alpha < 1/2$ cells are squamous ($h_{eq}/a_{eq} < 1$) whereas at $\alpha > 1/2$ they are columnar ($h_{eq}/a_{eq} > 1$). The two regimes are separated by isometric, i.e., cuboidal cells with $\alpha = 1/2$. The relevant range of α is probably between 0 and about 5 because the aspect ratio of the tallest cells in an epithelium does not exceed about 10.

If the cell is deformed, its energy will increase and this is the origin of its elastic response. The above results allow us to evaluate the elastic modulus for a uniaxial pure-shear deformation where cell shape is stretched or compressed along the vertical axis such that its shape remains cuboidal but the aspect ratio at a given α departs from the equilibrium value. Let us parametrize the deformation by the deviation of the edge of the base from the equilibrium value a_{eq} defined by $\delta a = a - a_{eq}$. The dimensionless modulus is defined by the coefficient of the harmonic expansion of the energy around the minimum, which is plotted in Figure 4.12 using dashed lines:

$$\mathcal{E} = \frac{1}{2}\kappa\left(\frac{\delta a}{a_{eq}}\right)^2. \tag{4.17}$$

A one-line calculation gives

$$\kappa = 6(2\alpha)^{1/3} \tag{4.18}$$

which shows that the modulus increases quite dramatically at small reduced apical tensions. In a similar fashion, one can calculate the elastic moduli for various other deformation modes.

The aim of this discussion is to show that tissue elasticity can also be interpreted using ideas from the physics of liquids, thereby circumventing conceptual issues associated with models based on the elasticity of solids. The geometrical simplifications made (especially the assumption that cell faces are flat) allow one to understand the workings of the tension-based model as transparently as possible without altering any of the qualitative conclusions. By examining the equilibrium shape of the above cuboidal cell we learn that if apical tension is increased at a fixed lateral tension, the areas of the apical and the basal face (which both have the same tension) decrease and the cell becomes taller. The same logic applies to cells with dissimilar apical and basal tensions where the equilibrium cell shape is controlled by the reduced apical tension α and by the reduced basal tension

$$\beta = \frac{\Gamma_b}{\Gamma_l}. \tag{4.19}$$

In this case the cell is no longer necessarily prismatic because if the differential apico-basal tension $\alpha - \beta$ is large enough, it may prefer a cell shape with a small or even completely constricted face with the largest tension.

This effect is most easily analyzed in a two-dimensional version of the model. Here cells are represented by trapezoids which reduce to rectangles if α and β are the same and to triangles if their difference is large enough (Figure 4.13). Thus by controlling the two reduced tensions, one can make each cell itself prefer a shape characterized by a certain curvature just like in Lewis' brass-bar-and-rubber-band model (Figure 4.2). It is obvious that a suitable number of such cells arranged in a circle form a curved aggregate which may be viewed as a cross-section of an epithelial tube, and it is equally clear that in three dimensions the differential apico-basal tension $\alpha - \beta$ would drive the formation of shell-like structures.

The differential apico-basal tension may be regarded as one of the physical aspects of epithelial polarity, an umbrella concept referring to the various structural and functional differences between the apical, lateral, and basal domains of the cell membrane. In part, the origin of the apico-basal tension may be the contractile acto-myosin ring running around the circumference of the apical surface (Figure 1.8), which should be treated as a separate

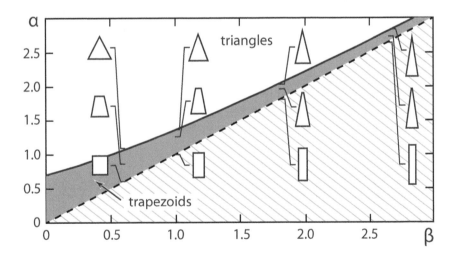

Figure 4.13 Equilibrium shape of an isolated cell in a two-dimensional version of the tension-based model. If the reduced apical and basal tensions (α and β, respectively) are identical, the cell is rectangular but if $\alpha > \beta$ then the apical side is shorter than the basal side and the cell is trapezoidal (shaded region) or triangular (white region). The hatched region below the diagonal contains basally constricted trapezoidal and triangular cells. (Adapted from Ref. [118].)

mechanical part of the cell in its own right [63, 119]. Although measuring the tensions of the three domains *in vivo* is difficult, their effect can be seen in epithelial patches detached from surrounding tissues. For example, explants of the *Xenopus* embryonic epithelium fold rather dramatically so as to form spherules or cylindrical shells and curl back when peeled off the ectoderm (Figure 4.14).

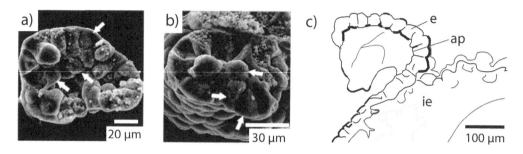

Figure 4.14 Spherules (a) and cylindrical wrinkles (b) formed spontaneously by *Xenopus* embryonic epithelial explants (reproduced from Ref. [120]). Panel c shows the epithelium (e) which has curled when peeled off the inner ectoderm (ie) so as to shrink its apical area (ap). (The drawing is a faithful representation of a micrograph from Ref. [120].)

The apico-basal tension may also give rise to corrugated epithelial morphologies although the mechanisms driving it are less intuitive than the area-mismatch theory, which is the tissue equivalent of the bimetallic strip. In a waveform consisting of cells of identical mechanical properties, all cells prefer the same curved shape at a given nonzero differential apico-basal tension but obviously not all of them can assume it because the waveform must contain both the groove and the crest or, in a more technical jargon, two segments of curvatures of opposite sign. Thus if the preferred shape is apically constricted, cells in the groove may approach it very closely whereas those in the crest must depart from it considerably (Figure 4.15). The deformation of the crest cells increases the overall energy but those in the grooves lower it compared to the flat state where the energy of each cell is larger than that of the preferred constricted shape. To see whether fold formation is favored, the tension-based model must be solved numerically by varying the wavelength and the number of cells in the waveform, which can be simplified by considering the two-dimensional version

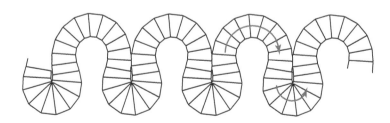

Figure 4.15 Cross-section of folded epithelium where the differential apico-basal tension is positive and large enough to cause the preferred cell shape to be apically constricted. The four or five cells at the bottom of the groove are in the minimum-energy state whereas the remainder of the waveform is not. (Adapted from Ref. [118].)

of the model representing the tissue cross-section. Indeed it turns out that folds are stable at intermediate differential tensions [118] whereas at small and large $\alpha - \beta$ they are not and the minimal-energy state is flat. As both the flat and the folded morphology contain cells which are very deformed compared to the preferred minimal-energy shape, the exact phase diagram depends on the choice of the mechanical model used. Thus we expect that by generalizing the basic tension-based model [Eq. (4.13)] by including additional terms, fold stability can be either enhanced or suppressed.

One possible extension is to include the basement membrane. So far, any role of the membrane was tacitly included in the effective basal tension but since the membrane is a network-like fibrous matrix it should have a nonzero bending rigidity. With this generalization, the folded minimal-energy solutions of the tension-based model can be divided into four distinct classes of folds: i) compact folds with tightly packed grooves and folds, ii) invaginated folds with deep and closed grooves, iii) evaginated folds with broad open grooves, and iv) wavy folds with a wavelength much larger than the cell size [121]. As shown in Figure 4.16, these classes of waveforms are indeed observed in several animal epithelia.

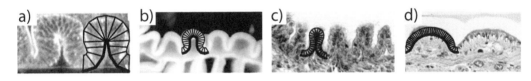

Figure 4.16 Theoretical compact (a), invaginated (b), evaginated (c), and wavy (d) longitudinal folds overlaid over epithelia of wing imaginal disc in *Drosophila* larva, *metridium* pharynx, cat lung bronchiole, and on the body surface of snail *Helix sp.*, respectively. (Adapted from Ref. [121].)

Within the tension-based model, folds are generated by the apico-basal tension which may be interpreted as an intra-epithelial shear prestress favoring identical wedge-like shapes for each cell. In this respect, the underlying logic differs from the area-mismatch theories where the epithelium is under an isotropic prestress due to cell division, which would give rise to an in-plane expansion of the tissue if it were not restricted by the elastic deformation of the stroma and the epithelium itself. The tension-based model does not rely on the stroma elasticity although the stroma is still needed so as to ensure that the folded epithelium is globally planar, which may be achieved by placing the epithelium on a large fluid drop.

4.1.3 Epithelial dysplasia

So far, the epithelium was treated as a solid layer, which is suitable for single-cell-thick tissues where strong cell-cell adhesion is needed to ensure the structural integrity of the tissue. Stratified epithelia, on the other hand, consist of several layers of cells and while here too cell-cell adhesion is strong, the number of neighbors and thus the binding energy of cells below the surface do not depend on the exact position of the cell within the tissue. If a cell is displaced a little to an equivalent position within the epithelium, it is still surrounded by other cells on all sides. In this respect, the stratified epithelium may be likened to a drop of a common molecular fluid where the molecules are free to move about as long as they stay in the bulk.

This analogy is very advantageous because the mechanics of simple fluids is well understood, and it can be used to interpret the instability observed in abnormal growth of epithelial tissues known as dysplasia. In this pre-cancerous condition, the epithelium invades the stroma in a characteristic finger-like fashion (Figure 4.17a). The morphology is remi-

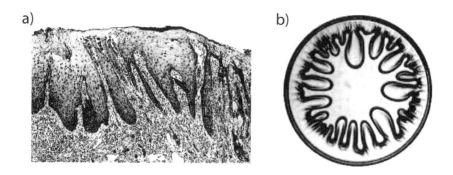

Figure 4.17 Dysplasia in vulvar epithelium (adapted from Ref. [122]). Panel b shows the Saffman-Taylor instability in a drop of silicon oil (image courtesy of J. Nase).

niscent of the Saffman–Taylor viscous fingering observed, e.g., in a drop trapped between horizontal plates. As the plates are pulled apart, the fluid-air interface becomes unstable, developing finger-like pockets which advance into the drop until it is detached from the top plate (Figure 4.17b).

Viscous stress and Navier–Stokes equation

In an ideal, inviscid fluid, Newton's second law $m\mathbf{a} = \mathbf{F}$ spelled out for per unit volume reads

$$\rho \frac{d\mathbf{v}}{dt} = -\nabla p, \qquad (4.20)$$

where p is the hydrostatic pressure. The right-hand side can also be written as the divergence of the hydrostatic stress tensor $-p\delta_{ij}$, i.e., $\sum_j \partial(-p\delta_{ij})/\partial x_j = -\partial p/\partial x_i$.

In a viscous fluid, stress includes both the hydrostatic term and a term that depends on the relative motion of nearby strata within the fluid. The viscous part of the stress tensor σ'_{ij} is constructed based on four conditions: i) it should depend on the gradient of the velocity $\partial v_i/\partial x_j$ often referred to as the strain rate; ii) it should vanish for rigid rotations which means that it must depend on the symmetrical part of the velocity gradient, i.e., on $\partial v_i/\partial x_j + \partial v_j/\partial x_i$; iii) to the lowest order, it should be linear in $\partial v_i/\partial x_j + \partial v_j/\partial x_i$; and iv) it must be isotropic like the fluid itself. Condition iii) defines the so-called Newtonian fluids and condition iv) implies that the viscosity coefficient for all diagonal components must be the same and this must also apply to all off-diagonal components. The most general form of viscous stress is thus

$$\sigma'_{ij} = \eta \left(\frac{\partial v_i}{\partial x_j} + \frac{\partial v_j}{\partial x_i} - \frac{2}{3}\delta_{ij} \sum_k \frac{\partial v_k}{\partial x_k} \right) + \zeta \delta_{ij} \sum_k \frac{\partial v_k}{\partial x_k}. \qquad (4.21)$$

Here η is the dynamic viscosity whereas ζ is the second or volume viscosity. In incompressible fluids where $\sum_k \partial v_k/\partial x_k = 0$, the divergence of Eq. (4.21) reduces to $\eta \nabla^2 \mathbf{v}$. If we include it on the right-hand side of Eq. (4.20) and spell out the material derivative we arrive at the Navier–Stokes equation

$$\rho \left[\frac{\partial \mathbf{v}}{\partial t} + (\mathbf{v} \cdot \nabla)\mathbf{v} \right] = -\nabla p + \eta \nabla^2 \mathbf{v}. \qquad (4.22)$$

The hydrodynamic theory of epithelial dysplasia can be elaborated using classical concepts from continuum mechanics [123]. In dysplasia, cells grow and divide faster than they die, and the net increase of epithelial volume per unit time is parametrized by the growth rate k_p. Because of rapid growth, cells must flow away from any small control volume within the epithelium, and the continuity equation relating velocity $\mathbf{v}$ and k_p reads $\nabla \cdot \mathbf{v} = k_p$ or

$$\sum_j \frac{\partial v_j}{\partial x_j} = k_p; \qquad (4.23)$$

here we assume that the epithelium is incompressible. The stress within the fluid-like epithelium contains the hydrostatic and the viscous term and reads $\sigma_{ij} = -p_e \delta_{ij} + \eta \left(\partial v_i/\partial x_j + \partial v_j/\partial x_i\right)$, where p_e is pressure and η is the viscosity of the epithelium. In the stationary state and in the small Reynolds number limit where the inertial terms in the Navier–Stokes equation describing the motion of the fluid vanish, this equation reduces to the condition for the force equilibrium, $\sum_j \partial \sigma_{ij}/\partial x_j = 0$, or

$$\eta \sum_j \frac{\partial^2 v_i}{\partial x_j^2} + \eta \frac{\partial k_p}{\partial x_i} - \frac{\partial p_e}{\partial x_i} = 0. \qquad (4.24)$$

A very similar equation describes the stroma approximated by an incompressible Hookean material with $\nu = 1/2$. In this case, the equation of mechanical equilibrium involving the displacement vector $\mathbf{u}$ [Eq. (4.5)] reduces to

$$\mu \sum_j \frac{\partial^2 u_i}{\partial x_j^2} - \frac{\partial p_s}{\partial x_i} = 0, \qquad (4.25)$$

where $\mu = Y/[2(1+\nu)]$ and p_s is pressure within the stroma; note that the term containing $\sum_j \partial^2 u_j/\partial x_i \partial x_j$ is zero because the stroma is incompressible $\left(\sum_j \partial u_j/\partial x_j = 0\right)$.

Equations (4.24) and (4.25) describe the velocity of the epithelium and the corresponding displacement of the stroma, and they are coupled by the boundary conditions at the interface involving interfacial tension and surface friction, which is parametrized by a coefficient ξ. These boundary conditions specify that the normal component of the stress is discontinuous at the epithelium-stroma and the epithelium-medium interfaces as specified by the Young–Laplace law involving the tensions at these interfaces denoted by Γ_{ES} and Γ_{EM}, respectively. The other boundary conditions co-determining the evolution of the epithelium pertain to tangential stresses at boundaries, continuity of displacement and velocity, and displacement at the bottom of the stroma [123].

The final element of the model is the dependence of the growth rate on position within the epithelium. The growth rate is controlled by the supply of nutrients and growth factors, and it is plausible to assume that it decreases with the distance from the epithelium-stroma interface which then leads to an ever smaller k_p. A suitable *ansatz* contains an exponentially decaying positive term describing cell growth and division combined with a constant negative term representing cell death:

$$k_p(z) = k \exp\left(-(z - L)/\ell\right) - k_0. \qquad (4.26)$$

Here k and k_0 are coefficients controlling cell growth and apoptosis,[5] respectively, and ℓ is the length scale characterizing the depletion of nutrients and growth factors with distance from stroma of thickness L.

---

[5] Apoptosis is a highly regulated activated pathway of reactions which leads to the programmed death of a cell.

A net production of cells creates flow. In a flat epithelium, cell velocity must point in the vertical direction as cells migrate from the bottom, nutrient-rich region to the top, depleted region where the net cell production rate is negative; in Figure 4.18 the two regions are separated by the dashed line. The corresponding velocity field can be calculated analytically by integrating Eq. (4.23) and the result is plotted in Figure 4.18. As expected, the epithelial velocity is largest at a height where net cell growth rate vanishes.

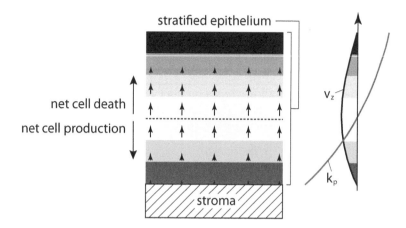

Figure 4.18 Vertical velocity field in the flat ground-state stratified epithelium where the net cell production is positive below the dashed line and negative above it. Grayscale levels represent the magnitude of cell velocity, which vanishes at the epithelium-stroma interface and at the free epithelial surface. Also shown are the profiles of the net cell production and the vertical tissue velocity (right).

For any given stroma thickness L and growth rate parameters k, k_0, and ℓ, the stationary thickness of a flat tissue is determined by the condition whereby its free surface does not move. After the stationary thickness has been found, a standard linear stability analysis can be performed, assuming that the epithelium-stroma interface is perturbed by a plane overdamped wave $\delta L(x,t) = \delta L_0 \exp(iqx + \omega t)$, where q and ω are the wavevector and the growth rate of the deformation, respectively. The perturbation gives rise to three eigenmodes and obtaining the dispersion relation from the boundary conditions is a little involved [123]. Depending on parameters, one of them may have a positive ω so that the mode is not overdamped but grows with time, spontaneously transforming the flat epithelium into a fingered structure. Figure 4.19 shows $\omega(q)$ in a 300 μm-thick epithelium supported by a 300 μm-thick stroma for a realistic choice of the seven material parameters involved.[6] At large enough q, the slowest mode indeed is unstable and its fastest growing wavelength is a fraction of a millimeter, which is reasonable—the "fingers" of epithelial tissue in vulvar dysplasia are typically separated by about 20 cell diameters.

Because of the many material parameters involved, it is worthwhile examining the approximate expression of the growth rate valid at large qs [123] where

$$\omega \approx k - k_0 - \frac{\mu}{\eta} - \frac{\Gamma_{SE}}{2\eta}q. \tag{4.27}$$

---

[6]These parameters include epithelium viscosity $\eta = 10$ MPa · s, stroma shear modulus $\mu = 100$ N/m^2, epithelium-stroma interfacial tension $\Gamma_{SE} = 10$ mN/m, epithelial apical surface tension $\Gamma_{EM} = 1$ mN/m, cell growth coefficient $k = 8.6$/day, cell production decay length $\ell = 200\,\mu$m, and epithelium-stroma friction parameter $\xi = 10^{10}$ Pa · s/m [123].

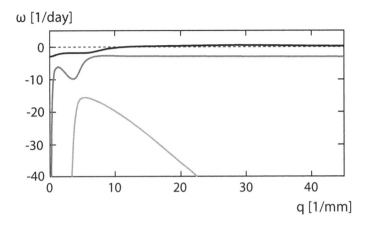

Figure 4.19 Growth rate ω of the three eigenmodes of the flat stratified epithelium as a function of the wavevector q for typical values of model parameters listed in the footnote on p. 118. One of the modes has a positive ω at large enough qs, indicating an instability. (Adapted from Ref. [123].)

In this expression, the last term represents the effect of the stroma-epithelium interfacial tension Γ_{SE}, which decreases the growth rate ω and hence stabilizes the epithelium provided that q is large enough. The necessary condition for the growth rate to be positive and thus for instability is

$$\eta(k - k_0) > \mu, \tag{4.28}$$

stating that dysplasia occurs when the viscous stress generated by the dividing cells at the epithelium-stroma interface exceeds the shear modulus of the stroma.

On long time scales, the assumption that stroma is an elastic solid no longer holds and replacing it by a viscous material is more appropriate. In the viscous-stroma model of epithelial dysplasia, the necessary condition for instability analogous to Eq. (4.28) is simply that at the interface the epithelium grows, i.e., that $k - k_0 > 0$. This is of course not sufficient and the typical dependence of the exact ω on the wavevector is qualitatively similar to Figure 4.19, peaking at a finite q. At a given profile of the net cell production, instability can be prevented by increasing the epithelium-stroma tension much like in the case of a solid stroma.

4.2 AGENT-BASED MODELS

In the above theories, cells are regarded as passive entities in the sense that their key mechanical properties that control buckling are determined by an external process, which must be of biochemical nature. But cells are not really passive. They respond to their local environment, adapting their state to physical and chemical conditions and adjusting their behavior so as to alter these conditions. For example, if the apico-basal tension model were generalized so as to include cell division, it would be natural to expect that the division rate should depend on the local stress which varies along the fold because a part of the waveform is close to the preferred constricted minimal-energy state but the rest is not. The stress-dependent division rate provides a feedback mechanism and in the stationary state, the form of the buckled epithelium would generally be different from that in absence of division.

This is just one possible example of agent-based models of tissues, which typically consist of a mechanical model and a set of rules describing scenarios of cell growth, division, death, and remodeling. All models of intra-epithelial structure in Chapter 3 that include cell division are, in fact, agent-based, and using similar concepts to understand the shape of epithelia is a natural step toward more realistic theories. At the same time, we will see that the mechanical component of agent-based models is usually simplified, sacrificing the detailed description of cell shape and cell-cell interaction.

An early implementation of the agent-based approach was proposed in the early 1980s to describe the invagination of the embryonic epithelium. In this key morphogenetic movement, a part of the epithelium infolds so as to form a cavity within the embryo, and the internalized tissue later develops into various organs—this process is discussed in detail in Chapter 5. In the spirit of Lewis' brass-bar-and-rubber-band model in Figure 4.2, invagination can be interpreted as a localized contraction of a certain segment of cells (Problem 5.2), and the novelty introduced by the agent-based view is a simple rule-based mechanism of cell excitation leading to the initiation and propagation of the contraction wave [124].

At the heart of this approach is the view that stresses exerted on a given cell are not only the cause of a deformation from a given reference state but also a way of changing the reference state itself. This idea can be illustrated by a spiral spring from a ballpoint pen. If stretched a little it readily returns to the initial unstretched length L_0' once it is released, but if the extension is too large this does not happen: instead it assumes a new, larger reference length L_0 after release but a further application of a small force still leads to an elastic response described by Hooke's law.

In this example, a large extensional force results in an increase of the unstretched length of the spring due to the plastic deformation. In an epithelium, the elastic response of cells to stress is largely due to cytoskeletal networks including the apical network consisting of actin filaments, which is an important structural feature of the apical side of cell and defines its resting diameter as well as its elastic modulus. This network may be characterized by an inverse response where a large extension results in a smaller reference area. A possible microscopic mechanism for this effect involves an influx of calcium ions into the cell after cell stretching [124], which is thought to de-crosslink the network. This should allow the actin filaments to contract due to the action of myosin molecular motors so that after the initial calcium level is re-established and the network is re-formed, its projected area is smaller than at the beginning. Given that epithelial cells exert forces on each other, an excited cell should stretch the apical surfaces of its neighbors. In turn, stretching may trigger the transition from a larger to a smaller reference apical surface area in these cells, thereby initiating an excitation wave (Problem 5.2). Eventually a finite segment of cells will be apically constricted, which should then produce an invagination.

The excitable epithelium was first discussed within a mechanical framework fairly similar to Lewis' brass-bar-and-rubber-band model. Each cell was represented by its cross-section consisting of six Kelvin–Voigt viscoelastic units of a spring and a dashpot connected in parallel (Figure 4.20): one at the apical and at the basal side, two along the diagonals, and two at the lateral sides, the latter being shared by adjacent cells. The dynamics of each unit is described by the inertia-free equation of motion given by

$$\eta \frac{\mathrm{d}L}{\mathrm{d}t} = -k(L - L_0) + F_{\text{load}}, \tag{4.29}$$

where η is the viscous damping parameter, k is the spring constant, L_0 is the rest length of the spring, and F_{load} is the total lengthwise force exerted by all other spring-and-dashpot units attached to both ends of the unit in question.

This equation stated that at any nonzero F_{load}, the length of the unit will gradually approach its equilibrium value of $L_{\text{eq}} = L_0 + F_{\text{load}}/k$ starting from the initial length $L(t = 0)$,

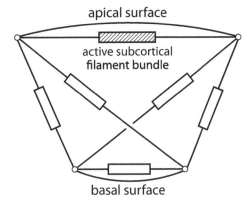

Figure 4.20 Model cell of the excitable epithelium, which consists of five passive and one active apical Kelvin–Voigt spring-and-dashpot units [124].

and the solution of the equation is

$$L(t) = L_{\text{eq}} - [L_{\text{eq}} - L(t = 0)] \exp\left(-\frac{kt}{\eta}\right). \tag{4.30}$$

In all spring-and-dashpot units except the apical unit, the rest length of springs L_0 is constant. This represents the passive response of the cell because irrespective of their deformation in the past, these units come to the same rest length once the forces are removed. The apical unit is special because its rest length depends on its instantaneous length. Initially, the rest length is fixed at some value and it remains unchanged until the deformation of the unit exceeds a given threshold. When this happens, the rest length is decreased irreversibly to a smaller value so that the cell becomes apically constricted. This process is described by another differential equation of the form $dL/dt = G(L, L_0)$, where $G(L, L_0)$ is a function of both instantaneous unit length and the rest length. The precise form of $G(L, L_0)$ can be spelled out mathematically [124] but it is really not important as long as it ensures that the unit fires as described.

The excitable response of the apical spring-and-dashpot unit makes the whole cell undergo a deformation from the initial cuboidal shape to the final apically constricted columnar shape (Figure 4.21a) provided that the many parameters of the model are chosen properly.[7] This, in turn, may stretch the apical units of the two neighbors beyond the threshold. As a result, they fire and constrict themselves, triggering a wave of excitation propagating away from the first cell. Thus an active constriction of a single cell may drive the deformation in a whole patch of cells. In Figure 4.21b such a wave is seen in the sequence of simulated cross-sections of a gastrulating sea-urchin embryo. The first cell to undergo the constriction is in the middle of the bottom part of the embryo and the wave propagates in both directions, making the invaginated section grow in width.

The mechanical form of active apical constriction is really not too different from the passive, minimal-energy framework proposed by Lewis. The viscous components of the Kelvin–Voigt units are important primarily if the viscosities of the different units are not the same: in this case, they do affect the propagation of the contraction wave but otherwise they

---

[7]In Figure 4.21a, the apical and the basal sides are upside down compared to Figure 4.20. This is so because we focus on the invaginating part of the epithelium usually shown at the bottom of the embryo [125] and because the apical side is on the external boundary of the epithelium, which is in contact with the fibrous membrane enveloping the embryo. Thus the apical side of the invaginating cells is at the bottom whereas the basal side is on top.

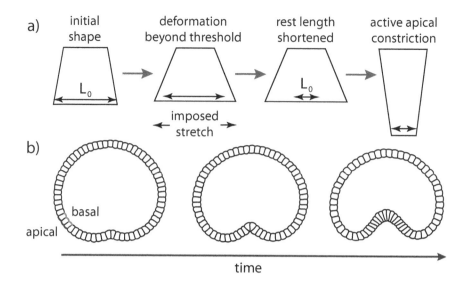

Figure 4.21 Schematic of active apical constriction in the excitable epithelium: the initially trapezoidal cell with a large apical rest length L_0 is stretched beyond firing threshold, and L_0 is decreased (a). Panel b shows the initial stages of a simulation of the contraction wave in gastrulating sea urchin embryo. (Adapted from Ref. [124].)

merely provide a plausible way of relaxation towards equilibrium. Yet including active constriction is a considerable conceptual step ahead in that it represents a microscopic cell-level process known from smooth muscles [126] and believed to also be relevant in other tissues. If stretched at a large enough velocity, the membrane of smooth muscles becomes depolarized, setting up a train of action-potential waveforms which leads to an increased concentration of calcium ions both by release from internal stores and by influx across the membrane. This, in turn, gives rise to muscle contraction [126,127].

This particular type of mechanical feedback has been used to describe epithelial invagination in embryos of various animals [124] in processes which happen rather quickly so that one can presume that the dynamics may be important. Here we described it as a simple yet far-reaching and mechanically sophisticated generalization of the theoretical framework behind the minimal-energy models such as Lewis' brass-bar-and-rubber-band model, the area-mismatch model (Section 4.1.1), or the tension-based model (Section 4.1.2). The much more stationary epithelial morphologies such as those in Figure 4.1 probably rely on a different type of coupling between tissue shape and any processes that generate it, one possibility being a differential net growth rate.

4.2.1   Curvature-dependent division in renewing epithelium

The main result of the area-mismatch buckling theory in Section 4.1.1 is that the growth of epithelium counterbalanced by the elasticity of the stroma and of the epithelium itself may lead to folded morphologies. Yet the harmonic analysis employed to estimate the wavelength of the instability cannot yield the detailed shape of the final waveform, and it does not distinguish between the evaginated villus-like morphologies and the invaginated crypt morphologies. A rather simple and physically meaningful way of introducing this distinction is the apico-basal polarity of the epithelia mentioned in Sections 4.1.2 and 4.2. Because of polarity, cell division rate may depend on the local shape of the epithelium and in principle, it may either be larger in the invaginated domains (grooves or crypts) compared to the evaginated domains (crests or villi) or *vice versa*; we note that while theoretically possible, the latter option is inconsistent with experiments which show that in many folded epithelia cells are typically produced in the grooves and then they migrate toward the tips of the crests where they are shed.

Introducing this coupling thus allows one to view the epithelium as an inherently dynamic rather than as a static, energy-minimizing system, and cell flow is an essential element of the process. Like in Section 4.1.3, a positive net production of cells where the local division rate k_d is larger than the apoptosis rate k_a generates an outflow of cells from the region in question, whereas domains with a negative production act as sinks. Mathematically, this effect is expressed by the continuity equation for incompressible materials

$$\nabla \cdot \mathbf{v} = k_d - k_a, \tag{4.31}$$

where $\mathbf{v}$ is the local cell velocity.

Imagine now a flat epithelium characterized by a reference negative surface tension $-\Gamma$. In this state, there can be no in-plane migration of cells, simply because otherwise two neighboring regions of the tissue would not be equivalent; in more technical terms, because of translational invariance. Thus cell division and cell death must locally cancel each other exactly ($k_d = k_a$) and the net cell production must vanish. Should such an epithelium buckle, the tension will no longer be uniform, but we may expect that it will not differ very dramatically from the reference value. In turn, the spatially modulated part of the tension $-\delta\Gamma(\mathbf{r})$ will give rise to an imbalance between k_d and k_a. We may assume that the imbalance should be proportional to $-\delta\Gamma(\mathbf{r})$ so that $k_d - k_a = -\xi\delta\Gamma(\mathbf{r})$, where ξ is positive so that a more negative surface tension results in an increased net cell production (Figure 4.22; Problem 4.10).

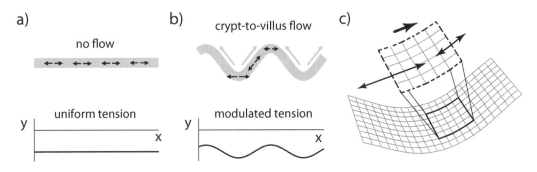

Figure 4.22 Flat epithelium with uniform surface tension which does not generate any cell flow (a) and a buckled epithelium with a non-uniform tension giving rise to a crypt-to-villus cell flow (b) as well as to a net force (thick arrow) on any part of the epithelium (dashed box in panel c).

As mentioned above, this is not the only mechanism affecting the imbalance of cell division and death rates: we also need to include the curvature-dependent term to explicitly distinguish between invaginations and evaginations. In the Monge representation (Box on p. 104), curvature is given by $\nabla^2 w(\mathbf{r})$ and thus finally

$$\nabla \cdot \mathbf{v} = -\xi\delta\Gamma(\mathbf{r}) + \alpha\nabla^2 w(\mathbf{r}), \tag{4.32}$$

where $\alpha > 0$ so that cells are produced at a faster rate in the invaginated regions (where $\nabla^2 w > 0$) compared to the evaginated regions.

So far, this reasoning only revolved around the mass or volume conservation. But the spatially non-uniform effective surface tension also means that there is a net in-plane force on any given part of the tissue (Figure 4.22c). This force equals $-\nabla\delta\Gamma$ and in absence of inertial effects, it can only be balanced by the friction between the epithelium and the basement membrane which is proportional to $\mathbf{v}$:

$$\zeta\mathbf{v} = -\nabla\delta\Gamma. \tag{4.33}$$

Here ζ is the friction coefficient. This equation allows us to eliminate $\mathbf{v}$ from Eq. (4.32) and to directly connect the shape of the tissue represented by $w(\mathbf{r})$ to the modulation of the surface tension $-\delta\Gamma(\mathbf{r})$. It is easy to see that the relative magnitude of the two terms containing $-\delta\Gamma(\mathbf{r})$ is $\lambda^2\zeta\xi$, where λ is the wavelength of the waveform. In the large-friction regime, this dimensionless product is much larger than 1 and

$$-\delta\Gamma(\mathbf{r}) \approx -\frac{\alpha}{\xi}\nabla^2 w(\mathbf{r}), \tag{4.34}$$

whereas in the small-friction regime

$$-\delta\Gamma(\mathbf{r}) \approx \alpha\zeta w(\mathbf{r}). \tag{4.35}$$

Qualitatively, these two limiting cases describe the same behavior. In the invaginated regions of the epithelium both $\nabla^2 w(\mathbf{r}) > 0$ and $w(\mathbf{r}) < 0$ so that in either case $-\delta\Gamma(\mathbf{r}) < 0$ consistent with an increased pressure within the tissue. On the other hand, in the evaginated regions $\nabla^2 w(\mathbf{r}) < 0$ and $w(\mathbf{r}) > 0$ so that $-\delta\Gamma(\mathbf{r}) > 0$ in both large- and small-friction regime.

This generalization allows us to spell out the dynamics of the epithelium based on the dependence of the in-plane surface tension on its shape. To this end, one must again resort to the classical elasticity of solid plates [23] and extend it by including i) friction, which leads to overdamped dynamics, ii) in-plane expansion of the epithelium due to cell division, and iii) stretching of the basement membrane and stroma deformation as two stabilizing mechanisms [114]. The final equation of motion for $w(\mathbf{r})$, the epithelial displacement from the (x, y) plane, is not particularly illuminating but much can be learned from the inspection of the numerically obtained stationary states.

Figure 4.23a shows the buckled epithelial morphology for vanishing coupling between cell division and curvature, i.e., for $\alpha = 0$. In this case, all of the epithelium is characterized by the same value of surface tension and there is no net crypt-to-villus cell flow typical for a renewing epithelium. This morphology consists of both crypt-like invaginations and villus-like evaginations arranged in a checkerboard fashion. The height profiles of invaginations

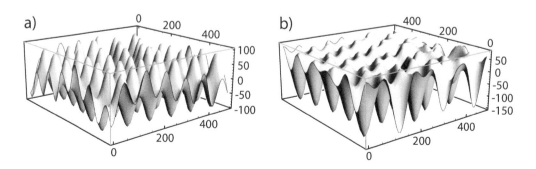

Figure 4.23 Stationary height profiles of the non-renewing epithelium with $\alpha = 0$ (a) and a crypt-like renewing morphology with $\alpha > 0$ (b). (Adapted from Ref. [114].)

and evaginations are fairly symmetric and approximately sinusoidal as expected. In contrast, the renewing $\alpha > 0$ morphology shown in Figure 4.23b is almost devoid of protrusions and consists essentially only of deep crypt-like recesses from the base level.

The similarity of these two morphologies with the mammalian small intestine and colon epithelium (Figure 4.24) is very impressive despite the Monge-representation implementation of the model, which does not allow too steep a shape or even overhangs simply because the shape of the surface is a single-valued function of x and y coordinates. In part, this is the reason why the exact reproduction of, e.g., the test-tube-like form of the crypts is beyond the scope of the model.

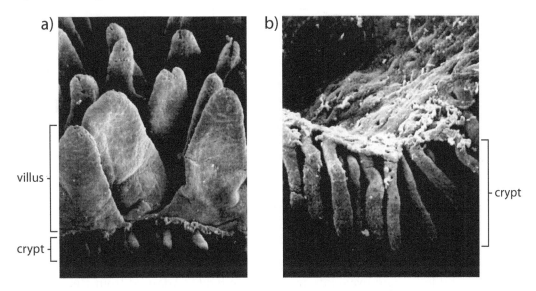

Figure 4.24 Murine duodenal epithelium with large leaf-like villi and small crypts (a) and crypts in the colon epithelium (b). (Adapted from Ref. [128].)

Apart from the in-plane surface tension due to homeostatic pressure, the main parameter controlling the tissue morphology within the renewing-epithelium model is the cell-division–curvature coupling constant α. As it is customary in physics, not all of the many material constants matter by themselves. It can be shown that in the low-friction limit the main dimensionless parameters are $\alpha\zeta h_b/\Gamma_c$ and Γ/Γ_c, where h_b is the thickness of the basement membrane and Γ_c is the critical surface tension equal to $(3D_cY_s^2)^{1/3}$ as mentioned in Section 4.1.1 immediately below Eq. (4.9). The phase diagram in the space spanned by these two quantities has a rather simple topology (Figure 4.25). At small $\alpha\zeta h_b/\Gamma_c$, the buckled epithelium consists of the more symmetric alternating invaginations and evaginations either of the crypt-villus type in or of the groove-crest type characteristic of longitudinal folds, which may give rise to the herringbone-like or a labyrinthine pattern depending on Γ/Γ_c (Figure 4.25). On the other hand, a large enough cell-division–curvature coupling constant suppresses the crypt-villus symmetry and promotes a crypt morphology. Since the only topographic feature of this morphology are recesses from the base level referred to as the intercypt table, it is quite natural that they should form a hexagonal lattice.

The predictive power of this mechanically sophisticated model is impressive, and it is possible that a technically even more advanced implementation of the mechanisms involved could provide a still better description of real epithelia, especially when extended to large deformations. At the same time, the model includes several physiologically relevant effects, some of which are intertwined, and it is not easy to see how essential each of them is for

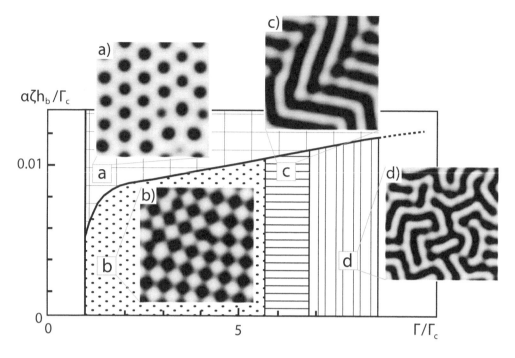

Figure 4.25 Stationary states of the renewing-epithelium model include crypt-villus (b) as well as herringbone (c) and labyrinthine (d) fold morphologies at small $\alpha\zeta h_b/\Gamma_c$ and crypt-only hexagonal morphology at large $\alpha\zeta h_b/\Gamma_c$ (a). In the grayscale-coded snapshots of these morphologies, invaginations are dark whereas the intercrypt table (a) and evaginations (b–d) are light. (Adapted from Ref. [114].)

the morphologies obtained. It may be that to some extent, the precise shape of villi, crypts, and folds are decoupled from the processes that drive epithelial buckling. So it could well be advantageous to study the precise shape of the crypts separately from the onset of buckling instability itself so as to more directly associate the cause and the effect.

This could be done within a suitably simplified framework, and one possible simplification is to exclude the elastic stroma. As shown in Figure 4.7, stroma is needed within the area-mismatch theory so as to introduce a finite buckling threshold and thus prevent the epithelium from buckling at small negative surface tensions. Apart from this, stroma does not qualitatively affect the tissue morphology, which also holds for the theory of epithelial dysplasia in Section 4.1.3. Moreover, in the polarity-induced-folding model (Section 4.1.2) the buckling threshold is well-defined although the stroma is fluid-like rather than elastic.

4.2.2 Shape of crypts

Such an approach has been explored within a cell-based model of a single crypt in colon [129], and the main conclusion of the model is that both spontaneous curvature at the crypt base and two cell populations, a proliferative and a non-proliferative one, are needed so as to reproduce both the test-tube-shaped crypt cross-section and their small spacing.

In this discrete computational model, cells are represented by their centers as if they were point particles, and cell-cell forces are described by Hookean springs acting between the centers (Figure 4.26). In turn, cells themselves are defined by the Voronoi tessellation (Box on p. 83) used to compute their respective areas.

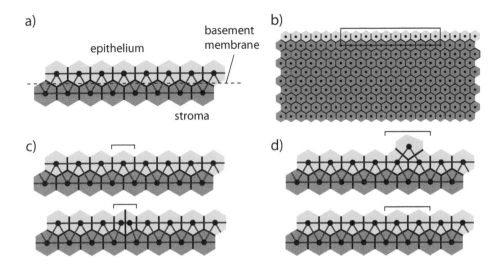

Figure 4.26 Point-particle representation of a supported epithelium (a). Centers of neighboring cells are connected by springs, and cells are defined by the Voronoi construction. Also shown is the basement membrane. The epithelium of a few dozen cells is supported by stroma consisting of a few 100 cells and the model is solved using periodic boundary conditions; the bottom boundary is rigid (b). During division, the mother cell is replaced by daughters a small distance apart (c) and cell death takes place when a cell loses contact with the basement membrane (d). (Adapted from Ref. [129].)

The two-dimensional model consists of a single-cell-thick epithelium resting on the stroma and the motion of epithelial cells is restricted to the surface. Cells in the stroma, on the other hand, may move past each other as specified by the forces exerted on them, exchanging neighbors as needed. Thus the stroma is really represented by a fluid rather than by a solid. Also included in this model is the basement membrane described by an infinitely thin curve running halfway between the centers of epithelial cells and the stromal cells immediately adjacent to them. The basement membrane is an elastic body with a finite bending rigidity and a certain spontaneous curvature, which is nonzero at the base of the crypt and zero otherwise. The basement-membrane restoring force is added to the direct cell-cell forces on the epithelial cells.

Much like in the renewing-epithelium model of Section 4.2.1, the model dynamics is overdamped so that the velocity of each cell center is proportional to the total force exerted on it by the neighbors. This allows the system to reach the stationary state generated primarily by the division and death of epithelial cells; the stromal cells neither divide nor die. In the discrete model, these two processes can be implemented by a set of rules describing cell cycle and cell death as follows:

- **Cell growth** is described by a time-dependent rest length of springs connecting daughter cells during mitosis, which increases from 10% to 100% of the length in a mature cell.

- **Oriented cell division** is ensured by replacing the center of the mother by centers of daughter cells within the epithelial plane a small distance apart (Figure 4.26c).

- **Cell death** takes place when a cell moves too far from the basement membrane and no longer adheres to it—whether this is the case is determined by absence of direct cell-basement membrane contact in the Voronoi construction (Figure 4.26d). If so, the cell is removed from the epithelium.

- **Inhibition of mitosis** due to too large a local density is modeled by an area threshold below which cell division does not take place. Cells that do not reach the threshold stay in their non-dividing state until they are large enough. The mechanical feedback is thus a regulator of the position-dependent tissue growth.[8]

These events are implemented within a realistic model of the cell cycle including the Gap 1 phase of a somewhat randomized duration and the fixed-duration DNA synthesis phase, Gap 2 phase, and mitosis phase; the simulation was run for 40 to 50 rounds of cell division after the stationary state has been reached.[9] The whole framework contains many parameters and their precise choice [129] does play a role. Yet the main qualitative conclusions of simulations can be summarized in Figure 4.27 showing the final stationary state of the unit-cell containing a single crypt under various conditions. In all three cases, the central portion of the epithelium lies on the basement membrane with a given spontaneous curvature. In Figure 4.27a, the width of the spontaneously curved domain is 50% and all cells divide. While the model crypt is invaginated, it is too broad and too shallow compared to real crypts shown in Figure 4.24b. A much more test-tube like profile is obtained if the size of the spontaneously curved domain is reduced to 20% of the tissue (Figure 4.27b) where stroma size has been enlarged too so as to accommodate all of the crypt. A comparison of these two cases suggests that i) cell division is needed in order to generate a large enough crypt area so that the epithelium can buckle and that ii) a certain preferred spontaneous curvature in the tissue serves as a locus of the buckling instability, a little like creases on origami paper. From the latter perspective, a small spontaneously curved domain is preferred simply because it leaves a larger portion of the epithelium shapewise unbiased so that it can adapt more readily to forces and processes in it and around it.

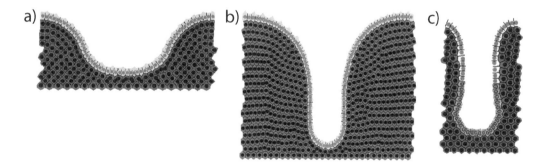

Figure 4.27 Three stationary crypt shapes: crypts consisting of a single population of dividing cells with a broad 50% (a) and a narrow 20% spontaneously-curved domain (b), and a crypt with a proliferating base and non-proliferating top section (c; dark and light gray cells, respectively). (Adapted from Ref. [129].)

---

[8]A similar mechanical feedback can also be implemented in a continuum model [130].

[9]The cell cycle is typically divided into four phases. In the G1, or Gap 1 phase, the cell synthesizes mRNA and proteins and prepares for mitosis. In the S phase, the cell DNA is replicated. In the G2, or Gap 2 phase, the cell rapidly grows until it enters the M (mitosis) phase in which the growth stops and the cell divides into two daughters. The resting phase, in which no mechanisms preparing the cell for mitosis have been triggered, is called G0 or Gap 0 phase.

The test-tube-shaped crypt in Figure 4.27b is more similar to those seen in real epithelia but its collar is too broad and open, and the spacing of the crypts is too large. These features can be adjusted by restricting the size of the proliferating epithelial domain to the bottom part of the crypts (dark gray epithelial cells in Figure 4.27c) so that cells in the top part and in the intercrypt table (light gray epithelial cells in Figure 4.27c) are merely pushed upwards by the proliferating bottom part. This scenario leads to a fairly accurate description of the crypt shape even in the two-dimensional model representing the cross-section of the epithelium.

At the same time, the model crypts in Figure 4.27c may well be regarded as a cross-section of the longitudinal folds in the stomach of *Styela clava* discussed in Figure 4.1b and c known to include a proliferating groove population and a non-proliferating crest population. This is evidenced by autoradiographs in Figure 4.28 showing how the dark labeled cells migrate from the grooves to the crests [107]. One hour after injection of tritiated thymidine, a radioactive nucleoside commonly used in cell proliferation assays which is incorporated into genetic material at the time of cell division, the labeled cells are located at the bottom of the groove but in 10 days time they migrate halfway towards the crest, and 15 days after injection they are spread across the whole fold.

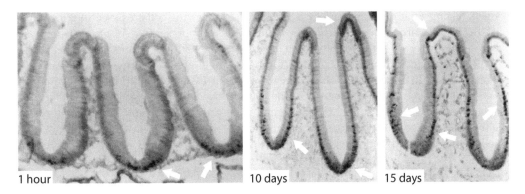

Figure 4.28 Autoradiograms of stomach grooves in longitudinal folds in *Styela clava* 1 hour, 10 days, and 15 days after exposure to radiolabeled tritiated thymidine showing that the labeled dark cells (pointed at by the arrows) migrate from grooves to crests. (Reproduced from Ref. [107].)

This experiment shows that there indeed exist two different cell populations but it also demonstrates that the time scale of cell migration from the grooves to the crests is of the order of 10 days. This is a rather slow migration, and assuming that the dynamics is subdominant seems a reasonable approximation provided that one can account for the differences between the proliferating and the non-proliferating populations by treating them as materials with distinct mechanical properties. For example, the spontaneous curvature of the basement membrane at the base of the crypt seems to play a very similar role as the differential tension of Section 4.1.2 in ensuring that a part of the epithelium prefers a curved shape—except that the latter is the same in all of the tissue. Should the differential tension in the polarity-induced folding model be allowed to vary from the grooves to the crests, an even broader variety of fold morphologies than that presented in Figure 4.16 may be expected.

The main advantage of the above agent-based model is that by tracking the spatio-temporal evolution of each individual cell, the various events determining its fate can be described in considerable detail. On the other hand, the very mechanics of the cells and their shape is simplified by representing the cells as point particles connected by springs, which is quite close to how one typically visualizes inanimate matter such as atoms or molecules in a solid or in a liquid. Given that the stromal cells in this section can move past each other, the stroma may be likened to a classical liquid, and the epithelium can then be viewed as a solid plate lying on it.

In this respect, the models based on individual cells are really not too different from considering both the epithelium and the stroma as elastic or fluid continuous media as suitable, provided that their material properties averaged over a large enough patch of cells and a long enough time are well defined. This, in turn, also implies that the typical time needed to establish the stationary state must be much longer than the cell cycle or any other process involved so that once the stationary state is reached, the material coefficients such as the effective Young's modulus and viscosity no longer change with time. If this condition is not fulfilled, then the dynamics itself may well be the dominant factor dictating the shape of the tissue.

4.2.3   Rapid folding

A theory of folding dominated by cell division can also be explored within an individual-cell model of a somewhat different variety [131]. Instead of viewing cell cross-sections as polygons constructed by Voronoi tessellation, they may well be represented simply by spheres of a given rest diameter. When two cells come in contact, they adhere to and press on each other, and these two effects can be accounted for by a single elastic-energy term modeled by a Hookean spring. In Section 4.2.2 the extension or compression of the springs is unconstrained because the cell-cell boundary is determined a *posteriori* by the tessellation. But in reality very large extensions never take place because then the cells are typically no longer nearest neighbors and large compressions are unphysical because of the intrinsic incompressibility of cells themselves. To enforce the integrity of both each cell and of the tissue, it is thus convenient to truncate it for compressions and extensions beyond a certain length as shown in Figure 4.29a. The two characteristic lengths σ and δ may be interpreted in two different ways: one is that the rest diameter of the sphere corresponding to the minimal energy is $\sigma + \delta$ and that the rest length of the spring is 0, and the other one is that the rest diameter of the sphere is σ and that the rest length of the spring is δ; here δ is the maximal allowed deformation of the spring. Given that the spheres themselves are an idealization, these details are unimportant—in fact, the exact form of the nearest-neighbor interaction does not really matter either as long as it ensures that the neighboring spheres neither approach each other too closely nor become too distant.

The other important type of cell-cell interaction is related to the tissue bending energy. In a two-dimensional model, the bending energy only depends on the local curvature defined by the radius of the arc passing through the centers of the sphere and their neighbors (Figure 4.29b). This radius may vary both because the centers are not aligned and because the center-to-center distances of the spheres are not $\sigma + \delta$ but typically the latter effect is less pronounced. The nearest-neighbor interaction and the bending energy each carry a certain energy scale given by the depth of the parabolic potential ϵ in Figure 4.29a as well as by the product of the bending rigidity and sphere diameter, but since spring deformation is restricted the shape of the tissue is controlled primarily by the bending energy. Given enough time, the tissue is expected to reach an equilibrium state of minimal bending energy depending, of course, on the boundary conditions.

Yet if the tissue proliferates rapidly, growth may prevent it from reaching the global equilibrium. In the sphere representation of cells, the natural manner of introducing cell growth is to let it deform into two overlapping spheres (Figure 4.30). During simulation, the distance between their centers is gradually increased provided that the neighboring

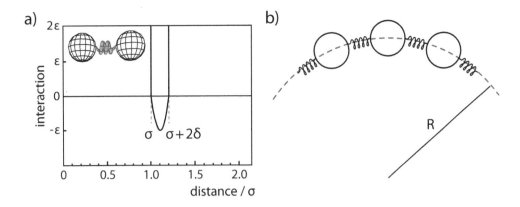

Figure 4.29 Nearest-neighbor interaction of model spherical cells (a) [131] and their bending energy depending on the local radius of curvature R (b).

cells are not too close to the dividing cell [131]. The cell is also forced to maintain an in-plane orientation tangential to the local contour, thereby mimicking the behavior of real epithelia where the cleavage plane is usually perpendicular to the plane of the tissue so that the daughters stay in the tissue. The Monte-Carlo simulation protocol also includes cell translation and reorientation moves, and the key control parameter is the growth rate represented by n_g, the number of attempted translation and reorientation moves between two subsequent growth attempts. In any event, n_g must be considerably larger than 1 so as to allow the model tissue to explore the energy landscape; as expected the duration of the cell cycle τ_0 increases with n_g.

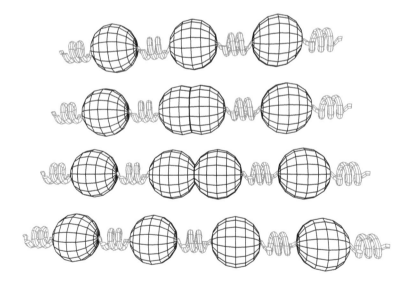

Figure 4.30 Cell division within the rapid-folding scenario. A spherical mother is deformed into a dumbbell-like body consisting of two overlapping spheres, and the distance between the centers is gradually increased. Once it exceeds the sphere diameter σ, division is completed.

Figure 4.31 shows the initial undeformed state of a model tissue of 100 cells forming a straight line, its endpoints being fixed. If it is allowed to grow until the number of cells is increased to 110, this tissue will evolve into a convoluted buckled contour[10] to accommodate the increased length between the endpoints. The characteristic wavelength depends on the duration of the cell cycle τ_0. Although there is a considerable variability between the individual waveforms in the two buckled snapshots in Figure 4.31, it is still evident that at the smaller τ_0 their wavelength is about 0.1 of the end-to-end distance whereas at a longer cell cycle time of $4\tau_0$ it is almost doubled.

initial state   folded states   minimal-energy state

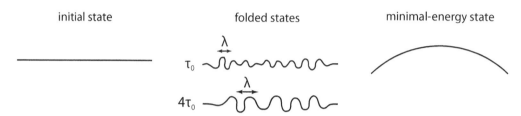

Figure 4.31 Results of the rapid-folding scenario: the initial 100-cell state, the folded 110-cell states with a short and a long cell cycle τ_0 and $4\tau_0$, respectively (adapted from Ref. [131]), and the single-arc minimal-energy state.

None of these buckled contours reaches the global energy minimum, which corresponds to a circular arc spanned between the endpoints. This state is never reached by the rapidly dividing tissue because the orientation of the cleavage plane of a given cell is determined by its nearest neighbors, reducing the bending energy locally by increasing length and thus the radius of curvature. To appreciate why the folded state is still metastable, consider a simple toy-model contour consisting of two semicircles of radii r_1 and r_2 (Figure 4.32a). At a fixed number of cells, the length of the contour $\ell = \pi r_1 + \pi r_2$ is fixed too, and so the distance between the endpoints $2(r_1 + r_2) = 2\ell/\pi$ is constant. The bending energy of the contour $E_{\mathrm{bend}} = (\pi D/2)\left(r_1^{-1} + r_2^{-1}\right)$, where D is the bending rigidity. These two equations can be

a)   b)   c)

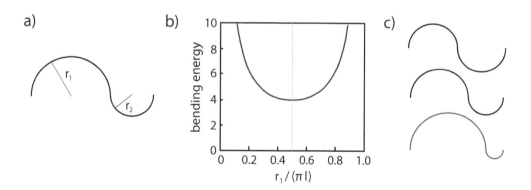

Figure 4.32 Toy model of a fixed-length fold waveform consisting of two semicircles of radii r_1 and r_2 (a) and the bending energy of a waveform (b) which is minimal in symmetric folds where $r_1 = r_2$. In the sequence of waveforms in panel c, the bending energy increases from top to bottom.

---

[10]The line buckles so to remain in plane, as the simulations are performed in two dimensions.

combined, giving

$$E_{\text{bend}} = \frac{\pi D}{2}\left(\frac{1}{r_1} + \frac{1}{\ell/\pi - r_1}\right).$$ (4.36)

The bending energy is minimal at $r_1 = \ell/2\pi$, that is, for identical r_1 and r_2 representing a symmetric fold (Figure 4.32b). This conclusion is consistent with the locally fairly symmetric contours in Figure 4.31 where the radius of curvature varies smoothly from turn to turn.

Although the numerically obtained contours do not consist of circular arcs, this model still explains why reaching the global minimum is not easy. Should a given locally symmetric contour unfold, it must pass through a sequence of states where half of the waveform grows at the expense of the other half (Figure 4.32b) which increases the total energy.

The rapid folding mechanism itself requires no additional mechanical features such as the stroma, apico-basal tissue polarity, inhibition of division, or cell death, which may of course be included so as to construct a more realistic model. With an imbalance between cell division and cell rearrangement, this mechanism is not relevant for the epithelial morphologies mentioned at the beginning of this chapter where division is rather slow, but it may well play a role during embryonic development or in non-embryonic dividing tissues with a fast cell cycle including certain pathological conditions that involve tumor growth. In addition, it may also be initiated by external stimuli such as x-ray irradiation or a growth factor [131].

In the rapid-folding scenario, the mechanics of cells is much simpler than in most other models in this chapter. A step further in the abstraction of the tissue as a physical entity are models where mechanics is incorporated in the rules that specify the cell cycle, division, migration, apoptosis, and differentiation as well as the interaction of cells with their neighbors. These models are known as cellular automata [132], the most famous of them being Conway's Game of Life. In this automaton cells live, die, or reproduce on a lattice, depending on the number of their neighbors, and these processes lead to dynamic patterns. These patterns include stationary but spatially non-uniform states which could represent, at least conceptually, a dynamic tissue with localized apoptosis and cell division and the associated flow of cells between the different regions, say from crypts to villi. Rule-based models or tissues are very versatile in encoding the different possible behaviors of cells, which rely on molecular control, regulation, and signaling mechanisms.

Such models have been explored in various contexts including, e.g., growth of epithelial cells in monolayer culture [133] or tumor growth [134], where the shapes and patterns of tissues emerge primarily from the social behavior of cells. One of the advantages of this approach is that it can be used to simulate an ensemble of many interacting cells, allowing one to investigate phenomena at the scale of organ parts such as epithelial villi or crypts. In one of such studies involving liver lobules described at the beginning of Chapter 6, a few thousand hepatocyte cells modeled as proliferating adhesive elastic spheres were employed to theoretically examine various scenarios of liver regeneration after intoxication, and then compare them to observations [135]. The best agreement was obtained in the scenario that includes the alignment of daughter cells along the neighboring blood capillary known as the sinusoid. Obviously, this conclusion could not have been reached without a model that can accommodate an accurate enough representation of the three-dimensional hepatocyte architecture and the interpenetrating vascular network, which inevitably involves a large number of cells.

Cells in a human body die in billions every day, although not all at the same rate. While brain is a fairly fixed structure so that the neurons can stabilize consciousness throughout the lifetime, erythrocytes live a couple of months, leukocytes about a year, intestinal epithelial cells only a few days... Except for neurons, most cells in the body are replaced many times so that the body itself is more like a blueprint than like a fixed arrangement of building blocks. In this sense, a renewing tissue is closer to a stream than to a rock—the shapes of both a stream and a rock are reasonably stable in time but the stream flows and whirls and the water in it is never the same whereas the rock is static rather than stationary, its grains locked in a networks of contacts and cracks. The models of epithelia that we present in this chapter are between the stream and the rock. In all of them the tissue shape depends on some form of mechanical energy and virtually all of them contain a dynamic element, sometimes packed in an effective parameter such as the negative surface tension due to cell division in Section 4.1.1 and sometimes more explicit like in the rapid folding theory in Section 4.2.3. Thus most of these models pertain to a stationary state.

Life is however stationary only when fully developed and even then just for a period of time. An animal starts out from a single cell, and after many rounds of division, differentiation, and various shape deformations this cell develops into the final stationary form. The many details of this marvelous process and its comprehensive physical interpretation are beyond the scope of any single book. Yet in some transformations of simple tissues, especially during embryonic development, the existing theoretical insight is quite elaborated as sketched in the next chapter.

PROBLEMS

4.1 Consider an elastic cylinder made of a thin sheet pressed between two parallel plates. Assume that the profile of the deformed cylinder is a combination of planar pieces, which are in contact with the plates, and cylindrical surfaces [136]. Use this profile to variationally determine the energy of the deformed cylinder. How does the force required to press the plates change with the displacement of the plates?

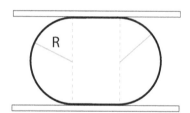

4.2 Assume that the number of segments in the numerical representation of Lewis' brass-bar-and-rubber-band model is large and express the energy [Eq. (4.1)] in terms of its position-dependent spontaneous curvature and bending rigidity.

4.3 Estimate the critical force for Euler buckling of a rod with one end fixed as shown in Figure 4.5 by analyzing the energy of a compressed rod and the energy of a bent rod with a constant radius of curvature [Eq. (2.67)]. *Hint:* Compare the energies at identical displacement of the tip of the rod δ. Show that the compression energy is proportional to δ^2 whereas the bending energy is proportional to δ. (This calculation predicts the correct functional dependence and an approximate value of the critical buckling force; the exact solution is $F_c = \pi^2 Y I / L^2$ [112].)

4.4 A bimetallic element is made from two strips of different metals, glued together and cut to the same length L at room temperature. The thicknesses of both strips are the same and denoted by h. As the temperature is changed by ΔT, the bimetallic element bends. Show that the angle ϕ is given by

$$\phi = \frac{L}{h}(\alpha_2 - \alpha_1)\Delta T, \tag{4.37}$$

where α_1 and α_2 are the coefficients of linear thermal expansion of the two metals. Express the energy of the bimetallic element in terms of Young's moduli of the two metals Y_1 and Y_2.

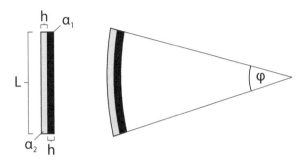

4.5 Within the model represented by Eq. (4.13), calculate the moduli for a simple shear and a bend deformation where the deformed cell is shaped as right rhombic prism and as a right frustum, respectively.

4.6 In the unconstrained Lewis' brass-bar-and-rubber-band model shown in Figures 4.2 and 4.3, each segment is in its respective minimal-energy state, its shape completely independent of the neighbors. This is not the case in the periodic profiles shown in Figure 4.4. Elaborate on this by studying the profiles and the shapes of segments in different regions of each profile.

4.7 Dysplasia often occurs in patches: anomalous and fingered epithelial tissue is surrounded by normal, flat epithelial tissue. Discuss how this can be explained within the framework of the model described by Eqs. (4.24)–(4.26).

4.8 In Section 4.1.3, epithelial dysplasia was explained with a dynamic model—cells divide and die, and growth depends on the supply of nutrients and on growth factors. However, a dynamic process produces stationary states, which, in case of dysplasial instability, have a significantly increased contact area between the stroma and the epithelium. Discuss this within the framework of a static model in which the two tissues are modeled as liquids with surface tensions (and interfacial tension Γ_{SE}). Discuss the formation of epithelial "drops" and their migration into the stroma, which is experimentally observed.

4.9 A strip is made of three pieces: the middle one of length l_2 and bending rigidity D_2 flanked by two identical pieces of length l_1 and bending rigidity D_1. The middle piece

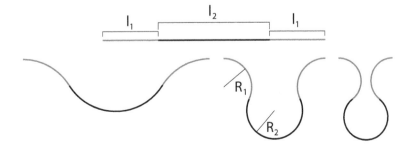

has a spontaneous curvature c_0 whereas the spontaneous curvature of the left and right piece is zero. Assume that $D_2 \gg D_1$ and determine the equilibrium shape of the strip. Find the variational solution for periodic boundary conditions by representing the shape as a union of three circular arcs, two of radius R_1 and one (in the middle) of radius R_2. Show that

$$R_2 = \frac{1}{c_0}, \qquad R_1 = 2R_2\frac{l_1}{l_2}, \qquad \text{and} \qquad E = \frac{D_1 l_2}{4}\frac{l_2}{l_1}c_0^2, \qquad (4.38)$$

where E is the variational estimate of the energy of the strip.

4.10 Consider a cylindrical drop with a patterned surface consisting of lengthwise strips of identical width and alternating positive and negative surface tension. The width of each strip is fixed and so is the length of the cylinder. Use the Young–Laplace law to discuss how the shape of the cylinder changes depending on the pressure difference across the surface of the drop. Show that there are non-vanishing, in-plane forces acting across the borders of the strips.

4.11 Imagine a disk of tissue where only cells in a given narrow sector divide, producing new cells in the circumferential direction (panel c in Figure 1.4). The new cells induce stress in the tissue if the disk is restricted to a plane, and the elastic energy can be decreased considerably if the disk is allowed to buckle out of the plane. This growth mode can be represented as an insertion of a circular sector (new tissue) into the disk (existing tissue). A similar shape may be expected under more general circumstances where growth is strongly anisotropic and more pronounced in the circumferential direction than in the radial direction. The physics sketched may be associated with the shape of sympetalous flowers—their petals are fused into a single sheet of plant tissue (these include the well-known *Brugmansia* and sweet potato flowers, but also hedge binweed). Discuss the general features of the shape of the buckled disk and estimate its bending energy.

4.12 Calculate the energy barriers separating the folded states in the simple two-dimensional two-radii model described in Section 4.2.3 from the minimal-energy state using Eq. (4.36) by assuming that the shortest relevant wavelength is of the order of cell size.

4.13 Growth of an elastic material in confinement produces structures with an organic feel. In organisms, confinement is induced by neighboring tissues that surround a growing organ or structure. Consider fat elastic closed tubes of radius R_t growing within a cylinder of base radius R_c and height $2fR_c$. Assume that growth results only in the elongation of the tube whereas its radius remains fixed. Discuss the shapes of the growing tube which minimize its elastic energy [137] for different ratios of tube volume and the volume of the cylindrical confinement v. Discuss the formation of helical structures in cases when the tubes are not closed, i.e., when their ends are free to move independently. Compare these shapes to compact fluorescent tubes.

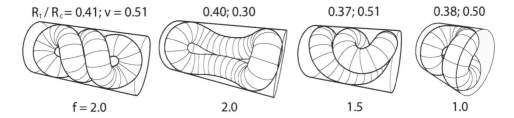

| $R_t/R_c = 0.41; v = 0.51$ | 0.40; 0.30 | 0.37; 0.51 | 0.38; 0.50 |
| $f = 2.0$ | 2.0 | 1.5 | 1.0 |

CHAPTER 5

Morphogenesis

I N EARLY ANIMAL DEVELOPMENT, the embryo consists of a mass of dividing cells and a yolk, both contained within an envelope known as the vitelline membrane. As cells divide, they are gradually spatially organized so as to form compartmentalized structures, which then develop into organs. The shaping of tissues and of the organism itself is referred to as morphogenesis, and it takes place together with cell growth and differentiation.

Morphogenesis is, of course, not restricted to the embryonic stage of life; it is also present in adult animals as well as in tumors and other types of cell aggregates such as cultures and organoids. Each of these instances is interesting in its own right—understanding tumor growth is of enormous importance from the medical perspective and cell cultures are essential for tissue engineering—but from the theoretical point of view, embryonic development is rather special. During this stage, morphogenesis is easier to interpret simply because it involves fewer components either as active elements which drive tissue transformation or as supporting structures. Nevertheless, constructing a phenomenological mechanical model of the various morphogenetic processes is still an arduous challenge.

In an embryo, cells divide so as to become separate entities but they are held together so that the organism can be formed; sometimes they must move past each other and sometimes they need to form a solid-like aggregate such as an epithelial sheet. These functional dychotomies are accommodated by relying on several types of tissues. Embryos include both epithelial tissues, which consist of cells that are strongly bound to each other, and fluid-like tissues known as mesenchymal. Mesenchymal cells are embedded in a gel-like matrix and can migrate through it rather smoothly. The two types of tissue can transform into each other in processes referred to as the epithelial-mesenchymal and mesenchymal-epithelial transition. This is only the roughest division, which emphasizes how dramatically the mechanical properties vary across tissues. In fact, even variations within a single tissue are often considerable.

Without appreciating these differences it is hard to visualize how embryos may undergo the elaborate transformation needed to shape them. Figure 5.1 shows examples of a few basic deformation modes in three animals and in an organoid. At the blastula stage, the acorn worm embryo is a spherical shell formed by an epithelial monolayer reminiscent of a raspberry (Figure 5.1a). During gastrulation, the 300 μm-diameter blastula develops a dimple that then deepens and transforms into a tube which grows and eventually reaches the opposite side. The *Drosophila* gastrulation begins with the formation of ventral furrow, a lengthwise closed groove on the ventral side of the ellipsoidal embryo seen in Figure 5.1c. Also visible is the cephalic furrow separating the future head from the trunk. Neurulation in vertebrates (Figure 5.1b) is geometrically similar to the formation of ventral furrow in that it produces a subsurface tube but it takes place at a later stage of development and is mechanically more sophisticated, involving hinges and partial delamination [139]. The

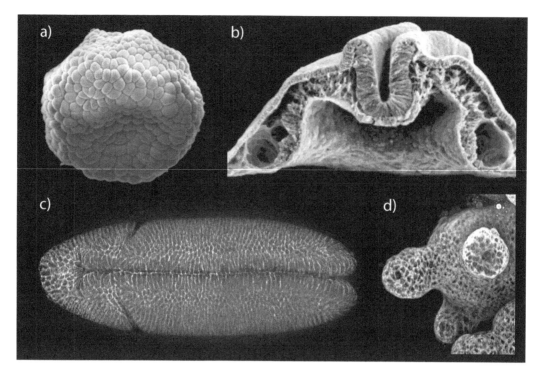

Figure 5.1 Simple examples of morphogenesis: onset of gastrulation in acorn worm *Saccoglossus kowalevskii* (a; adapted from Ref. [138]), cross-section of chick embryo at the neural-groove stage of neurulation (b; adapted from Ref. [139]), *Drosophila* embryo after formation of the ventral and the cephalic furrow (c; image courtesy of M. Rauzi), and buds in a murine small intestinal organoid (d; image courtesy of J. Heuberger).

last example in Figure 5.1 is an organoid rather than an animal or its part. Organoids are small bud-like structures grown *in vitro* from stem cells that are morphologically and functionally very similar to real organs including heart and liver [140], and they are used as experimental models for disease and as a source of tissue for therapy or transplantation. The murine intestinal organoid in Figure 5.1d is composed of epithelial cells that self-organize in the *in vivo*-like crypt-villus structures with stem cells located in the crypt-like buds which continuously produce new cells.

These four examples of geometrically straightforward deformations involve a few tissues. The complete picture is much more complex as illustrated by the sequence of micrographs of the *Drosophila* embryo (Figure 5.2) which covers the period from early embryogenesis until the end of segmentation when the trunk is divided into fourteen segments. In order to understand such a sophisticated transformation, it is best to break it up into morphogenetic movements as the basic deformation modes that shape the tissue. This chapter describes the basic morphogenetic movements, starting with the simpler processes and then turning to the more complex ones.

Any mechanical theory of these movements must take into account that they take place autonomously, that is, without the action of a force generated outside of the embryo. At least one of the tissues or structures involved in a movement must consume the stored chemical

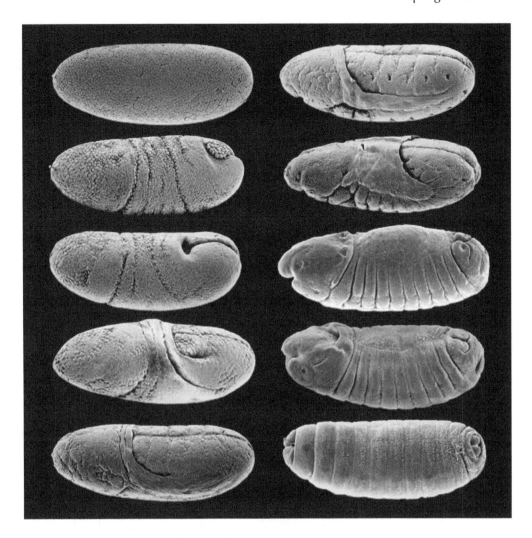

Figure 5.2 Side view of the *Drosophila* embryo in the early embryogenesis, gastrulation, and segmentation stage (reproduced from Ref. [141]).

energy and use it so as to move, deform, and exert forces; its motion and deformation also depend on the physical constraints enforced by yolk, vitelline membrane, etc. The epithelial, mesenchymal, and other embryonic tissues should generally be regarded as the prototypical active matter (Box on p. 8), and any model of a given movement must include one or more elements that drive the process.

The active nature of these elements can be either explicit or implicit depending on the representation of the energy input. An explicit theory of an active tissue includes the energy balance whereas an implicit description does not and relies instead on, e.g., a prescribed kinematics of deformation like in the continuum-mechanics model of ventral furrow formation in *Drosophila* [142], on a prescribed temporal variation of forces or tensions that cause this deformation [143], or on a sophisticated theory of excitable tissues where the active deformation of cells in the invaginating furrow is triggered by stress [124]. These ideas were tacitly included in the models of formation of epithelial folds and other processes discussed in Chapter 4. For example, the area-mismatch theory of folds involves cell growth which obviously requires energy—but in a mechanical theory focused primarily on the form of the folds, growth can be represented by a negative surface tension as an effective macroscopic parameter.

The explicit and the implicit descriptions of active deformation may seem like two fairly different approaches, the former being more complete. Yet preferring one over another is often merely a matter of convenience, much like the motion of a swinging pendulum can be parametrized either by the total energy or by amplitude provided that the mass and length are known. At the same time, we must bear in mind that any mechanical cell- or tissue-level model is inevitably phenomenological rather than based on first-principle molecular origin of the physical forces generated within and by the cells. From this perspective, the division is less dramatic than at first sight.

This chapter discusses six basic morphogenetic movements which include (Figure 5.3):

- **Invagination**, where an epithelial sheet forms a fold- or a finger-like cavity protruding inside;

- **Convergent extension**, where an epithelial patch extends in a given direction and contracts in the perpendicular direction while the cells themselves elongate in the perpendicular direction;

- **Intercalation**, where cells in an epithelium undergo topological rearrangement much like cars on a highway when the number of lanes is reduced;

- **Epiboly**, which refers to the spreading and thinning of a sheet-like tissue;

- **Involution**, where a cell sheet forms a fold like in a pleated fabric, typically advancing in a tank-treading fashion; and

- **Ingression**, where some cells withdraw from the epithelium and transform into motile mesenchymal cells.

Not all of these movements are equally complex. Ingression, for example, pertains to individual cells (unless we also think of their neighbors, which patch the hole left behind) whereas involution is typically a coordinated deformation of a multilayer tissue. On the other hand, intercalation constitutes a well-defined tissue deformation *per se* but is often also a part of convergent extension and epiboly. As such, the illustration depicting convergent extension in Figure 5.3 also shows the mediolateral intercalation mode where cells intercalate in an in-plane direction as opposed to the radial mode where they intercalate in the direction perpendicular to the epithelial sheet.

The insight into the mechanics of these movements is uneven, some being understood much better than others. We present them in a comparative manner by combining experime-

---

Endoderm, mesoderm, and ectoderm

During gastrulation, three distinct tissues referred to as germ layers are formed by cell movements from the blastula epithelial sheet facing the external medium: the endoderm, the mesoderm, and the ectoderm. The endoderm is the innermost layer and develops into gut and gut derivatives (lungs in vertebrates). The mesoderm lies between the endoderm and the ectoderm and produces muscular, vascular, and reproductive systems as well as the connective tissues. The ectoderm is the outermost layer of the embryo and forms epidermis and the nervous system. These three layers are present in the higher animals which are known as triploblasts. In diploblasts such as medusae and hydra the true mesoderm is absent.

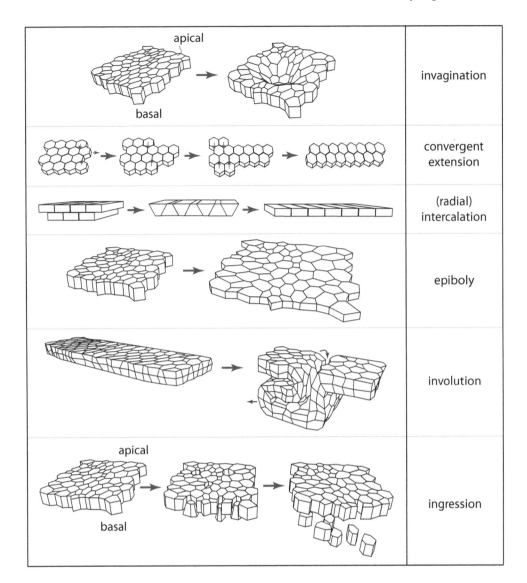

Figure 5.3 Basic morphogenetic movements.

ntal facts with some of the available theories and by describing the processes in several tissues and animals. The main emphasis is on invagination and convergent extension, where the theoretical understanding is rather advanced. Also discussed are some aspects of mesenchymal morphogenesis.

5.1   INVAGINATION

The inward buckling of an epithelium which creates a subsurface cavity is referred to as invagination (Figure 5.4a). During embryogenesis, the first process involving invagination is gastrulation, that is, the transformation of a single-layer shell-like blastula into a more complex gastrula featuring three layers of cells, each developing into a specific system or part of the body. The formation of the ventral furrow in *Drosophila* (Figures 5.1c and 5.4)

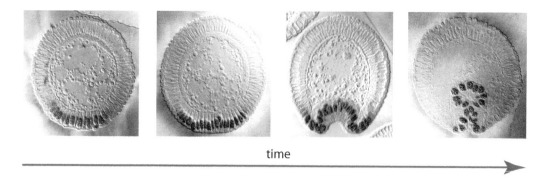

time

Figure 5.4 Transverse cross-section of the *Drosophila* embryo during the formation of the ventral furrow, the final stage showing the fully invaginated epithelium (images courtesy of M. Leptin).

is an example of invagination which may be geometrically likened to the buckling of a piece of paper held by hands and then compressed; naturally, the similarity is merely geometrical because the mechanics at work in the tissue is much more involved.

The ventral furrow is a channel-like cavity with the axis running parallel to the tissue surface. In other cases, the invaginated part of the epithelium may also be oriented perpendicular to the surface. The dimple in the acorn worm embryo in Figure 5.1a is the initial stage of this type of invagination and later evolves into a well-like protrusion. In some cases, the epithelium buckles outward into the lumen rather than inward, which is referred to as evagination.

Despite many similarities and shared features, the different instances of invaginations are not driven by the same mechanism. By examining and comparing the possible physical theories, some of which are quite distinct from each other, we will see that the same final form of a tissue may be attributed to more than a single process. In view of the robustness expected to be built into any kind of a developmental plan of an organism, it is actually more likely that several processes are engaged simultaneously, all of them working toward the same goal. Considering more than just one possible scenario is thus much more than just an exercise, especially if one examines how the conceptually similar mechanical theories compare to the observed behavior in different embryos. Here we will look into selected theories of gastrulation in sea urchin and in *Drosophila*, some chosen for their historical importance and others because of their faithfulness.

One of the earliest and the simplest models of gastrulation in sea urchin builds on assumptions qualitatively reminiscent of the area-mismatch theory of epithelial folding described in Section 4.1.1. In this scenario, the deformation of the initially spherical blastula to an invaginated shape is caused by a pressure difference, much like a table-tennis ball buckles if the external pressure is too large. This possibility was first proposed as early as in 1902 [144] but then abandoned because it was shown that when cut out of the embryo, the vegetal plate itself deformed spontaneously much as if it were still a part of the embryo [145].

Since then, the mechanics of gastrulation was generally interpreted by associating the forces driving invagination to a specific part of the tissue rather than to the blastula as a whole [146]. Among the various mechanisms potentially responsible for the process, those involving the vegetal plate as the active element include i) apical constriction of cells in the vegetal plate, ii) their contraction in the apico-basal direction, and iii) the differential swelling of the extracellular matrix deposited on the apical side of the tissue, which appears to be instrumental for invagination [147]. In addition to these three scenarios, the vegetal plate may also buckle because it is being pushed on by a ring of cells around it, which is referred to as the cell tractor model, or because of the contraction of a cytoskeletal cable running around the perimeter of the vegetal plate [146].

Before describing these modes of localized active deformation of the vegetal plate and the adjacent tissue, it is nonetheless instructive to consider the possibility that the primary gastrulation is assisted by a global process, partly because the simple buckling scenario requires little signaling and differentiation which is advantageous.

Gastrulation in sea urchin

At the 128-cell stage, the sea urchin embryo forms a hollow sphere known as the blastula. The cells are arranged in a single-cell-thick shell forming a fluid-filled cavity called the blastocoel. As cells continue to divide, a dimple similar to that in Figure 5.1a develops at the vegetal pole; this is the primary invagination. During secondary invagination, the dimple grows in depth and transforms into a tube. This tube is referred to as the archenteron and the rim of the opening is known as the blastopore. At this stage, cells are added to the archenteron epithelium so that its volume and area are increased. The archenteron then fuses with the opposite side of the blastula wall, creating the future digestive tract.

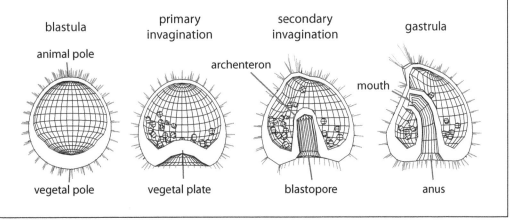

5.1.1 Drying-drop model

The interest in the global scenario of primary gastrulation in sea urchin relying on buckling due to a pressure difference was revived by a tabletop soft-matter experiment involving millimeter-size drops of aqueous latex solution placed on a superhydrophobic substrate and left to dry [148]. These drops retain a spherical shape despite gravity and their deformation upon drying depends on the concentration of latex particles. Drops containing a more concentrated solution remain more or less spherical after drying but the more diluted drops gradually develop a depression at the top (Figure 5.5a). In the extreme case the depression collapses and reaches the bottom, connecting to the small air pocket trapped between the drop and the substrate (Figure 5.5b). This transforms the drop into a toroidal shape.

During drying, the drop initially merely shrinks. At this stage the concentration of latex particles at the surface of the drop increases until a gel-like skin is formed; the pores in the skin contain solvent. In drops of concentrated solutions skin formation is faster than in the more diluted ones, and the skin itself is thick and rigid. Upon further drying, the drop size no longer decreases as the rigid skin is hard enough to withstand deformation due to low intra-drop pressure caused by evaporation; instead, a pocket of air is formed at the bottom.

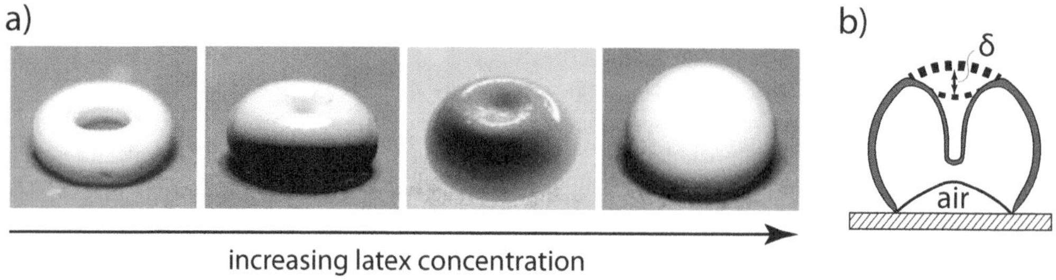

a)   b)

increasing latex concentration

Figure 5.5 Drying latex drops on a superhydrophobic substrate (a) and a schematic cross-section showing a well-developed depression protruding from the top deep into the drop as well as the air pocket between the substrate and the drop (b). (Adapted from Ref. [148].)

In the diluted drops, the skin is thin and elastic and it does deform such that the initial shrinkage is followed by the flattening and buckling of the top.

The deformation of the skin can be interpreted by equating the work expended by the pressure difference across it to either in-plane compression or bending. The pressure difference is determined by Darcy's law describing evaporative discharge through the porous skin:

$$p = \frac{\eta V_E h}{A_d k}. \tag{5.1}$$

Here η is the viscosity of the fluid, V_E is the discharge rate (that is, the volume of evaporated water per unit time), h and k are shell thickness and permeability,[1] respectively, and A_d is drop area. The work associated with the flattening of the top of the drop is $W = p\Delta V$, where ΔV is the volume of the depression which is, up to a numerical factor of order 1, given by the product of the indentation δ and the area of the flattened zone A. Thus

$$W \sim pA\delta. \tag{5.2}$$

In a flattened-top drop, this work is expended in straining the shell. Assume for simplicity that the flattened zone can be viewed as the base of a spherical cap. Then the relative area decrease of the whole shell is of the order of δ^2/R^2, where R is the drop radius. The corresponding elastic energy is obtained by integrating the elastic energy due to area decrease over the volume of the shell, which gives

$$E_{flat} \sim Yh\delta^2, \tag{5.3}$$

again up to a numerical factor of order 1; here Y is Young's modulus. In equilibrium, the in-plane compression energy must be equal to the work done by the pressure difference and this condition determines the magnitude of indentation.

The invaginated regime is characterized by a deformation localized in the bent strip along the rim of the depression. The deformation includes both shell bending in the meridional direction and in-plane compression in the azimuthal direction, the width of the strip determined by minimizing their sum. The detailed analytical discussion due to Pogorelov [149] is concisely presented in Ref. [23] and the final result is

$$E_{bend} \sim Y\frac{h^{5/2}\delta^{3/2}}{R}. \tag{5.4}$$

---

[1] The permeability of the skin measures its capacity to allow the fluid flow through it once a pressure difference across the skin is established. Permeability depends on the porosity of the skin, with larger porosities resulting in larger discharge rates V_E at a given pressure difference p.

Here the elastic energy is proportional to a smaller power of indentation δ than in the flat-top regime, which means that at small δ the flat-top shape is preferred whereas at large δ the drop invaginates as seen in experiments.

From the physical perspective, this scenario of the onset of gastrulation in sea urchin is interesting because it relies on a rather generic and, above all, robust mechanism. Appealing as it may be, it must still be interpreted with care because the theoretical model does depart from the actual embryo in various ways. The sea urchin blastula is hardly a shell of uniform thickness. In fact, its thickness at the vegetal pole where the embryo invaginates is considerably larger than at the opposite animal pole. If the epithelium could indeed be modeled as a solid elastic shell, the thicker vegetal part should have a larger bending modulus than the rest of the tissue because the bending rigidity of a solid plate is proportional to h^3 (Box on p. 51). At the same time, adhesion between cells at the vegetal pole is reduced so as to facilitate ingression and formation of the primary mesenchyme. A weaker cell-cell adhesion implies a smaller effective Young's modulus and thus a smaller bending rigidity of the tissue, which may partly cancel the effect of the increased thickness.

Despite these reservations, it is worthwhile noting that the drying-drop analogy is consistent with experiments where the osmotic pressure within the blastocoel was varied by putting the embryos in seawater containing sucrose [150]. By controlling the concentration of sucrose during gastrulation, the embryos were either shrunken or swelled. The degree of invagination in the shrunken embryos was somewhat larger but generally rather similar to that in control embryos, whereas the swelled embryos did not invaginate as much as the controls.

5.1.2 Spontaneous-curvature model

In the drying-drop model, the epithelium is regarded as a homogeneous solid-like shell although it really consists of several layers. The external, apical side of the cell monolayer is covered by the hyaline layer (Figure 5.6), a filamentous extracellular matrix facing seawater and consisting primarily of hyalin, a high-molecular-weight glycoprotein. The hyaline shell around the blastula is structurally independent of the cell layer underneath and it retains its integrity even if the cell sheet is removed. Between the hyaline layer and the cell sheet is the another fibrous layer called the apical lamina. Cell microvilli reach across the lamina and through the hyaline layer so as to bond the three structures together. The basal side of the cell sheet is attached to the basal lamina.

These different layers are mechanically quite distinct. With a network-like structure, the hyaline layer must have a small bending rigidity but offers resistance to in-plane stretching. The most important mechanical feature of the apical lamina is probably its incompressibility, which becomes important during swelling due to hydrophilic granules secreted by the cell monolayer prior to gastrulation. This swelling leads to an increase of area of the apical lamina which may then bend the blastula wall [147]. On the other hand, the bending rigidity of the cell monolayer is expected to be considerably larger than that of the hyaline layer and the blastula wall as a whole may possess a certain finite in-plane shear rigidity. Finally, cells themselves are apico-basally polarized and thus they may assume a somewhat conical rather than a cuboidal or columnar shape, thereby imparting a certain nonzero preferred curvature to the monolayer.

Except shear rigidity, all of these mechanical features of the blastula wall are accounted for in the bending-elasticity theory originally developed for lipid vesicles. Here the deformation free energy of a closed shell of a given area and volume is given by

$$W = \frac{k_c}{2} \oint (C_1 + C_2 - C_0)^2 \mathrm{d}A + \frac{k_r}{2A_0} \left(\overline{C} - \overline{C}_0 \right)^2 . \tag{5.5}$$

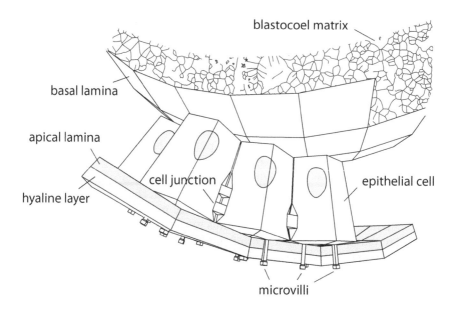

Figure 5.6 Structure of the sea urchin embryonic epithelium: the cell monolayer is covered by the apical lamina and the hyaline layer, and it is separated from the blastocoel matrix by the basal lamina.

The first term is the Helfrich expression for the local bending energy, which depends on the difference between the sum of principal curvatures $C_1 + C_2$ and the spontaneous curvature C_0 integrated over the neutral surface of the blastula wall; k_c is the bending modulus. This theory was originally developed for fluid membranes such as phospholipid bilayers where the molecules may freely rearrange within each leaflet, which is witnessed by the fact that it features the sum of the curvatures rather than C_1 and C_2 separately. Here the theory is seen as a part of a phenomenological description of the blastula wall a thousand times thicker than the phospholipid bilayer, much like both a concrete slab and a sheet of paper may be regarded as solid plates.

The second term in Eq. (5.5) represents the non-local bending energy associated with the difference in the reference areas of the hyaline layer, the apical lamina, and the cell sheet. Although the three layers are attached together, their preferred areas are generally not the same because they are controlled by specific mechanisms such as the degree of swelling of the lamina and the shape of the cells. The effect of the area difference is most transparently visualized by considering the layer and the lamina as a single entity. If this composite layer were rigidly attached to the cell sheet like two metal plates riveted together in a bimetallic strip, any area difference must be accommodated locally by a suitable curvature which depends on the distance between the layer and the sheet. Figure 5.7a shows that two parallel planar curves of lengths l_1 and l_2 fit together if the radius of curvature is

$$R = h\frac{(l_1 + l_2)/2}{l_1 - l_2}, \tag{5.6}$$

where h is the separation between the curves.

This type of deformation is already subsumed in the spontaneous curvature C_0. On the other hand, if the hyaline layer and the apical lamina can slide past the cell sheet then it is only the overall area difference of the extracellular matrix and the cell sheet that matters.

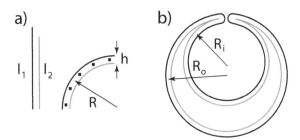

Figure 5.7 Two segments of different lengths riveted together at a fixed separation must curve locally (a; dots representing rivets). An invaginated closed contour (b) showing that in shapes formed by an inner and an outer line that may slide past each other, the length difference is a global rather than a local parameter.

This is best illustrated by an invaginated closed planar contour in Figure 5.7b. If the neck is neglected, the total length difference of the outer and the inner curve is $\Delta l = 2\pi h (R_o - R_i)$, where R_o and R_i are the radii of the convex and the concave arc, respectively. This example shows that the non-local area difference is compatible with shapes where the local curvature may vary quite considerably from point to point: the curvatures of the convex and the concave arc in the contour in Figure 5.7b even differ in sign.

The non-local bending energy associated with the global difference of areas of the extracellular matrix and the cell sheet is thus an additional term distinct from the local deformation energy which depends on the spontaneous curvature. The non-local term can be regarded as the energy associated with the relative stretching of the matrix and the cell sheet needed because their intrinsic, preferred area difference ΔA_0 is generally not identical to ΔA characterizing a given blastula shape. Instead of using ΔA and ΔA_0, the non-local bending energy may also be expressed in terms of the integrated sum of curvatures

$$\overline{C} = \oint (C_1 + C_2) \mathrm{d}A, \tag{5.7}$$

which is proportional to ΔA, as well as $\overline{C}_0$ as the preferred value of the integrated sum of curvatures. Thus one arrives at the second term in Eq. (5.5), where k_r is the non-local bending modulus and A_0 is the area of the blastula wall.

Like in the drying-drop model, the equilibrium shape is found by minimizing the elastic energy of the blastula wall given by Eq. (5.5). Instead of specifying the pressure as the parameter that controls the volume we may fix the volume itself, and since we assume that the main deformation mode of the wall is bending rather than stretching we also fix its area as the second extensive quantity describing the size of the blastula. The volume and the area can be combined in the dimensionless reduced volume

$$v = \frac{6\sqrt{\pi}V}{A^{3/2}} \tag{5.8}$$

defined as the ratio of the volume of a given body and the volume of the sphere whose area is the same as that of the body. As such, v measures how full a shape is with a fixed surface area. In a sphere, $v = 1$ and in any other shape it is smaller than 1.

The spontaneous-curvature model can be used to construct a plausible sequence of shapes transforming a spherical blastula into an invaginated shape. This is done by computing the equilibrium shapes at each reduced volume v and reduced integrated curvature

$$\bar{c} = \frac{\overline{C}}{\sqrt{16\pi A}} \tag{5.9}$$

at fixed ratio of local and non-local elastic constant k_r/k_c, fixed reduced spontaneous curvature $c_0 = C_0\sqrt{A/(4\pi)}$ and fixed reduced preferred integrated curvature $\bar{c}_0 = \overline{C}_0/\sqrt{16\pi A}$. A plausible shape-deformation scenario should meet two criteria: i) it should consist of a continuous sequence because experimentally no abrupt shape changes are seen, and ii) all shapes involved should have a large reduced volume because during the primary invagination the overall shape of the blastula remains rather spherical—except, of course, for the invaginated part itself. These restrictions are met, e.g., in the sequence of axisymmetric shapes shown in Figure 5.8 where $k_r/k_c = 100, c_0 = 2$, and $\bar{c}_0 = 1.054$. In the initial part of the sequence (shapes 1–3), the reduced volume is decreased while the reduced integrated curvature is slightly increased, which leads to the flattened oblate shape 3. The second stage contains shapes 4–8 obtained by decreasing the reduced integrated curvature at constant v, which makes the shell invaginate.

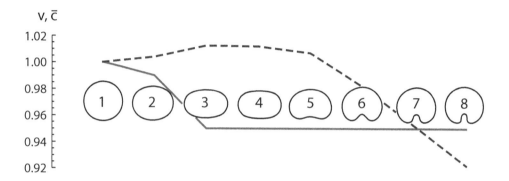

Figure 5.8 Continuous sequence describing the evolution of an initially spherical shape to an invaginated shape within the spontaneous-curvature model. All shapes are axisymmetric about the vertical axis. Solid and dashed lines show the reduced volume and the reduced spontaneous curvature, respectively. (Adapted from Ref. [151].)

In the sequence in Figure 5.8, small variations of the model parameters of the order of a few percent are sufficient to drive invagination. On the other hand, the intermediate oblate shapes (2–5 in Figure 5.8) are not consistent with the observations and the elongated, finger-like invaginations that can be reproduced using this theory are all rather shallow. Of course, the model parameters can be varied so as to make the invagination deeper but this is necessarily accompanied by a round, more or less spherical invagination, whereas the observed invagination in the sea urchin is tube-like as seen in Figure 5.9 which shows two sections of the early-prism stage embryo. The spontaneous-curvature model is nonetheless interesting because it associates the invagination with tensions within the blastula wall rather than with a pressure difference across the wall like the drying-drop model. In addition, it shows that the process may take place even if all of the wall were a homogeneous shell, suggesting that a part of the forces causing invagination may be provided by a global deformation of the blastula as a whole.

This model contains the spontaneous curvature and the preferred integrated curvature as two shape-generating features of the blastula wall, and they need not act in concert. Instead they could well compete with each other: for example, a negative c_0 favors invagination whereas a positive $\bar{c}_0$ is compatible with an outward-budded shape. At the same time, in an isolated patch of a wall with $c_0 < 0$ the preferred integrated curvature $\bar{c}_0$ is immaterial, and

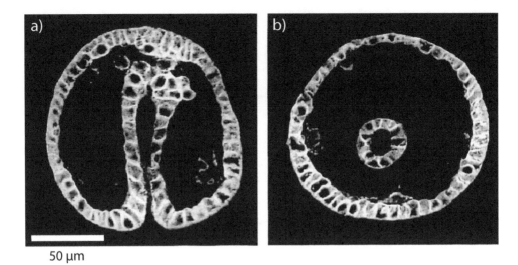

50 μm

Figure 5.9 Saggital (a) and transverse section (b) of the early-prism stage sea urchin embryo showing the fully developed tubular archenteron. (Reproduced from Ref. [150].)

thus such a patch should invaginate just like an excised vegetal plate as seen in experiment mentioned at the beginning of Section 5.1 [145].

5.1.3 Localized active deformation

The spontaneous invagination of the vegetal plate can be reproduced by invoking an active deformation of the plate itself or a specific force exerted by the adjacent tissue. Generated in the plate or in its immediate vicinity, these deformations and forces operate independently of the more distant parts of the embryonic epithelium. In this respect, they are very different from the mechanisms of the drying-drop and the spontaneous-curvature models which rely on the deformation of the whole embryo.

There exist five localized modes with distinct microscopic, cell-level mechanisms that deform the epithelial cell monolayer, the apical lamina, or the hyaline layer [146]:

- **Apical constriction** refers to the decrease of cells' apical area caused by the contraction of microfilaments located at the apical side (Figure 5.10a). This is accompanied by an increase of the basal area, most plausibly due to the conserved cell volume. As a result, each cell is deformed from a prismatic to a keystone shape and this leads to invagination.

- **Apico-basal contraction** is a process whereby the cells in the vegetal plate shorten along the direction perpendicular to the tissue (Figure 5.10b). Since the volume of the cells remains unchanged, this shortening results in an in-plane dilation of the plate. As a result, the vegetal plate then buckles inward if held in place around the perimeter by the adjacent lateral tissue.

- **Gel swelling** is caused by secretion of hydrophilic granules into the apical lamina, which makes the lamina volume increase by water uptake. The hyaline layer, on the other hand, does not swell and the differential projected area (Figure 5.10c) of the two layers bends the blastula wall.

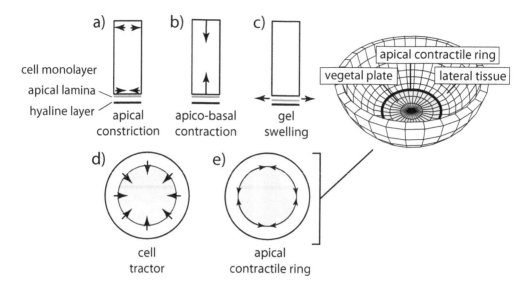

Figure 5.10 Localized active deformations of the sea urchin embryo during gastrulation represented either by the generated or the driving forces. In the apical constriction, apico-basal contraction, and gel swelling models these forces are located within the vegetal plate shown in cross-section in panels a–c, respectively. In the cell tractor (d) and the apical contractile ring models (e) they are exerted by the lateral tissue and a cytoskeletal cable running around the plate, respectively; panels d and e show the plate from top as indicated in the inset.

- **Cell tractor** mode relies on the lateral cells around the vegetal plate, which pull themselves toward the center of the plate by extending contractile protrusions. This creates a radial in-plane stress which makes the vegetal plate buckle (Figure 5.10d).

- **Apical ring constriction** postulates the existence of a cytoskeletal cable on the apical side of the lateral cells immediately adjacent to the vegetal plate. Running around its perimeter, the contractile ring exerts an azimuthal purse-string force on the plate, again giving rise to buckling (Figure 5.10e).

These scenarios were examined using a numerical model where the cell monolayer, the apical lamina, and the hyaline layer were represented as isotropic elastic materials, each with a different Young's modulus [146]. Geometrically, the model was restricted to the vegetal hemisphere and the thickness profile of the cell monolayer accurately represented the actual sea urchin embryo prior to gastrulation, the vegetal plate being considerably thicker than the lateral cells (Figure 5.11a). Each of the five deformation modes was implemented by including a specific local deformation or a force-generating structural element as appropriate. Apical constriction was encoded by an imposed strain gradient within the cells in the vegetal plate such that the areas of their apical and basal sides were decreased and increased, respectively. Apico-basal contraction was driven by a number of springs reaching from the apical to the basal side and initially stretched so as to produce a contractile stress within the vegetal plate. To study the gel swelling hypothesis, the volume of the apical lamina in the vegetal-plate region was increased. In the cell tractor model, cells around the perimeter of the vegetal plate were equipped by prestretched springs lying at the apical sides and pointing towards the center of the plate, and the apical contractile ring was represented

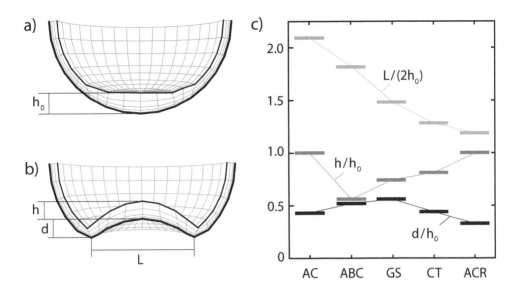

Figure 5.11 Hemispherical geometry of the model sea urchin blastula used to study localized active deformation modes (a) [146]. Panel b introduces the quantities used to parametrize the deformed shapes, and panel c shows the dimensionless invagination diameter $L/(2h_0)$, the vegetal plate thickness h/h_0, and the invagination depth d/h_0 for the example deformations induced by the five modes [146]. Lines are guides to the eye; AC, ABC, GS, CT, and ACR stand for apical constriction, apico-basal contraction, gel swelling, cell tractor, and apical contractile ring, respectively.

by a prestretched circular spring. After each of these elements was activated, the reference stress-free blastula wall spontaneously deformed so as to minimize the total elastic energy.

As expected, all five modes do produce the primary invagination but the detailed shape of the deformed vegetal plate and the depth of invagination depend on the parameters involved. Nonetheless, the examples pictured in Ref. [146] are quite telling and the differences between them can be succinctly presented in terms of the three lengths parametrizing the deformed blastula shape: the thickness of the vegetal plate h, the depth of invagination measured from the rim d, and the diameter of the invagination L (Figure 5.11b). In Figure 5.11c we plot them in dimensionless units for the example shapes, scaling h and d with h_0 but L with $2h_0$ for presentation purposes. This diagram shows that in the example shapes, the ratio of the largest and the smallest value across the modes is about two for any of the three lengths, emphasizing the distinct deformation pattern arising from each mode.

In particular, the invagination produced by apical constriction is generally characterized by a thick vegetal plate, its thickness being essentially unchanged by buckling. The depth of the invagination decreases with increasing elastic moduli of the apical lamina and the hyaline layer, which is quite intuitive. The apical-contractile-ring mode also predicts a thick vegetal plate, which, interestingly, may also slightly bend the tissue outward provided that the moduli of the apical lamina and the hyaline layer are large enough compared to that of the cell monolayer [146]. This non-physiological prediction is obviously a consequence of placing the ring at the apical side of cells under the two extracellular layers.

The other three active deformation modes lead to a somewhat thinner vegetal plate. This is explicitly included in the apico-basal contraction mechanism, which bends the plate inward only if the elastic moduli of the hyaline layer and the apical lamina are large. If this is not the case, the resistance of the hyaline layer and the apical lamina to the pushing in-plane stress generated by cell contraction in the transverse direction is simply insufficient. In this respect, the apical constriction model is more autonomous as the desired keystone shape of cells compatible with the invagination is produced directly by cells themselves rather than by a mechanism involving other structural components of the embryo.

The main detail of the invaginated embryo shapes in the cell tractor mode is the hinge-like deformation of the blastula wall around the perimeter of the vegetal plate where the protrusions of the lateral cells reach under the plate [146]. As a result, the invagination is not dome-like as in the other modes but blunt as if it were formed by a flat stamp, which is reflected in the small diameter of the invagination. The smallest-diameter invagination is, as anticipated, produced by the apical contractile ring which acts around the perimeter of the vegetal plate. Finally, the gel swelling mechanism seems to depend on the relative elastic moduli of the three layers very much, generating a large invagination only if the moduli of the apical lamina and the hyaline layer are rather similar and considerably larger than the modulus of the cell monolayer [146]. This mode also predicts a large outward bulging if the hyaline layer is soft compared to the apical lamina. In this case, the hyaline layer offers little resistance to the expanding apical lamina and since it is considerably thinner than the cell sheet the wall develops an evagination rather than an invagination.

To decide which of these modes are relevant, one should measure Young's moduli of the cell sheet, the apical lamina, and the hyaline layer, and then compare them with the numerical results which, interestingly, suggest that most combinations of the moduli that produce a deep enough invagination are associated with just one of the five localized active deformation modes [146]. The moduli of the extracellular matrix and of the cells were estimated using the diametral-compression test [152], providing an upper limit for the former and a lower limit for the latter. These estimates rule out the apical constriction and the apical contractile ring modes. Yet another aspect where the five scenarios differ from one another is the overall volume of the whole blastula including the blastocoel. Unlike the spontaneous-curvature model from Section 5.1.2 where the constant-volume and constant-area constraints are the key elements restricting the range of possible shapes, the localized active deformation modes disregard these effects, assuming that the fluid within the blastocoel can be squeezed out. Taking these constraints into account may help distinguishing between the three remaining modes.

5.1.4 Excitable epithelium

All of the models described above postulate that the mechanical state of the blastula including its shape is determined by the minimum-energy principle and that its development is quasistatic and controlled by a temporal variation of the external parameters. An alternative and, in fact, historically precedent approach to generate complex deformation patterns is based on the excitable-epithelium model described in Section 4.2. This model was developed primarily to provide an interpretation of the various types of invaginations [124] including gastrulation in sea urchin, the main underlying mechanism being the conditional apical constriction triggered by a large enough stretching of the apical side. In the two-dimensional implementation, the cross-section of the blastula is represented as a ring of quadrilateral cells, each contain six Kelvin–Voigt spring-and-dashpot units as shown in Figure 4.20. The model consisted of a total 64 cells, of which 28 were excitable and 36 were non-excitable. The area confined by the cell ring, which represents the blastcoel, was not conserved because the blastula wall was allowed to leak. The model included about a dozen other material parameters including spring constants, viscous damping constants, etc. [124].

The excitable-epithelium model relies on actual overdamped dynamics of cells explicitly included in Newton's equation of motion so that within this representation the sequence

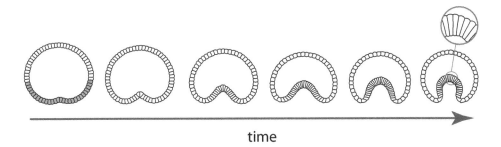

time

Figure 5.12 Sea urchin cross-section during primary gastrulation within the two-dimensional excitable-epithelium model. In the leftmost panel, the 28 excitable cells are shaded gray whereas in the rightmost panel the inset zooms in on the apically constricted cells in the invagination. (Adapted from Ref. [124].)

of snapshots in Figure 5.12 can be regarded as a physically consistent temporal evolution of the epithelium. In the simulation, the excitation wave is initiated by triggering apical constriction in the central cell at the vegetal pole. As expected, the 28 constricted cells located at the bottom of the model blastula eventually form an invagination characterized by a well-defined curvature. By close inspection of cell shapes, one can easily distinguish between the excitable cells which underwent apical constriction and the rest of cells which did not do so. Also apparent is that the enclosed area was decreased by about 50%, which is considerable. In this particular case, invagination ended after all excitable cells apically constricted but this is not the only option. It is conceivable that for a somewhat different choice of parameters the opposite ends of the blastopore would touch and push against each other, and this too could terminate the deformation. Yet another possibility is that the process would stop because of a global constraint due, e.g., to the incompressibility of the blastocoel which could render extreme invaginations costly or even impossible.

Although they feature a deeper invagination, the final stages of the sequence in Figure 5.12 are quite similar to shapes 5–8 in Figure 5.8 illustrating the predictions of the spontaneous-curvature model. It is possible that a three-dimensional variant of the excitable-epithelium blastula would be quantitatively closer to the actual embryo for geometrical reasons. In two dimensions, only the cells at the tip of the invagination are close to their resting, stress-free shape whereas those close to the blastopore are not: the simulation snapshots clearly show that their apical sides are only a little shorter than their basal sides, and this is the reason why a long invagination is not energetically favored (Figure 5.13a). In three dimensions, all excitable cells arranged along the invagination would be close to their apically constricted resting shape (Figure 5.13b), those in the hemispherical tip even more so. Thus we may expect that a geometrically more realistic, three-dimensional version of the excitable-epithelium model (as well as other models that rely on apical constriction) could readily provide a more accurate description of gastrulation.

Another important organism where invagination is being studied in considerable detail is the *Drosophila*. As shown in Figure 5.2, the *Drosophila* embryo is roughly ellipsoidal with the length-to-diameter ratio of about 2.2. The embryo consists of a single-cell thick epithelial layer enveloping the yolk; in turn, the epithelium is contained within the vitelline membrane which may be viewed as a semihard shell. In Figure 5.14 which shows the transverse cross-section of the embryo, the epithelium is seen as an annulus of cells.

About 2 hours 50 minutes after fertilization, cellularization is completed and the epithelium consists of approximately 6000 cells. At this point gastrulation begins, initiating

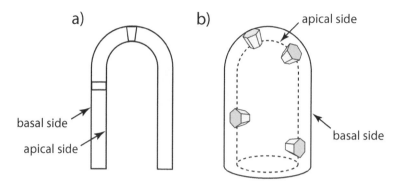

Figure 5.13 Schematic of the invagination in two dimensions where the cells at the tip of the archenteron are keystone-like whereas those arranged around the tubular section are not (a). In three dimensions, the apico-basal area difference is nonzero along all of the tubular section as well as in the hemispherical tip (b).

many morphogenetic processes. The first one is the formation of ventral furrow, a lengthwise invagination on the dorsal side of the embryo seen in the rightmost panel in Figure 5.1b. This process is completed in just about 15 minutes.

The ventral furrow formation is among the best-studied morphogenetic transformations and one of the earliest mechanical theories proposed was in fact the excitable-epithelium model [124]. In this case, considering a two-dimensional cross-section is certainly more justified than in the sea urchin. The chief additional element of the model needed to produce an overall more circular shape of the embryo cross-section and a closed invagination as seen in experiments is the fixed-yolk-volume constraint represented by the fixed enclosed area within the ring of cells. Compared to the sea urchin simulation in Figure 5.12, the model parameters were a little modified too—the total number of cells was the same but there were only 16 excitable cells, the reference shape of cells was more columnar, and the choice of elastic constants and damping constants was modified too. However, the vitelline membrane around the epithelium was not included.

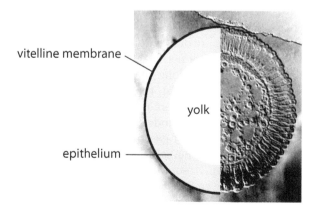

Figure 5.14 Cross-section of a *Drosophila* embryo at approximately 50% egg length at the cellular blastoderm stage just before gastrulation. Cell nuclei and cell membranes are readily visible. (Micrograph courtesy of M. Leptin.)

Gastrulation in *Drosophila*

The development of the *Drosophila* embryo includes many morphogenetic movements. Gastrulation is the transformation of the single-cell-thick shell called the blastoderm into the gastrula consisting of three germ layers, and it includes three events: ventral furrow formation, which marks the beginning of the process, and anterior and posterior midgut invagination. The ventral furrow (here drawn interrupted to show cross-section) is a lengthwise infolding that develops into the mesoderm which later on forms muscles. During the anterior and posterior midgut invagination, two disjoint parts of the blastoderm epithelium transform into endoderm, the tissue that then develops into the gastric tract. Simultaneously with these three invaginations, several other movements take place, the germband extension and the formation of cephalic furrow seen in Figure 5.1c starting shortly after the ventral furrow has been completed.

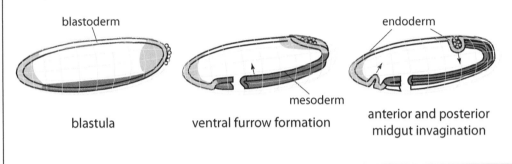

Not all of the ingredients of this elaborate model are equally important. On one hand, the first half of the sequence in Figure 5.15a where the excitable cells did not quite complete the apical constriction fails to predict the flattened vegetal plate seen in experiments [154];

a)

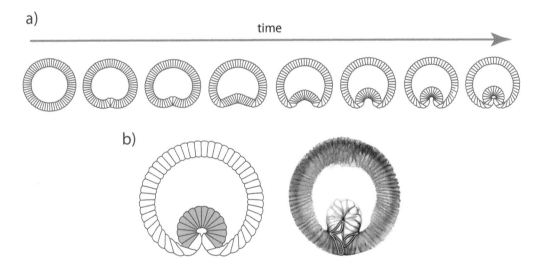

Figure 5.15 Sequence of theoretical excitable-epithelium cross-sections of the *Drosophila* embryo during the formation of ventral furrow (a; adapted from Ref. [124]). A magnified final state with internalized furrow where the 16 excitable cells are shaded gray compared to observed embryo cross-section (b; electron micrograph adapted from Ref. [153]).

instead, they feature a wide V-shaped invagination with a clear notch in the center. The cross-section itself is elliptical rather than circular, which is due to the absence of the vitelline membrane. However, as the apical constriction of the excitable cells continues, the embryo cross-section is progressively more and more circular, consistent with the epithelium minimizing the external perimeter at a constant enclosed area which represents both the yolk and the internalized part of the epithelium.

The large features of the final cross-section are reasonably close to those of the experimentally observed furrow (Figure 5.15b). In the magnified final shape, the radial arrangement of the shaded 16 excitable cells in the furrow is better visible. The two cells with arc-like apical surface at the opening of the furrow are not excitable; these two have in fact undergone a basal contraction so as to allow the furrow to completely internalize. Cells adjacent to the furrow are tilted from the epithelium normal, which is consistent with experiments as seen in the highlighted cell contours in the micrograph in Figure 5.15b.

5.1.5 Global active deformation

Many of these fine details are important but the main lesson learned from the excitable-epithelium model is that the propagation of mechanical signal from cell to cell can too produce a sequence of *Drosophila* embryo cross-sections quite similar to those seen in experiments without a prescribed variation of, say, spring constants. On the other hand, given that in the simulations of invagination in both sea urchin and *Drosophila* all of the excitable cells did eventually undergo apical constriction, the elaborate firing scheme should not be very important provided that the deformation is slow enough. In this case, the underlying idea of active deformation specified by the apically constricted shape of cells can also be captured by imposing a given preferred cell shape and then letting the tissue further deform so as to minimize the elastic energy [142]. Since the magnitude of cell shape change involved is typically considerable, one should resort to finite-deformation elasticity based on the deformation gradient tensor F defined by

$$F_{ij} = \frac{\partial x_i}{\partial X_j}, \tag{5.10}$$

where x_i and X_j are the current and the initial position of a point in a body, respectively. We divide the total deformation gradient into an active part F^a and an elastic part F^e:

$$\mathsf{F} = \mathsf{F}^a + \mathsf{F}^e, \tag{5.11}$$

where the active part of the deformation gradient is a kinematic quantity describing the preferred strain in the material.

The epithelium consists of the mesoderm and the endoderm, each characterized by a different type of active deformation (Figure 5.16). The two modes considered include apical constriction and apico-basal contraction/elongation represented by local coordinate transformations given by

$$x = (1 + \tau_1 Y)X \qquad \text{and} \qquad y = Y \tag{5.12}$$

and

$$x = (1 + \tau_2)X \qquad \text{and} \qquad y = \frac{1}{1 + \tau_2}Y, \tag{5.13}$$

respectively. Here X is the coordinate along the azimuthal direction and Y is the radial coordinate. The magnitudes of the two modes are controlled by dimensionless parameters τ_1 and τ_2. In the apical constriction, a square domain of edge length a is transformed into an isosceles trapezoid of base lengths $(1 + \tau_1 a/2)a$ and $(1 - \tau_1 a/2)a$ and the apico-basal

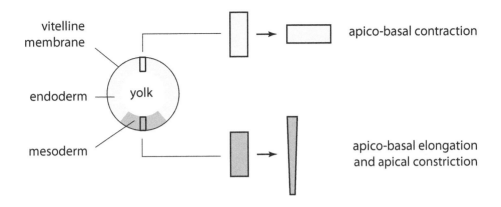

Figure 5.16 Two-dimensional active-elastic model of *Drosophila* gastrulation: the epithelium is divided into the mesoderm undergoing apical constriction as well as apico-basal elongation and into the endoderm where the active deformation mode is apico-basal contraction. The vitelline membrane and the yolk are represented by a rigid frictionless ring and an incompressible liquid, respectively.

contraction/elongation is an example of a pure-shear deformation which changes the aspect ratio of a rectangular cell, turning, e.g., a square domain into a rectangle of sides $(1 + \tau_2)a$ and $a/(1 + \tau_2)$. In both modes the area of the epithelium is locally preserved.

In this model, the endoderm undergoes apico-basal contraction so that each cell is flattened and its apical diameter is increased whereas the active deformation of the mesoderm consists of both apical constriction and apico-basal elongation, deforming cells into tall trapezoids (Figure 5.16). The whole endodermal segment is characterized by a single value of τ_2 and the same holds for the mesodermal segment where τ_2 is negative and equal in magnitude to that in endodermal cells. If the arc lengths of the endoderm and mesoderm were the same, this would imply that the net perimeter change of the whole epithelium due to active apico-basal contraction/elongation is zero. However, the mesodermal segment is much smaller than the endoderm, the arc-length ratio in the initial undeformed state being about 22%. As a result, the active deformation alone does give rise to an increased total perimeter of the epithelial ring.

After the active deformation with a given τ_1 and τ_2 is applied, the equilibrium shape of the epithelium is computed by minimizing the neo-Hookean elastic energy with the energy density of the form

$$f = \frac{\mu}{2}\left(I_1^e - 3 - 2\ln J^e\right) + \frac{\lambda}{2}\left(\ln J^e\right)^2, \tag{5.14}$$

where $I_1^e = \mathrm{tr}\left((\mathsf{F}^e)^{\mathrm{T}}\mathsf{F}^e\right)$ is the first invariant of the right Cauchy–Green deformation tensor $(\mathsf{F}^e)^{\mathrm{T}}\mathsf{F}^e$ and $J^e = \det \mathsf{F}^e$; μ and λ are the Lamé constants. Selected shapes in Figure 5.17 clearly show that both active modes are needed for a deep enough and closed furrow similar to that seen in the embryo (Figure 5.15b). If the endoderm and the mesoderm undergo apico-basal contraction and elongation, respectively, and the mesoderm apical constriction is absent, the mesoderm buckles inward yet forms a flattened invagination rather than a furrow (Figure 5.17a). Thus within this model, pushing and pulling the epithelium towards the ventral region alone is not sufficient and apical constriction must also be involved in furrow formation. This conclusion is supported by the shapes in Figure 5.17b and c where apical constriction of the mesoderm is active. A small constriction bends the mesoderm inward quite considerably (Figure 5.17b) and a more pronounced constriction produces a

apical constriction	absent	small	large	large
apico-basal deformation	large	large	large	absent
	$\tau_1 = 0$	0.66	1.2	1.2
	$\tau_2 = 0.39$	0.33	0.3	0

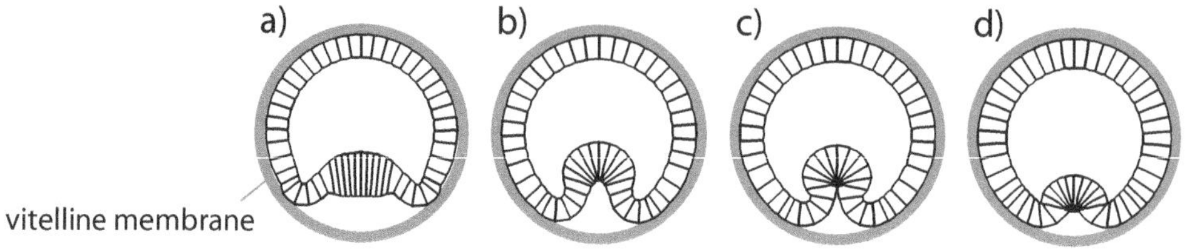

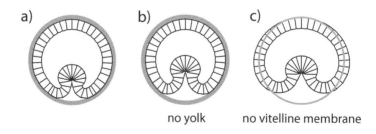

a) b) c) d)

vitelline membrane

Figure 5.17 Final shapes of the deformed *Drosophila* epithelium within the global active deformation model for four pairs of magnitudes of active apical constriction τ_1 and apico-basal deformation τ_2 (adapted from Ref. [142]). The shape in panel c is closest to the observed invagination shown in Figure 5.15b. The quadrilateral domains mimicking cells facilitate visualization of displacements and relative deformation.

deep and closed furrow (Figure 5.17c). On the other hand, apical constriction alone leads to a shallow furrow as shown in Figure 5.17d, which suggests that the endoderm too must participate in the furrow formation.

These results agree with the excitable-epithelium model in highlighting that while local active deformation modes within the mesoderm are important, the formation of furrow involves all of the embryonic epithelium rather than just the invaginating part. At the same time, equally important for both proper furrow shape and proper overall shape of the embryo are the constraints imposed by the yolk and by the vitelline membrane which ensure a more round overall embryo shape from the inside and from the outside, respectively. The effect of the two physical constraints is presented in Figure 5.18. As expected, relaxing the constant-yolk-volume constraint leads to a somewhat deeper yet less closed furrow, which can be seen by comparing the yolk-free embryo shape (Figure 5.18b) and the reference

a) b) c)

no yolk   no vitelline membrane

Figure 5.18 Effect of physical constraints on furrow formation illustrated by a comparison of cross-sections of the embryo with the yolk and the vitelline membrane (a) to those without the constant-yolk-volume constraint (b) and without the rigid-membrane constraint (c). In all panels $\tau_1 = 1.2$ and $\tau_2 = 0.3$. (Adapted from Ref. [142].)

shape (Figure 5.18a). If this constraint is retained but the vitelline membrane is removed, the outer contour of the embryo departs from a circle and the furrow itself is even more open than in the no-yolk case (Figure 5.18c).

The global active deformation model assumes that the epithelium can be described as an elastic continuum, and one may argue that this approximation is not consistent with the structure of cells and the liquid-like rather than solid-like cytoplasm (Chapter 2). Thus a discrete model where the cytoplasm is described by an incompressible liquid and the elasticity of the cell originates in the membrane like in the excitable-epithelium model may seem like a more justified choice. On the other hand, in view of the different approaches discussed so far one may expect that a discrete model of this type should be able to capture the same effects as the continuum elasticity theory, and any agreement found would then point to the robustness of the processes involved.

In the surface-tension-based discrete model, the idea that furrow formation is a collective rather than a local active process is taken an additional step further [143]. In a two-dimensional implementation of this approach introduced in Section 4.1.2, cells are represented by quadrilaterals with an apical side, a basal side, and two lateral sides. The surface tensions of the three types of cell sides, Γ_a, Γ_b, and Γ_l, are all different and the preferred shape of cell is controlled by the reduced apical and the reduced basal tension defined by Eqs. (4.15) and (4.19):

$$\alpha = \frac{\Gamma_a}{\Gamma_l} \tag{5.15}$$

and

$$\beta = \frac{\Gamma_b}{\Gamma_l}. \tag{5.16}$$

As shown in Figure 4.13, the preferred cell shape may be altered by changing α and β. For example, the apico-basal contraction of a cell is induced by decreasing the magnitude of both α and β, and the apical constriction can be triggered either by increasing α or by decreasing β. At each α and β which may vary from cell to cell, the equilibrium shape of the embryo is found by minimizing the total surface energy at fixed cell and yolk cross-section area and with a semihard vitelline membrane around the embryo.[2]

Now that we know that both apico-basal contraction and apical constriction are needed for furrow formation, we can make an additional simplification and disregard the distinction between the mesoderm and the endoderm. This may seem a bold step or even a step back, but the logic behind it is based on two expectations. Firstly, in the global active deformation model the overall increase of the epithelium perimeter is due to the apico-basal contraction of the endoderm, which generates a stress that pushes the tissue toward the ventral region where the buckling takes place. Secondly, the force-based description of active deformation may well operate in a softer fashion than the kinematic description where the preferred shape of cells in mesoderm and endoderm are specified directly. Here "softer" refers to the difference between cause and effect: If we specify the forces (more precisely, surface tensions) that drive a certain active deformation this does not necessarily imply that the deformation will take place. In particular, applying a large enough positive differential apico-basal tension $\alpha - \beta$ in all cells will promote apical constriction globally but with the many constraints at work, constriction may still be localized.

The non-segmented, non-differentiated epithelium consisting of identical discrete cells indeed invaginates much like real embryos. As expected, at large reduced apical and basal tensions α and β the cross-section is circular [143], which corresponds to the blastula prior to gastrulation. On the other hand, at small enough α and β the embryo cross-section does

---

[2]Note that the constraints of the infinitely hard vitelline membrane and exact conservation of yolk volume cannot be applied simultaneously.

buckle. In fact, the sum of the two tensions $\alpha + \beta$ matters much more than each of them separately, which is not too surprising. If the combined tension on the apical and basal side is decreased then the cell height-to-diameter ratio decreases too and cell shape changes from columnar to cuboidal (Figure 4.13) just like in the apico-basal contraction of the endoderm in the active-elastic model (Figure 5.16). In turn, a decrease of the combined tension $\alpha + \beta$ leads to an increase of the perimeter of the epithelium and since the tissue is confined within the vitelline membrane, it must buckle inwards (Problem 5.4).

The detailed shape of the invagination depends on the reduced differential tension $\alpha - \beta$. This is illustrated in the three equilibrium embryo cross-sections in Figure 5.19 which all have the same value of $\alpha + \beta = 3.8$ but different reduced differential tensions. The depth of the furrow is virtually identical in these three cases (Problem 5.6) but the fine details of its shape are not. A positive $\alpha - \beta$ promotes apical constriction and makes the tip of the furrow a little more fan-like, and thus the furrow in the $\alpha - \beta = 1.8$ model embryo in Figure 5.19c consists of a slightly larger number of cells than those at $\alpha - \beta = 0.2$ and $\alpha - \beta = 1$ in Figure 5.19a and b, respectively. In addition, the cells in the furrow are taller because of their preferred constricted shape.

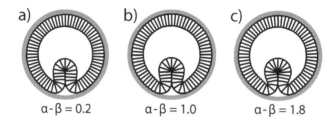

a)   b)   c)

$\alpha - \beta = 0.2$ $\alpha - \beta = 1.0$ $\alpha - \beta = 1.8$

Figure 5.19 Three invaginated shapes of the *Drosophila* embryo in the tension-based discrete model at different reduced differential tensions $\alpha - \beta$. The number of cells in the cross-section is 80, which is approximately consistent with the real embryo; in all cases $\alpha + \beta = 3.8$. (Adapted from Ref. [143].)

The simplest variant of the tension-based model does not include the apico-basal elongation of cells in the mesodermal, invaginating tissue. Within a non-differentiated model of the epithelium where all cells are equivalent, invagination may take place at any arbitrary location; no specific part of tissue is preferred over another. This spontaneous symmetry breaking is not expected in a real embryo and there should exist a way of ensuring that the furrow is formed at a predefined spot. A possible mechanism which could take care of this without the need to include position-dependent tensions is provided by the slight asymmetry of the embryo. As seen in Figure 5.2, the embryo is not exactly rotationally symmetric about the long axis and its cross-section in Figure 5.20a shows that prior to gastrulation, the mesoderm is about 20% thicker than the endoderm on the dorsal side. At the same time, the apical diameter of the mesoderm cells is larger by about 10% (Figure 5.20b), which means that their cross-section area exceeds that of the dorsal endoderm cells by as much as 30%.

It is possible that this variation of cell size could amplify the effect of differential tension even if the apical and the basal tensions were identical in all cells. This scenario can be tested by increasing the cross-section area of the 18 ventral cells by 30%, which indeed makes them more prone to invagination than the rest of cells and ensures that the furrow is formed in the ventral region (Figure 5.20c). In fact, a much smaller cross-section area difference of about 5% would serve the purpose equally well.

By disregarding the differences between the endoderm and the mesoderm in its basic

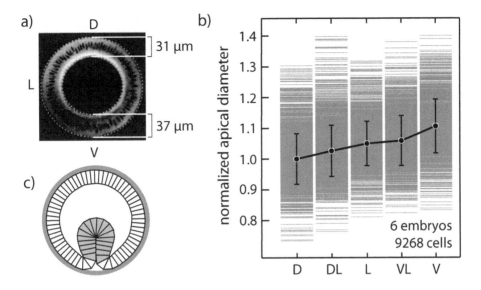

Figure 5.20 *Drosophila* embryo cross-section showing the spatial variation of tissue thickness (a). Panel b shows the apical diameter of the dorsal, dorsolateral, lateral, ventrolateral, and ventral cells (D, DL, L, VL, and V, respectively). The equilibrium shape of an 80-cell model embryo with $\alpha = 2.8$ and $\beta = 1.0$ and a 30% increase of cross-section area in the 18 shaded cells (c). (Adapted from Ref. [143].)

version, the tension-based model emphasizes that some of the mechanical details may not be easily manifestable in a system as restricted by the external structures as the *Drosophila* embryonic epithelium during the formation of ventral furrow. In addition, the predictions of this model are quite similar to those obtained within the active-elastic model—for example, the cross-sections in Figure 5.17c and Figure 5.19c are both equally close to the observations—and thus we cannot decide which of them is more reasonable solely by looking at still images.

5.1.6   Measuring forces

Existing experimental techniques allow us to go beyond comparison of morphometric features. One of the methods that provides a quantitative insight into the forces at work is known as video force microscopy [155]. The idea behind this technique is based on the assumption that the morphogenetic movements are overdamped so that in Newton's equation of motion for any given small part of the embryo

$$m\mathbf{a} = \mathbf{F} - \gamma\mathbf{v} \tag{5.17}$$

the inertial term $m\mathbf{a}$ can be neglected. In this case, the external force acting on the part in question $\mathbf{F}$ is equal to the viscous force $\gamma\mathbf{v}$; here γ is a suitable effective viscosity and $\mathbf{v}$ is the local velocity. The velocity can be measured based on the displacements read off a sequence of micrographs, and Newton's law then allows one to relate it to the force. In an embryo, one may monitor several dozen or more registration points—a little like in motion capture acting used in the film and video game industry. The natural choice of registration points are cell vertices so that the forces between a given pair of points can be interpreted as a tension, but one may also choose to skip some of the vertices or to introduce additional points so as to obtain a more refined description of cell displacement and deformation.

Another method of quantifying the forces within the tissue is the mechanical inverse approach [156] where one assumes that the velocity is negligible so that the tissue is in mechanical equilibrium at all times and the total force on any given vertex is zero. In this respect, the approach resonates with most models discussed above, although it can readily be generalized so as to include dynamics. The starting point of the analysis is that the force on a given vertex depends both on tension along the cell-cell contact as well as on the pressure difference across it, assuming that cells can be approximated by liquid drops (Section 2.3.1). The approach can be readily used to study the effective in-plane forces in the tissue which can be extracted from the external view of the apical side of the epithelium.

In two dimensions, we choose a local coordinate system such that the x axis points along the edge shared by cells a and b (Figure 5.21) and the y axis is perpendicular to it. The force on vertex i exerted by a curved edge between cells a and b points along

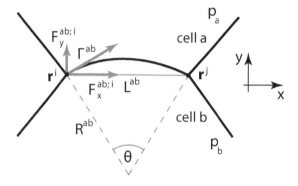

Figure 5.21 Geometry of a curved edge connecting vertices $\mathbf{r}^i$ and $\mathbf{r}^j$ shared by cells a and b [156]. Given by the Young–Laplace law, the normal component of the force depends on the pressure difference and on curvature of the edge whereas the tangential component is equal to the line tension.

the edge itself, and if θ is the angle corresponding to the edge then the force is given by $\mathbf{F}^{ab;i} = \left(\Gamma^{ab}\cos(\theta/2), \Gamma^{ab}\sin(\theta/2)\right)$, where Γ^{ab} is the tension of the edge. In terms of edge length and radius of curvature R^{ab}, the angle $\theta = L^{ab}/R^{ab}$ so that in the limit of slightly curved edges

$$\mathbf{F}^{ab;i} = \left(\Gamma^{ab}, \frac{1}{2}(p_a - p_b)L^{ab}\right), \qquad (5.18)$$

where we used the Young–Laplace law [Eq. (2.3)]; p_a and p_b are the two-dimensional pressures in cells a and b, respectively. This result shows how each of the forces exerted on a given vertex by one of the edges that meet at it depends in the line tension and length of the edge and the pressure difference.

The next step is to compute the edge tensions and pressure differences from the known positions of vertices and the assumption that the total force on every edge vanishes. A little bookkeeping shows that this is not so easy because the number of unknown tensions and pressures is larger than the number of force-balance equations. If we denote the number of vertices by V, then the number of force-balance equations is $N_e = 2V - 3$ because in two dimensions, the position of each vertex is specified by two coordinates and because rigid translation and rigid rotation of the tiling (parametrized by two and one degree of freedom, respectively) do not depend on the internal forces and must be subtracted. The numbers of tensions and pressures are equal to the number of edges E and cells C, respectively. In a closed tiling where cells form a finite patch, the region outside the patch can be viewed as an additional face so that the total number of faces is $F = C + 1$. Such a tiling lies on a topological sphere, and thus E and C are related by the Euler formula [Eq. (3.5)], which gives $V - E + C + 1 = 2$. In a tiling of three-valent vertices, $3V = 2E$ because each edge connects two vertices so that finally the combined number of tensions and pressures to be computed is $N_p = E + C = 2V + 1$, which exceeds the number of force-balance equations N_e by 4. Thus the system of linear force-balance equations is mathematically underdetermined, and further assumptions must be introduced to solve it. This can be done by introducing additional constraints by, e.g., setting the pressure in all cells to the same value or by expressing the tension along an edge as a sum of two terms, one corresponding to each cell. This makes the system overdetermined so that instead of solving it exactly one may seek the least-square solution [156].

The mechanical inverse method was used to study the tensions in the ventral part of the *Drosophila* embryo epithelium during the formation of the ventral furrow immediately prior to invagination. Panels a and b in Figure 5.22 show cell tilings in the same region of the ventral epithelium 4 and 2 minutes before invagination, respectively. By comparing them, it is easy to see that the tissue did elongate along the anteroposterior axis and contract along the dorsoventral axis in just 2 minutes, and a more detailed analysis shows that the tensions of the edges along the anteroposterior axis are larger than those along the dorsoventral axis at the end of this interval. To demonstrate this, edges were binned into two classes depending on whether their orientation relative to the anteroposterior axis was less or more than 45°. The histograms of edge tensions in the two bins are essentially indistinguishable 4 minutes before invagination but 2 minutes later the tensions of edges along the anteroposterior axis is increased whereas in the edges pointing along the dorsoventral axis they are decreased as can be seen by comparing panels c and d in Figure 5.22.

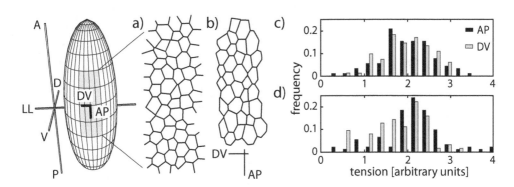

Figure 5.22 Tilings obtained from the micrographs of the ventral part of the *Drosophila* embryo 4 and 2 minutes before ventral furrow formation (a and b, respectively); the anteroposterior (AP) axis is vertical and the dorsoventral (DV) axis is horizontal. Histograms of tensions in edges pointing along the AP and DV axes at these two times, with c and d corresponding to panels a and b, respectively. (Adapted from Ref. [156].)

This process is consistent with the apical constriction in the ventral region as one of the key elements of the invagination but it also shows that while the apical sides of cells do shorten in the dorsoventral direction they also elongate in the anteroposterior direction. This elongation obviously cannot be directly included in two-dimensional models of the furrow formation where the embryo is represented by the cross-section alone, and thus the qualitative message of the analysis of tensions is that a full three-dimensional representation of the embryo is still needed and that the various effects at work during ventral furrow formation are not exhausted by the mechanisms discussed in Sections 5.1.4 and 5.1.5.

In fact, the mechanics of embryonic epithelium of *Drosophila* is even more complex. With the modern experimental techniques such as selective-plane illumination microscopy known under the acronym SPIM [157], one can monitor the progress of invagination with very fine spatial and temporal resolution so as to obtain an accurate map of cell displacements in all of the embryo [158]. These displacements can be represented by the movements of the apical sides of cells, which are best appreciated by unrolling the epithelium so as to obtain a cylindrical projection of the elongated embryo (Figure 5.23a). The projection allows one to view the dorsal, lateral, and ventral sides in the same image like that in Figure 5.23b where the dorsal region is shown both at the top and at the bottom. In this representation, the simultaneous movements and deformations of cells are more readily monitored and quantified than in the full three-dimensional view.

To show how cells move, one can plot a given section across the cylindrical projection

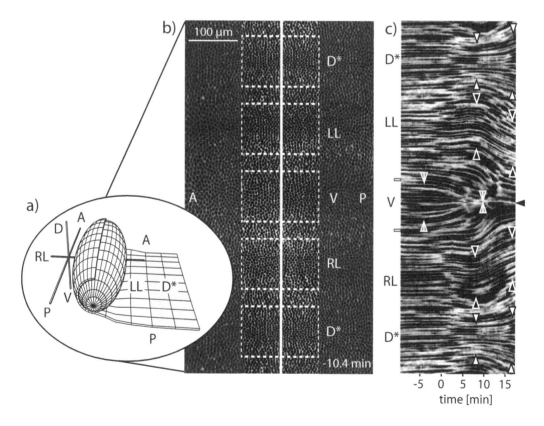

Figure 5.23 The apical surface of the *Drosophila* embryo may be represented by a cylindrical map (a) drawn such that the dorsal region is plotted twice (b; D, RL, V, and LL standing for dorsal, right lateral, ventral, and left lateral region, respectively). Kymograph of the central section indicated by the vertical white line in panel b, with pairs of arrowheads emphasizing the specific deformation modes of the different regions (c). (Adapted from Ref. [158].)

at different times. In the thus obtained kymograph, each time section consists of white and dark points representing cell-cell edges and the interior of cells, respectively. The main feature of a sequence of such sections are the streak-like trajectories of cells somewhat reminiscent of pictures of hydrodynamic flow obtained by introducing tracers into the fluid. The kymograph in Figure 5.23c represents the deformation of the section in the middle of the embryo. Until the onset of invagination at $t = 0$, all streaks are fairly horizontal, which means that none of the cells really moves. At $t = 0$ the converging streaks in the ventral region indicate that the mesoderm migrates towards the location of the future furrow. At ≈ 5 minutes after the onset of invagination when the ventralmost streaks coalesce into a single horizontal line, mesoderm internalization begins and the furrow is being formed. This process is emphasized by a pair of arrowheads located approximately at the edge of the ventral region in the center of the kymograph. After invagination, the distance between these arrowheads is zero, clearly showing that the mesoderm has internalized.

The displacement and the deformation of the lateral and the dorsal tissue start at a later time, about 8 minutes after the onset of invagination. Over the next 9 or so minutes, both left and right lateral tissues move toward the ventral region but they do so without stretching as if they were unaffected by the forces which have by then already deformed the ventral tissue, i.e., the mesoderm. This rigid displacement is readily seen by observing that the arrowheads that mark the boundaries of the left and the right lateral region move toward the center of the kymograph but the distance between them remains unchanged. The dorsal tissue, on the other hand, is dilated by as much as about 50% as shown by the large increase of the distance between the arrowheads close to the boundaries of this region at the top and at the bottom of the kymograph.

The most important conclusion of this experiment is that the dorsal, the lateral, and the ventral tissues are very different and that the detailed mechanics of invagination is more complex than previously thought. The key role of apical constriction in the mesoderm located in the ventral region has been recognized for a long time but the distinction between the lateral and the dorsal endoderm was not appreciated. The stark contrast between these two tissues suggests that former is considerably stiffer than the latter, and this is further confirmed by cauterization experiments where a part of the tissue was mechanically isolated from the rest [158]. In particular, these experiments ruled out the possibility that the large deformation of the dorsal tissue is driven by an active process such as the apico-basal contraction assumed in the active-elastic and in the tension-based model in Section 5.1.5. If so, an isolated dorsal region should buckle but this is not observed in cauterization experiments, which means that in a wild-type embryo the dorsal tissue does not push on the lateral tissue but is instead being stretched by it. This leaves the apical constriction of the ventral region as the main and possibly the only cause of invagination. In turn, this process could not take place as it does without a soft dorsal region.

5.2 CONVERGENT EXTENSION

In the previous section, we described the primary invagination in sea urchin embryo in considerable detail. This stage of gastrulation leads to the formation of a shallow depression which then develops into a long tube called the archenteron (Box on p. 145). The archenteron grows in length, eventually reaching the opposite side of the blastula and fusing with it. Thus the embryo is transformed into a topological torus. If archenteron growth were devoid of a preferred in-plane direction, the primary invagination would develop into a round bud rather than into a tubular structure, much like soap bubbles and most toy balloons are spherical rather than elongated.

Archenteron elongation involves an anisotropic deformation known as convergent extension, which is an important mechanism of directional growth. Here a patch of a cell sheet is transformed such that it extends along a given axis and contracts in the perpendicular direction. During convergent extension cells in the epithelial layer are polarized and oriented perpendicular to the extension axis, and rows of cells along this axis intercalate. This is nicely seen in the germband extension in *Drosophila*, another textbook example of this morphogenetic movement. In this process which takes place soon after the formation of ventral furrow, two bands of endoderm tissue in the ventral and lateral regions—one on either side of the embryo—stretch so as to roughly double in length. Initially, the germbands extend from the cephalic furrow to the posterior end but during the extension, they roll past the end into the dorsal side of the embryo (Figure 5.24). At the end of the process,

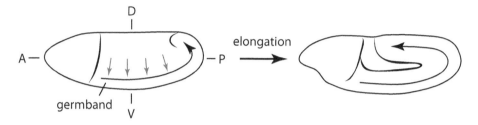

Figure 5.24 Germband extension in *Drosophila* embryo with the lengthwise arrows showing the extent and the direction of displacement of the germband; also indicated is the accompanying shrinkage of the tissue in the dorsoventral direction.

the width of the germband is only half of the initial width so that to a good approximation, the apical area of the germband remains unaltered. As cells themselves do not deform nearly as much as the tissue they must rearrange, and the main mode of rearrangement is mediolateral intercalation.

In more quantitative terms, the extent of germband deformation is visualized in Figure 5.25 where the average position of germband cells initially located at four distances from

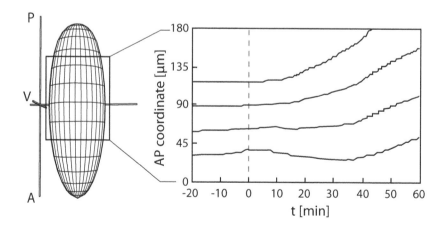

Figure 5.25 Cell displacement during germband extension depicted using the cell coordinate along the anteroposterior axis vs. time; $t = 0$ corresponds to the onset of the extension. The trajectories represent the average location of cells in four stripes of the germband initially located at different distances along this axis. (Adapted from Ref. [159].)

the anterior end is plotted against time; the data are adapted from an in-depth study of cell rearrangement during the process [159]. The figure covers the period from 20 minutes before the beginning of the extension and during the first 60 minutes of the extension, showing how each of the four stripes of germband cells moves toward the posterior end. Specifically, cell displacement seems to consist of two stages: the first one from $t \approx 10$ min to $t \approx 40$ min is marked by a fan pattern indicating that the tissue stretches whereas during the second one starting at $t \approx 40$ min the tissue merely moves toward posterior end. This is witnessed by the fact that the bottom three trajectories are parallel to each other so that the distances between them do not change with time.

Convergent extension is a rather generic morphogenetic process involved in various stages of embryonic development of many animals from ascidians to mammals [160]. Apart from archenteron growth in sea urchin and germband extension in *Drosophila*, we also discuss convergent extension during notochord elongation in *Xenopus*.

5.2.1 Anisotropic differential adhesion model

A simple theory of cell intercalation and rearrangement behind convergent extension can be constructed by assuming that cell-cell interactions can be reduced to a surface energy and by relying on the differential adhesion hypothesis (DAH; Section 2.4). Within this theory, the orientational order of cells can be reproduced by an anisotropic surface tension in a population of identical anisometric cells. Given that the cells are elongated, it is possible or perhaps even necessary to distinguish between their short and long edges; here we switched to the two-dimensional representation where we only consider the apical surfaces of the cells. Then we may postulate that contacts between two long edges, two short edges, and a long and a short edge each carry a different interfacial tension [161]. Let us denote these three tensions by Γ_{ll}, Γ_{ls}, and Γ_{ss}, respectively, and assume that the shape of cells is fixed. Depending on the three tensions, an aggregate of cells may either prefer an ordered structure where the long axes all point in the same direction or a disordered aggregate (Figure 5.26). Here we tacitly disregard any possible packing issues since the fixed rectangular shape of cells as the main underlying assumption is a big simplification hardly representative of real cells in any sense other than elongation itself.

Within the anisotropic-differential-adhesion model, the energy of a finite tissue patch includes four terms, three representing the interfacial energies of all long-long, long-short,

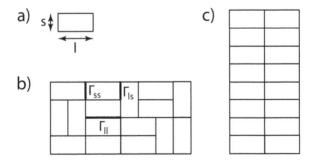

Figure 5.26 Anisotropic-differential-adhesion model of convergent extension: a rectangular elongated cell of sides l and s (a) and an orientationally disordered and ordered patch of tissue (b and c, respectively). In panel b, the bold lines highlight the three types of cell-cell contacts; also indicated are the corresponding interfacial tensions.

and short-short cell-cell contacts and the fourth one corresponding to the cell-medium interfacial energy. All of these terms depend on the total length of edges engaged in a given
type of contact. In a patch of N rectangular cells each measuring $l \times s$, the total length of all
long edges, $2Nl$, is divided into the lengths of the long-long, long-short, and long-medium
contacts L_l, L_{ls}, and S_l, respectively:

$$2Nl = 2L_l + L_{ls} + S_l. \tag{5.19}$$

Here we took into account that there are two long edges in every long-long contact. An
analogous argument shows that the total length of short edges is

$$2Ns = 2L_s + L_{ls} + S_s, \tag{5.20}$$

where L_s and S_s are the total lengths of short-short and short-medium contacts, respectively.
The surface energy of an aggregate reads $E = \Gamma_l L_l + \Gamma_s L_s + \Gamma_{ls} L_{ls}$ where we assume for
simplicity that the tension of the cell-medium contacts vanishes. The lengths of long-long
and short-short contacts can be expressed in terms of L_{ls} and the other extensive parameters
using Eqs. (5.19) and (5.20). This is particularly convenient because in an aggregate with
perfect orientational order, L_{ls} will be zero whereas in a disordered tissue it will be finite,
which means that L_{ls} may be viewed as an order parameter. Eventually one finds that the
total interfacial energy can be cast as [161]

$$E = \left(\Gamma_{ls} - \frac{\Gamma_{ll} + \Gamma_{ss}}{2} \right) L_{ls} + (\Gamma_{ll}l + \Gamma_{ss}s)N - \frac{\Gamma_{ll}S_l + \Gamma_{ss}S_s}{2}. \tag{5.21}$$

The second term is independent of the structure and the shape of the aggregate whereas the
third one involves only the total lengths of the long-medium and short-medium contacts so
that it scales as the perimeter of the tissue proportional to $\sqrt{N}$. In a large enough patch of
tissue, the third term is much smaller than the first one whose magnitude depends on the
total length of long-short contacts proportional to the number of cells N. In view of these
arguments, the structure of the tissue is controlled solely by the first term in Eq. (5.21).
The nature of the minimal-energy state depends on the sign of

$$\Delta\Gamma = \Gamma_{ls} - \frac{\Gamma_{ll} + \Gamma_{ss}}{2}, \tag{5.22}$$

which may be regarded as the energy change upon a 90° rotation of a cell within an ordered
patch.[3] If $\Delta\Gamma > 0$, the energy is minimal at $L_{ls} = 0$, which corresponds to an ordered state.
On the other hand, if $\Delta\Gamma < 0$ then long-short contacts are favored over long-long and short-
short contacts and the tissue will be disordered. The reader will notice that Eq. (5.22) is
merely a variant of the argument behind Eq. (2.53) describing cell segregation and dispersal
within the differential adhesion hypothesis (Section 2.4).

The prediction that the anisotropic-differential-adhesion model may, under suitable conditions, induce a disorder-order transition where cells in an initially isotropic tissue become
aligned is the first nontrivial albeit anticipated result. The other important finding pertains
to the shape of a finite patch. At this point, we have to make a further assumption concerning the interfacial tensions. Given that in a tissue cells are held together, it is reasonable

---

[3]This designation is technically correct only for square cells where a rotation of a single cell is geometrically possible but the lengths of all sides are the same. In elongated cells, one must rotate a square block
of cells instead (e.g., any two adjacent vertical or horizontal cells in Figure 5.26b that are paired along
the long edge) which can be done only if cell aspect ratio l/s is a rational number. In turn, the energy
change depends on l/s. Despite this reservation, it is still instructive to associate $\Delta\Gamma$ with a rotation of a
cell because the lattice model of the tissue is a rough approximation itself.

to expect that Γ_{ll}, Γ_{ls}, and Γ_{ss} must all be negative so that the cell-cell interfacial energy is really a binding energy. In this case, the third term in Eq. (5.21) is positive and may be viewed as an effective surface energy of the whole tissue patch. Now consider the case $\Delta\Gamma > 0$ so that $L_{ls} = 0$ and the patch is orientationally ordered: If the total length of long-short contacts L_{ls} vanishes, the long and short sides of each cell in the bulk can only be in contact with the long and short sides of its neighbors, respectively, which means that the long axes of cells all point in the same direction. In this case, the only remaining degree of freedom of the tissue is its shape. For simplicity, we compare the energy of rectangular patches of a fixed number of cells but different aspect ratios. In a patch of $N_\perp$ rows and $N_\parallel$ columns of cells (Figure 5.27), $S_l = 2N_\parallel l$ and $S_s = 2N_\perp s$ so that Eq. (5.21) reads

$$E = const. - \left(\Gamma_{ll} N_\parallel l + \Gamma_{ss} \frac{N}{N_\parallel} s \right), \tag{5.23}$$

where *const.* stands for the first and the second term in Eq. (5.21) which do not depend on the shape of the patch. We also used the fact that $N_\perp N_\parallel = N$.

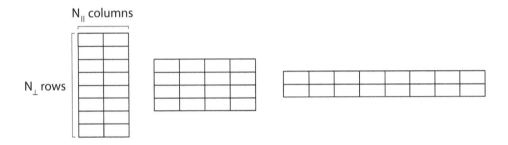

Figure 5.27 Three ordered tissue patches of 16 cells illustrating that the number of long-long and short-short contacts and thus patch energy depend on the shape of the patch, the ratio of numbers or rows and columns in the minimal-energy shape being given by Eq. (5.24).

By minimizing Eq. (5.23) with respect to $N_\parallel$ we find that the row-to-column ratio is

$$\frac{N_\perp}{N_\parallel} = \frac{\Gamma_{ll} l}{\Gamma_{ss} s} \tag{5.24}$$

so that if $|\Gamma_{ll}| > |\Gamma_{ss}|$ the patch is indeed elongated perpendicular to the long axis of cells, its aspect ratio being $(N_\perp s)/(N_\parallel l) = \Gamma_{ll}/\Gamma_{ss}$. Thus even if the lengths of cell sides are not very different, the patch will be anisometric provided that the interfacial tension is anisotropic enough. This is an important conclusion because in epithelia, the aspect ratio of cells' apical sides usually does not depart very dramatically from unity yet the in-plane polarization of cells (represented by, e.g., the different interfacial tensions) may be quite prominent.

The anisotropic differential adhesion hypothesis can also be implemented without the assumptions made above so as to demonstrate its workings. In the extended Potts model, cells are represented by clusters of contiguous sites on a regular (typically square) lattice, a single cell occupying many sites (Figure 5.28). By permitting a more detailed description of the cell shape, this lattice-based approach is a computationally efficient alternative to vertex-based models although cell-cell boundaries are jagged rather than smooth.

Within this scheme, cells may carry various types of energies, the most important one

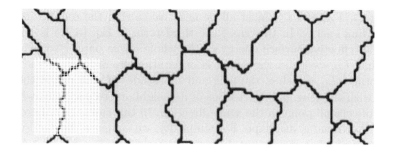

Figure 5.28 Snapshot of a few cells in the extended Potts model showing that in this representation, each cell contour is a jagged loop on a square lattice overlaid over the bottom-left part of the figure. (Adapted from Ref. [162].)

being anisotropic adhesion. Encoding the variation of the interfacial tension depending on the position of a given straight segment of cell-cell boundary relative to long axis of both cells in question is a little tricky [162]; the only other term needed is a phenomenological area-elasticity energy that stabilizes a preferred cell size [Eq. (3.32)]. After the parameters of the Monte-Carlo algorithm used to search for the minimal-energy structure have been fine-tuned, the snapshots observed during relaxation are indeed fairly reminiscent of cell rearrangement seen during convergent extension, reproducing the intercalation along the horizontal axis and the elongation perpendicular to the average cell long axis (Figure 5.29).

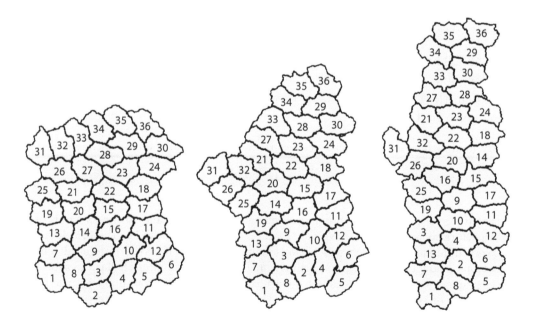

Figure 5.29 Sequence of snapshots obtained during relaxation within the extended-Potts-model implementation of the anisotropic differential adhesion hypothesis. Cells are labeled so that their rearrangement can be tracked more easily. (Adapted from Ref. [162].)

5.2.2 Passive deformation

The snapshots in Figure 5.29 are fairly similar to the observed cell rearrangement in the epithelial explant from the back of the *Xenopus* embryo [162, 163] and it is tempting to conclude that the rearrangement is driven by anisotropic differential adhesion. However, one must take into account that this particular explant contains one or two layers of mesodermal cells on the epithelial layer and that in principle both types of tissue could drive the deformation. Yet in this particular case, it appears that it is not the epithelium but rather the mesenchymal cells in the mesoderm that cause the patch to converge and extend [164] and that the role played by the epithelium is passive. This conclusion is supported by a number of experiments. Firstly, the mere fact that an explant does converge and extend in culture excludes the possibility that the process be controlled by surrounding tissues. On the other hand, no convergent extension is seen in isolated epithelial sheets whereas in explants containing mesenchymal cells covered by a non-native epithelium transplanted from another part of the embryo it does take place [164]. Finally, the deformation and displacement of epithelial cells during convergent extension in *Xenopus* can be divided into two phases, each characterized by a different pattern. The epithelial cells are initially stretched along the axis of elongation of the whole tissue and they contract in the perpendicular direction, and only later do they intercalate and redeform so as to acquire their final shape. This sequence of events can be readily visualized as a response of a passive material to the force generated by the underlying mesenchymal cells [164].

In *Drosophila* germband extension, a qualitatively similar behavior is seen when comparing the relative contributions of epithelial cell shape change and cell intercalation to the overall tissue deformation [165]. To this end, the displacement and shape of a few 100 cells in the tissue were tracked, and the total tissue deformation was divided into a term describing the deformation of individual cells and a term arising from cell intercalation. Both processes contribute to net tissue elongation but not in the same amount at all times during the 50-minute convergent extension, which can be divided into two phases. In phase I which lasts about 10 minutes, the contributions of cell deformation and intercalation to the total tissue extension are roughly the same. The beginning of phase II is marked by the maximum of the cell deformation strain rate. From this point on, this rate gradually decreases and eventually even becomes negative, which signals an elongation of cells in the direction perpendicular to the tissue elongation. As a result, the total strain rate during phase II is dominated by cell intercalation.

By comparing tissue deformation in wild-type embryos to that in *Krüppel⁻* mutants one arrives at a surprising fact. In these mutants, cells fail to intercalate so that eventually the germband does not extend properly. During phase I, the total tissue elongation remains the same as in wild-type embryos because cell deformation compensates for the suppressed intercalation. If both cell deformation and intercalation were active processes, it is difficult to imagine that the latter would exactly compensate for the former when it is suppressed. But if germband is being stretched by an extrinsic rather than by an intrinsic force that is unaffected in these mutants, then the compensation can be rationalized more readily. In all, we are led to conclude that germband extension in *Drosophila* is a passive rather than an active tissue deformation [165].

Which force could possibly be responsible for the deformation? From the mechanical perspective, the lengthwise elongation of the germband could be caused either by a pulling force along the head-to-tail direction, that is along the anteroposterior axis, or by a pushing circumferential force acting in the dorsoventral direction. If the tissue did behave as an elastic material, the pulling and the pushing scenario would each be characterized by a distinct area-change fingerprint, which can be employed to distinguish between the two deformation modes.

Pushing vs. pulling

Consider a uniaxial deformation of a layer of Poisson's ratio ν to illustrate the difference between pulling and pushing as two alternative modes of deformation leading to the same strain ϵ_{yy} along the y axis.

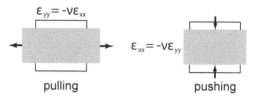

In the pulling mode, the layer is stretched along the x axis, the corresponding strain being $\epsilon_{xx} > 0$. Given that the strain characterizing contraction along the y axis is $\epsilon_{yy} = -\nu\epsilon_{xx}$, the relative projected area change of the layer is

$$\frac{\Delta A}{A}\bigg|_{\text{pull}} = \epsilon_{xx} + \epsilon_{yy} = (1 - \nu)\,\epsilon_{xx} = -\left(\frac{1}{\nu} - 1\right)\epsilon_{yy}. \tag{5.25}$$

At the same time, pushing on the other two sides of the layer so as to produce the same ϵ_{yy} as in the pulling mode results in the strain $\epsilon_{xx} = -\nu\epsilon_{yy}$ so that

$$\frac{\Delta A}{A}\bigg|_{\text{push}} = \epsilon_{xx} + \epsilon_{yy} = -\left(\nu - 1\right)\epsilon_{yy}. \tag{5.26}$$

In non-auxetic materials where $\nu > 0$, the projected area is increased on pulling and decreased on pushing, the magnitude of the relative area change being considerably larger on pulling than on pushing. The difference between the two modes arises from comparing the deformations at the same strain along one in-plane axis but not the other.

In wild-type embryos, cells intercalate and rearrange during germband extension, and hence their orientation and apical area are not controlled solely by the external force. On the other hand, in the *Krüppel*⁻ mutants intercalation is suppressed so that the tissue is more solid-like. Cells in the germband initially point in the dorsoventral direction but during the first half of deformation process this pattern changes. Cells in the central part of the tissue and close to the cephalic furrow do not have a preferred orientation but those in the posterior part are gradually elongated along the anteroposterior axis and their area is increased by as much as 20%. As suggested by the Box on p. 174, this deformation behavior is consistent with a pulling force acting in the anteroposterior direction [165].

In the second half of the process, the tension does not seem to be present any more and the cells shrink in area. The final apical areas of all cells are somewhat smaller than at the beginning and the dorsoventral orientation of cells is no longer seen. These observations suggest that after the pulling force is deactivated, germband cells in *Krüppel*⁻ mutants spring back as a passive elastic solid with some residual over-relaxation [165].

It is natural to expect that the pulling force should be generated by the morphogenetic transformations that take place shortly before germband extension or at the same time, which include mesoderm invagination, formation of cephalic furrow, and some other processes. The various pieces of evidence finally all point to the invaginating mesoderm [165]. This conclusion is conceptually very important because it emphasizes the cooperative nature of the morphogenetic behaviors. A given tissue may be deformed either by an active internal process or by an external force, and in turn the ensuing transformation can but need not affect the surrounding tissues, possibly involving some kind of feedback. The various levels of mechanical coupling between adjacent tissues thus lead to a spectrum of behaviors with different degrees of interdependence and active deformation.

5.2.3 Active hydrodynamic model

An appealing and efficient theoretical framework which captures cell rearrangement, extension, and alignment as the three essential qualitative features of convergent extension can also be constructed without considering individual cells as entities of interest, that is at a hydrodynamic level [166]. Here the local state of the tissue is described using coarse-grained variables such as the average orientation and elongation of cells as well as the velocity field.

In absence of topological rearrangements, any local change of cell elongation or orientation is due to a displacement of the cell centroid in the observed patch of the tissue, that is to hydrodynamic flow. If the in-plane area of each cell is fixed (which is usually true) then the divergence of the velocity field $\mathbf{v}$ must vanish just like in any incompressible fluid. In this case, flow is quantified by the traceless symmetrized gradient of the velocity

$$\tilde{v}_{ij} = \frac{1}{2}\left(\frac{\partial v_i}{\partial x_j} + \frac{\partial v_j}{\partial x_i} - \sum_k \frac{\partial v_k}{\partial x_k}\delta_{ij} \right), \tag{5.27}$$

which is zero for a rigid rotation to lowest order. A nonvanishing $\tilde{v}_{ij}$ implies a change of cell shape. Consider, for example, an incompressible flow with $v_{xx} = \partial v_x/\partial x = a$ and thus $v_{yy} = \partial v_y/\partial y = -a$; in this case $\tilde{v}_{ij}$ coincides with v_{ij} so that $\tilde{v}_{xx} = a$ and $\tilde{v}_{yy} = -a$. Let the centroids at $t = 0$ form a square lattice with unit cell of size $L_0 \times L_0$—in this case all cells in the tissue are identical and each unit cell of the lattice represents a biological cell. After a short time dt, the lattice is distorted due to flow of cell centroids so that the dimensions of the unit cell at time dt are $L_0(1 + a\,dt) \times L_0(1 - a\,dt)$; note that its area remains equal to L_0^2 for short enough times $dt \ll a^{-1}$. The square cell elongates more as time progresses, and the rates of elongation along the x axis and contraction along the y axis are given by $a = \tilde{v}_{xx}$ and $-a = \tilde{v}_{yy}$, respectively. More generally, a tensorial quantity $\mathbf{Q}$ encoding cell shape and orientation can be constructed such that its time derivative is the symmetrized gradient of the velocity field:

$$\frac{\mathrm{D}Q_{ij}(x,y)}{\mathrm{D}t} = \tilde{v}_{ij}(x,y). \tag{5.28}$$

The rate of change of the $\mathbf{Q}$ tensor must be computed such that it is insensitive to both net displacement and rigid rotation of the patch as a whole. This is ensured by using the convected corotational derivative $\mathrm{D}/\mathrm{D}t$, which is a generalization of the material derivative used in classical hydrodynamics [166,167]; instead of spelling it out, we refer the interested reader to an in-depth textbook on continuum mechanics [168]. The construction of the $\mathbf{Q}$ tensor is somewhat involved [169]. For our purposes, it suffices to say that $\mathbf{Q}$ is defined so that $\mathbf{Q} = 0$ represents isometric cell shapes such as regular hexagons or squares whereas a nonzero $\mathbf{Q}$ corresponds to elongated cells (Figure 5.30). In the above flow with $v_{xx} = a$ and

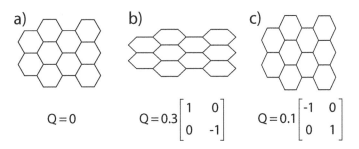

Figure 5.30 Examples of the $\mathbf{Q}$ tensor in isometric hexagonal cells (a) and in cells elongated along the x and y axis (b and c, respectively).

$v_{yy} = -a$ which starts from an isometric square-cell tissue, Eq. (5.28) gives $Q_{xx} = -Q_{yy} = a$ so that for short times, the diagonal components of the tensor encode the flow-induced relative change of cell lengths along the x and y axes: $\Delta L_x/L_0 = a$ and $\Delta L_y/L_0 = -a$.

Naturally, Eq. (5.28) also implies that cell shape changes that take place autonomously or in response to stress must result in flow. This is illustrated in Figure 5.31a where the initially elongated hexagonal cells contract along the x axis. As cells are incompressible, their centroids must be displaced during contraction, the displacements being indicated by arrows. However, if the tissue undergoes topological changes during cell deformation, DQ_{ij}/Dt will generally depart from $\tilde{v}_{ij}$. Each T1 transformation reconnects cells around a given edge, often such that the distribution of edge lengths becomes narrower. As illustrated in Figure 5.31, these transformations reduce cell elongation and affect the flow. This is evidenced by the cell centroid displacement patterns in Figure 5.31a and b which are very different although the initial and the final states of the cell shape tensor are identical:[4] in this case, T1 transformations considerably reduce the magnitude of flow.

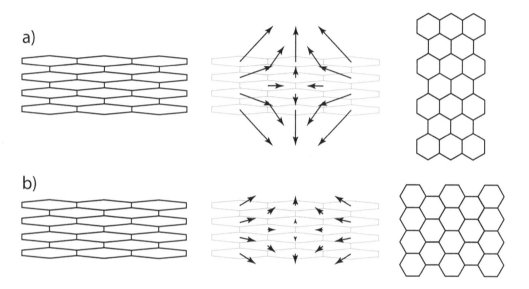

Figure 5.31 Displacement patterns of cell centroids in a tissue that is initially elongated along the x axis and undergoes contraction along this axis at fixed cell area without (a) and with T1 transformations (b). In the final state, the cells are isometric ($\mathbf{Q} = 0$) in both cases but the shapes of tissue patches are quite different.

In addition to T1 transformations, topological changes also include cell divisions and cell extrusions, some of the latter due to cell death. It is convenient to combine all of these terms in a single tensor referred to as shear rate due to topological rearrangements and denoted by R_{ij} [166]. Using R_{ij}, Eq. (5.28) is generalized to

$$\tilde{v}_{ij} = \frac{DQ_{ij}}{Dt} + R_{ij}, \tag{5.29}$$

where the dependence of the three tensors on spatial coordinates and time is dropped for clarity.

Apart from their average shape, cells are also characterized by an in-plane polarity

---

[4]Regular hexagons are described by $\mathbf{Q} = 0$ irrespective of their orientation as they are all fitted best with circles rather than ellipses (Box on p. 56).

arising from the distribution of any molecular entities or subcellular structures that control the mechanics of the cell. This distribution is independent of cell shape and described by another tensor

$$q_{ij} = p_i p_j - \frac{1}{2} \sum_k p_k p_k \delta_{ij}, \tag{5.30}$$

where p_i are the components of the cell polarity vector oriented along the direction of the concentration gradient of the molecules or structures in question within cells.

Equipped with the two tensorial order parameters of interest, we can now construct the constitutive equations of the epithelium by relating stress within the tissue σ_{ij} and the shear rate due to topological rearrangements R_{ij} to order parameters. The stress consists of an isotropic part arising from in-plane hydrostatic pressure P and of an anisotropic part $\widetilde{\sigma}_{ij}$, which is due to cell deformation and polarity:

$$\sigma_{ij} = -P\delta_{ij} + \underbrace{2KQ_{ij} + \zeta q_{ij}}_{\widetilde{\sigma}_{ij}}. \tag{5.31}$$

Here we restrict $\widetilde{\sigma}_{ij}$ to terms linear in Q_{ij} and q_{ij}, which can be seen as the lowest-order approximation; K and ζ are their respective moduli. In this form, the anisotropic stress depends on the instantaneous value of Q_{ij} and q_{ij}, much like force and extension of a spring are related by Hooke's law. The meaning of the two terms is however quite different. While the $2KQ_{ij}$ term can indeed be likened to Hooke's law because it relates stress and cell elongation, the ζq_{ij} term depends on cell polarization as an internal variable rather than on deformation. As such, it represents an active source of stress.

The dependence of R_{ij} on cell elongation and polarization is more involved as shown by measurements of R_{ij} in the *Drosophila* wing blade during pupal-stage morphogenesis [169]. As time progresses, R_{xx} decays to zero in an oscillatory fashion while Q_{xx} saturates at a finite value, which means that R_{xx} cannot depend solely on cell elongation; here the x axis points along the proximal-distal direction, that is from the trunk to the tip of the wing. This characteristic mode of relaxation is seen in wild-type, ablated, and even mutant wings so that it must be caused by an intrinsic mechanical feature of the tissue. In all studied cases, relaxation in the *Drosophila* wing can be interpreted using the same model [169] where

$$R_{xx}(t) = \frac{1}{\tau} \int_{-\infty}^{t} \frac{1}{\tau_d} \exp\big(-(t - t')/\tau_d\big) Q_{xx}(t') dt' + \lambda q_{xx}. \tag{5.32}$$

The first term states that the shear rate due to topological rearrangements R_{xx} at a given time t depends not only on the elongation Q_{xx} at this time but also on the value of Q_{xx} at earlier times t'. The memory kernel, sometimes referred to as the primary response function, decays with increasing $t - t'$ so that the effect of the distant-past states of the tissue is increasingly smaller. Here the memory kernel chosen is exponential and characterized by a relaxation time τ_d; the magnitude of the elongation-dependent term is controlled by τ.

The rate $R_{xx}(t)$ also depends on cell polarity as a variable akin to an external field, and this is represented by the second term in Eq. (5.32) where no memory effects are included. Like in the case of stress, the λq_{xx} term represents active cell rearrangements that are not caused by cell elongation.[5] Together with the equation of mechanical equilibrium $\sum_j \partial \sigma_{ij}/\partial x_j = 0$ [Eq. (2.29) in the Box on p. 29], Eqs. (5.29), (5.31), and (5.32) describe the shear flow and cell elongation as a function of time at a given cell polarity. To solve these equations, one also needs to include a suitable equation of state relating tissue in-plane pressure to cell area as well as to specify the boundary conditions.

---

[5]Although λ is a scalar, it is a complicated proxy-like quantity, relating internal cell polarization to cell shear flow caused by topological changes induced in the tissue by cell polarity.

Active forces vs. active motion

The notions of active forces and active motion can be illustrated using a toy model. Imagine a body of mass m suspended in a viscous fluid by a spring of stiffness k and equipped with an engine which allows it to move in the vertical direction at a speed v_0. The coordinate of interest $z(t)$ is measured from the bottom end of the unloaded spring and describes the position of the body as well as the extension of the spring. If the effective drag coefficient ξ is large, the inertial term md^2z/dt^2 can be neglected and Newton's law reads

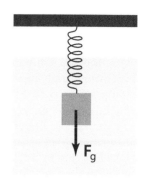

$$\xi\left(\frac{dz}{dt} - v_0\right) = -kz - mg; \qquad (5.33)$$

here mg is the effective force of gravity including buoyancy. The right-hand side is the net force $F(z)$ on the body, and the second term which is independent of extension z is its active part. In a similar fashion, v_0 is the active part of velocity because in the absence of both spring and gravity, $dz/dt = v_0$ and $z(t) = v_0 t + z_0$ where z_0 is the initial position of the body.

The general solution of the equation of motion is

$$z(t) = \underbrace{v_0\tau - \frac{mg}{k}}_{=z_\infty} + (z_0 - z_\infty)\exp\left(-\frac{t}{\tau}\right) \qquad (5.34)$$

where $\tau = \xi/k$ is the relaxation time. The first two terms together give the spring extension after a long time z_∞ where the effects of active motion and active force are indistinguishable. In the absence of gravity as the active force, a downward active motion with $v_0 = -mg/\xi$ leads to the same equilibrium extension as gravity without active motion. (This can be readily seen in the equation of motion itself.)

Despite its elaborated apparatus, there exists an insightful solution to this model. Consider a free unconstrained tissue with a fixed and uniform cell polarity field where $\mathbf{p}$ points along the x axis so that the $\mathbf{q}$ tensor is diagonal; we choose $q_{xx} = const. > 0$. Due to the cell polarity field, this tissue will reach a stationary state where R_{xx} is finite and thus the shear flow field given by Eq. (5.29) will be nonzero too. This means that in the stationary state, the tissue will indefinitely extend along the x axis and contact along the y axis or vice versa depending on whether the stationary value of $\tilde{v}_{xx}$ is positive or negative.

Let us now analyze this active process in more detail. In an unconstrained tissue, stress vanishes at the boundary and since Eq. (2.29) requires that in the absence of external forces it does not depend on position, stress must be zero everywhere. If we further assume that the in-plane pressure $P = 0$ then in the stationary state cell elongation reads

$$Q_{xx} = -\frac{\zeta}{2K}q_{xx}. \qquad (5.35)$$

Now we insert Q_{xx} in the first term in Eq. (5.32) and integrate it over time to find that $R_{xx} = [-\zeta/(2K\tau) + \lambda]\,q_{xx}$. Since in a stationary and homogeneous state undergoing uniaxial deformation $DQ_{xx}/Dt = 0$, Eq. (5.29) reduces to

$$v_{xx} = \left(-\frac{\zeta}{2K\tau} + \lambda\right)q_{xx}. \qquad (5.36)$$

These telling results involve two parameters that describe the active processes in the tissue, with ζ and λ measuring the contributions of active shear stress and active cell rearrangements, respectively. Equation (5.36) shows that either of these two processes can give rise to the shear flow so that based on an observed stationary-state v_{xx} alone one cannot tell whether this state is driven by active stress or by active topological rearrangements. Specifically, the pure active-stress case with $\zeta = \zeta_0$ and $\lambda = 0$ is indistinguishable from the pure active-rearrangement case with $\zeta = 0$ and $\lambda = -\zeta_0/(2K\tau)$. On the other hand, cell elongation Q_{xx} depends solely on the magnitude of active stress as shown by Eq. (5.35) and is thus zero in a tissue characterized only by active rearrangements.

The causal chains relating cell polarity and tissue deformation in the pure active-stress and in the pure active-rearrangement tissue are quite different and it seems appropriate to spell them out. In the former, cell polarity generates cell deformation by way of active stress, and cell deformation in turn gives rise to topological rearrangements which then result in tissue deformation. In the latter, the first step of this chain is skipped: cell polarity directly generates topological rearrangements which gives rise to tissue deformation (Figure 5.32).

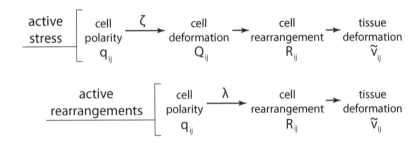

Figure 5.32 Causal chains in the pure active-stress and the pure active-rearrangement limits of the active hydrodynamic model in the stationary state.

The difference between these two extremes is illustrated in Figure 5.33 showing schematically the temporal evolution of an unconstrained active epithelium. In the pure active-stress case with $\zeta = \zeta_0 > 0, \lambda = 0$, and $q_{xx} > 0$, cells elongate along the direction perpendicular to the x axis according to Eq. (5.35), and elongation Q_{xx} is independent of time. Now recall that we assumed that the in-plane pressure is zero. In this case, cell area does not change so that $\nabla \cdot \mathbf{v} = 0$, and thus a steady shear-flow gradient with $v_{xx} < 0$ implies that $v_{yy} = -v_{xx} > 0$. After the active stress with $\zeta_0 > 0$ has been turned on, the tissue starts to contract along the x axis and elongate along the y axis. The cells too contract along the x and elongate along the y axis until they reach the final shape characterized by $Q_{xx} = -\zeta_0 q_{xx}/(2K)$. The pure active-rearrangement case with $\lambda = -\zeta_0/(2K\tau)$ differs from this scenario only in cell elongation, which is zero ($Q_{xx} = Q_{yy} = 0$).

Real epithelia are hardly unconstrained and hence the unlimited active deformation never really takes place. Instead, this process is restricted by confinement imposed by the surrounding tissues or structures. Confinement can be either hard or soft, the former referring to a rigid environment that cannot accommodate any tissue deformation of the tissue and the latter describing the case where some amount of deformation is allowed. Soft confinement can be represented by a tissue attached to a rigid frame by springs [166] so that in equilibrium, the buildup of stress due to tissue deformation is counteracted by extension or compression of springs.

In soft confinement, an active epithelium reaches mechanical equilibrium rather than a stationary state like the unconstrained active epithelium. In equilibrium, the shear rate vanishes and since the time derivative of Q_{xx} must vanish too, Eq. (5.29) implies that

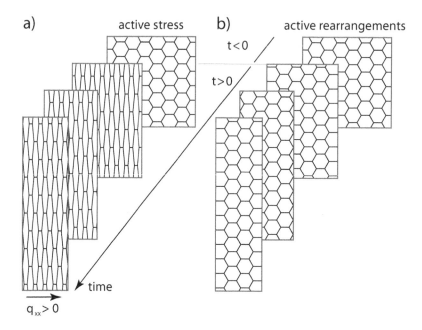

Figure 5.33 Evolution of the deformation of a free epithelium characterized by active stress (a) and by active topological rearrangements (b). In the active-stress tissue, cells are stretched in the same direction as the tissue itself whereas in the active-rearrangement tissue cell elongation is absent.

$R_{ij} = 0$. Like in the case of unconstrained epithelium, we integrate Eq. (5.32) over time with a Q_{xx} independent of time to find that

$$R_{xx} = \frac{Q_{xx}}{\tau} + \lambda q_{xx}. \tag{5.37}$$

Since R_{xx} is zero, cell elongation is now related to cell polarity by

$$Q_{xx} = -\lambda \tau q_{xx}. \tag{5.38}$$

Note that this result differs from its unconstrained-epithelium analog [Eq. (5.35)] in that it depends on the rate of active topological rearrangements rather than on the magnitude of active stress as witnessed by the presence of λ instead of ζ.

The extent of deformation of the epithelium is defined by the elasticity of the confining tissue, which determines the magnitude of shear stress in the epithelium. Thus in equilibrium,

$$\tilde{\sigma}_{ij} = 2KQ_{ij} + \zeta q_{ij} = const., \tag{5.39}$$

the constant depending on the elastic modulus of the confinement. After inserting Eq. (5.38), this condition reduces to

$$(-2K\lambda\tau + \zeta)q_{xx} = const. \tag{5.40}$$

so that both the pure active-stress tissue with $\zeta = \zeta_0 > 0$ and $\lambda = 0$ and the pure active-rearrangement tissue with $\zeta = 0$ and $\lambda = -\zeta_0/(2K\tau)$ give rise to the same final deformation. However, the two cases differ in the magnitude of cell elongation, which is zero in the pure active-stress case and finite in the pure active-rearrangement case. In this respect the confined active epithelium is qualitatively distinct from its unconstrained counterpart where cell elongation is driven by active stresses. Moreover, since λ corresponding to a tissue that contracts along the x axis and extends along the y axis is negative like in the unconstrained epithelium, Eq. (5.38) suggests that cells in the confined active-rearrangement tissue elongate along the x axis, that is perpendicular to the tissue elongation [166]. This pattern is indeed characteristic for many instances of convergent extension including germband extension in *Drosophila*.

The main features of the deformation of the confined active epithelium are schematically shown in Figure 5.34a and b. Also included are two details of the relaxation toward equilibrium; the transient elongation of cells in the active-stress tissue during the initial stages of deformation and the gradual development of cell elongation in the active-rearrangement tissue. These subtleties follow from an in-depth analysis of the model [166].

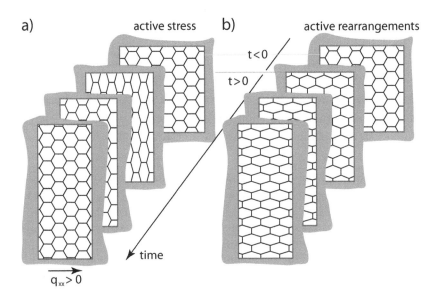

Figure 5.34 Relaxation of an active epithelium towards mechanical equilibrium in soft confinement represented by the gray area surrounding the tissue. In the active-stress tissue (a), cell elongation vanishes in the final deformed state whereas in the active-rearrangement tissue (b) cells are elongated in the direction perpendicular to tissue extension.

The active hydrodynamic model can also be explored in the limit of hard boundary conditions which ensure that the tissue as a whole does not deform. This case provides additional insight into the differential roles of active stresses and active topological rearrangements, and it is less artificial than it seems at first sight. In some animals, parts of the tissue can be immobilized by laser cauterization of cells along a given line or contour, which stitches them to an adjacent structure such as the vitelline membrane [158, 170]. As a result, the patch of the tissue contained within the cauterized boundary cannot move.

In a model active-stress/active-rearrangement tissue with rigid boundary conditions, the shear rate due to topological rearrangements $R_{xx} = Q_{xx}/\tau + \lambda q_{xx}$ [Eq. (5.37)] must vanish in equilibrium, and the shear velocity must vanish at all times. In this case, Eq. (5.37) shows that in the pure active-stress tissue with $\zeta = \zeta_0 > 0$ and $\lambda = 0$, cell elongation is zero $Q_{xx} = 0$ (Figure 5.35a). On the other hand, in the pure active-rearrangement tissue where $\zeta = 0$ and $\lambda = -\zeta_0/(2K\tau)$ the equilibrium cell elongation is finite and $Q_{xx} = -\lambda\tau q_{xx} = \zeta_0 q_{xx}/(2K)$. For cell polarization along the x axis where $q_{xx} > 0$, Q_{xx} is also positive. In

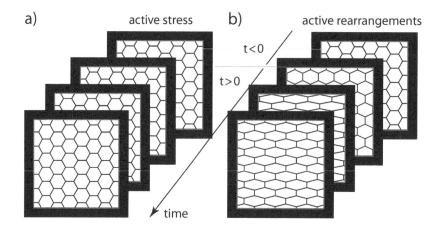

a) active stress   b) active rearrangements

t<0

t>0

time

Figure 5.35 Intercalation in the active tissue with rigid boundary conditions represented by the black frame around the tissue: in the active-stress case (a), no cell elongation takes place and the tissue remains unchanged, whereas the active-rearrangement tissue transforms such that the number of rows is increased which is accompanied by cell elongation along the horizontal axis (b).

this case, a patch of tissue consisting of roughly isometric cells transforms such that the number of rows is increased at the expense of cell elongation along the direction of rows, which is then accompanied by cell contraction in the perpendicular direction. This shows that a pure intercalation process cannot be driven by active stresses but it can (and must) take place due to active topological rearrangements.

The ideas behind the active hydrodynamic theory were investigated within microscopic, cell-level models well before they re-emerged in such a concise form. Here one starts with a suitable model of cell mechanics combined by rules governing the active elements of cell dynamics such as active intercalation. Any rules of this type are necessarily phenomenological and based on experimental observations, and one of the earliest theories of this kind was inspired by notochord formation in chordates as a dramatic example of convergent extension. The notochord is a flexible rod-like structure extending from head to tail along a considerable part of the embryo. One of its many functions is that it serves as a mechanical support to which muscles are attached, much like the spine in vertebrates. The notochord is also instrumental for the development of the neural tube.

In *Xenopus* the notochord consists of wedge-shaped cells stacked in a column, and the formation of this ordered structure involves several mechanisms acting in concert. Many of them were explored in a two-dimensional dynamical model of notochord morphogenesis [171] where the tissue is represented as a patch of adhering tiles with curved edges characterized by a given line tension. Any pressure difference across an edge gives rise to a normal force and during the simulation, cell vertices move in an overdamped fashion so that their velocity is proportional to the net force.

This rather general framework was then extended by several rules phenomenologically describing active cell motion. Active intercalation is introduced by a reduction of line tension at the protruding vertex so that the edges elongate due to the imbalance of pressure and tension, which makes the vertex advance along the boundary between adjacent cells. In addition, a vertex may advance because the adjacent cells contract and recede, which drags it into the freed space.

Another important element of cell motility is related to cell polarization. After one of the vertices of a cell has advanced, the vertices on the lateral edges are less likely to form protrusions themselves; the opposite vertex is not inhibited. This leads to bidirectional polarization and makes the cell elongated due to its protrusive activity, which mimics the behavior of many cells *in vivo* and *in vitro*. In addition, protrusions are inhibited along the boundary where the notochord tissue is confined by the neighboring somitic mesoderm so that the marginal cells may only extend towards the interior of the tissue, which is consistent with the experiments [171]. The last key ingredient of the model, again motivated by observations, is persistence of motion encoded by a probability that a given vertex continues to move along the same direction.

In the simulation, a vertex is chosen with a certain probability and the tensions of the neighboring edges are then reduced so that the vertex may advance according to the above rules. A comparison of the different scenarios shows that the three elements required to reproduce convergent extension in the notochord are contact inhibition of protrusion, polarized protrusions, and refractory boundary [171]. These predictions are qualitatively consistent with those obtained within the active hydrodynamic model (Section 5.2.3), and the results of the two approaches are very similar even in some details such as the gradual development of cell elongation seen in simulations [171] and evident from the exact analytical solution of the hydrodynamic model [166]. Given that both models were developed based on observations, the agreement should not be too surprising—yet it is still nice to see that a coarse-grained phenomenological theory works so well.

5.3   INTERCALATION

We already mentioned that intercalation is involved in convergent extension and in epiboly, but it also exists as a standalone morphogenetic movement [172]. In convergent extension, intercalation is an in-plane process and does not affect the number of layers in the tissue (Figure 5.3). The cells are characterized by an in-plane polarization and thus intercalation is directional. The proportions of the epithelial patch expressed in terms of the number of cells change such that the tissue elongates at the expense of its width while the thickness remains unchanged. This is referred to as mediolateral intercalation. On the other hand, epiboly typically involves a multi-layer tissue which expands areawise at the expense of the number of layers. Here cell intercalation takes place in the tissue cross-section (Figure 5.3), and this mode is known as radial intercalation.

In non-proliferating tissues, intercalation usually takes place as an active, autonomous process or in response to external uniaxial stress causing tissue deformation. Sometimes it also results from local fluctuations; in this case, cell rearrangement is reversible. On the other hand, topological changes in proliferating tissues are driven by cell division either in part or primarily. In Section 3.3.1 we learned that the disorder caused by division does affect the structure of the tissue considerably, and the mitosis-only model described in Section 3.3.2 emphasizes this fact very explicitly. Naturally, in absence of a preferred orientation of the cleavage plane, one should expect that the topological changes brought about by division are isotropic and so is any ensuing deformation of the tissue.

The extent of topological changes induced by division of a cell depends on how exactly the process proceeds. In the most common mode of cell division in epithelia, the daughter cells stay next to each other after division. Thus they do not disrupt the integrity of their immediate neighborhood in the sense that the ring of neighboring cells around the mother remains topologically unchanged; the ring will, of course, be remodeled as soon as one of the cells in it itself will divide. This type of division is seen in many proliferating epithelia: in the *Drosophila* imaginal discs, 94% of all cell divisions end such that the daughter cells are in contact [83]. Figure 5.36a shows an example of this process seen in chick embryonic epithelium disk in the fresh laid egg [173]. Apart from an additional vertex in

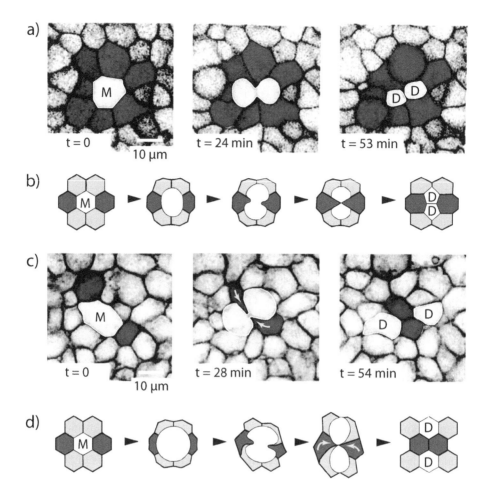

Figure 5.36 Part of stage-X epithelial disk in chick embryo showing cell division where daughter cells stay in contact (a; M and D stand for mother and daughter cell, respectively) and a part of the disk at stage 3 when the daughter cells separate (c). Panels b and d illustrate the two modes schematically. (Adapted from Ref. [173].)

the two cells that are in contact with both daughters, the seven neighbors surrounding the dividing cell remain topologically unchanged. The sequence of micrographs is schematized in Figure 5.36b, which emphasizes that in such divisions cell rearrangement is localized.

However, during the formation of the primitive streak in chick when the bilateral symmetry of the embryo is established, cell division takes place in a different manner. Instead of staying together, the daughter cells are separated such that a pair of neighbors located roughly in the cleavage plane intervene between them as illustrated in Figure 5.36c and d. The daughters in turn intercalate between the non-intervening neighbors, which causes a considerable topological rearrangement involving a larger region. In other words, the rate of rearrangements is faster than in divisions where daughters stay in contact.

In chick embryo during primitive streak formation, this mode is characteristic for as many as 90% of cell divisions, and it appears to be essential for a proper development of the embryo by allowing the epithelium to respond to stresses that drive the formation of the streak in a more fluid-like, energy-dissipating fashion [173]. In this case, cell intercalation is an important element of morphogenesis as a mechanism that determines the passive response of the tissue to forces generated outside it rather than as a force-generating process.

The illustration in Figure 5.36d is an example of a large-scale cell rearrangement involving more than just four cells like the T1 transformation (Box on p. 60). During germband extension in *Drosophila*, cells often form multicellular rosettes, that is, transient structures around a vertex with a large valence [174]. An example of such a rosette is shown in Figure 5.37a, the sequence spanning a period of about 20 minutes. In this process, two adjacent columns of cells oriented in the dorsoventral direction first deform such that the faces at the column-column contact constrict. In the second step, a large-valence vertex with a distinct wheel-like appearance is formed. Finally this vertex is dissolved along the anteroposterior direction, bringing in contact cells that were initially rather distant from each other including those at either tip of the columns. The number of cells in the rosette varies from six to about a dozen, the smaller rosettes being more frequent than the larger ones.

Rosette formation is evidently highly anisotropic and thus a clear sign of in-plane cell polarity, which is also confirmed by the strongly anisotropic distribution of myosin and

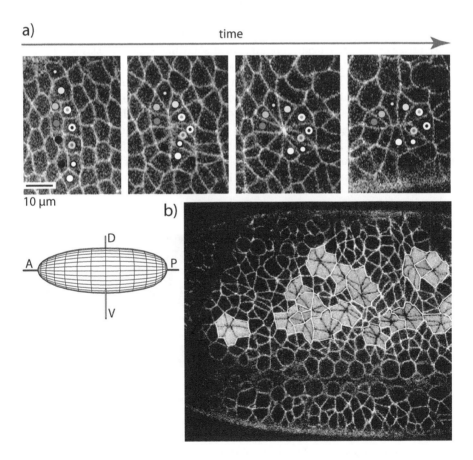

Figure 5.37 Sequence of images of an epithelial patch during the formation of a eleven-cell rosette in *Drosophila* germband (a), with each of the cells involved labeled by a different symbol. Side view of the embryo with highlighted rosettes simultaneously formed during germband extension (b); also visible are the cephalic furrow (left part of image) and the ventral furrow (bottom). (Adapted from Ref. [174].)

Bazooka/PAR-3 protein in cell-cell contacts. Myosin is enriched in contacts pointing in the anteroposterior direction whereas Bazooka, a protein involved in the establishment of adherens junctions and spindle orientation, is enriched in those oriented along the dorsoventral direction [174]. Moreover, the orchestrated simultaneous deformation of many cells is consistent with the picture of the acto-myosin network as a multicellular structure rather than just a ring running around the perimeter of a single cell. Some features of such a network can be subsumed in the polarity tensor q [Eq. (5.30)] used in the active hydrodynamic model in Section 5.2.3—where the nature of topological rearrangements need not be specified so that q can be thought to include T1 transformations as well as the more complex processes.

5.4 INGRESSION

During cell division in an epithelium, topological rearrangements are needed so as to accommodate an additional cell in an edge-to-edge packed tissue. Cell extrusion causes a converse effect, where a cell is squeezed out by its neighbors which then move in to fill space. This too is usually associated with a change of topology but it is also interesting from the perspective of the driving forces behind the process. In morphogenesis and embryonic development, the most interesting instance of cell extrusion is the transformation of epithelial cells, which are closely attached to their neighbors and form solid-like sheets of tissues, into mesenchymal cells which are motile and propel themselves by exerting traction forces, typically by filopodia (Section 2.3.2). This transformation is known as the epithelial-mesenchymal transition (EMT) and is initiated by the detachment of cells from the epithelium, which involves several cell-level processes including loss of adhesion to the epithelium. In the late blastula stage, some cells at the vegetal pole of the sea urchin embryo undergo ingression and become internalized, producing a lump of primary mesenchyme (Figure 5.38) which is easily distinguished from the more compact epithelial tissue. As a result of ingression, the epithelium becomes thinner which is also clearly visible in the cross-section.

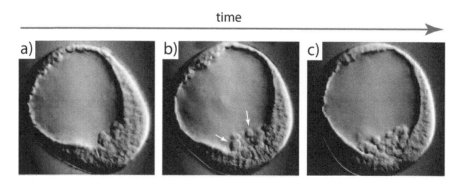

Figure 5.38 Ingression illustrated by the formation of primary mesenchyme during the late blastula stage of the sea urchin; in the cross-section, the mesenchymal cells are seen as the loosely attached lump at the bottom of the blastocoel (arrows in panel b). (Image courtesy of D. R. McClay.)

The EMT and the underlying ingression of cells are very important because mesenchymal cells can easily relocate and disperse but also aggregate in suitable condition. As such, they contribute to morphogenesis in a very different manner than epithelial cells. In addition, once relocated they may transmute back into epithelial cells, forming a new epithelial structure. This process is known as the mesenchymal-epithelial transition (MET).

It is widely accepted that the main reason for ingression from the epithelium is a reduced adhesion between a prospective mesenchymal cell and its neighbors. The simplest framework to quantify this is the differential adhesion hypothesis (DAH; Section 2.4). A toy model may go as follows. Assume for simplicity that the apical and basal surface tensions of cells in an epithelium are the same (Γ_a) and the cell shape is a square prism. If we denote cell height by h and the edge of the base by a, the surface energy of the cell is $E = 2\Gamma_l ah + 2\Gamma_a a^2$. Here we took into account that the energy of each cell-cell contact $\Gamma_l ah$ is divided between two cells, and Γ_l is the surface tension of the lateral sides. By minimizing this energy at fixed cell volume $a^2 h = V_0$ (Problem 5.9), we find that the equilibrium binding energy of the cell is

$$E_{\text{bound}} = 2(2^{1/3} + 2^{-2/3})\Gamma_a^{1/3}\Gamma_l^{2/3}V_0^{2/3}. \tag{5.41}$$

On the other hand, the surface energy of a cube-shaped detached cell in contact only with the medium is

$$E_{\text{detached}} = 6\Gamma_m V_0^{2/3}, \tag{5.42}$$

where Γ_m is the surface tension of the cell-medium interface, which is generally different from Γ_a.

Apart from the numerical prefactors which depend on the assumed shape of cells, the detached state is preferred over the bound state as soon as

$$\Gamma_a^{1/3}\Gamma_l^{2/3} \gtrsim \Gamma_m. \tag{5.43}$$

In this result, Γ_l is the effective surface tension of the lateral sides, which can be written as a sum of a positive cortex-tension term and a negative adhesion term like in Eq. (3.49). If adhesion strength is reduced, Γ_l will increase and thus at some point, E_{bound} will exceed E_{detached} and ingression will take place. This analysis may be generalized in several ways so as to more accurately account for the detached cell shape (which is certainly not cuboidal after ingression) and for its position immediately after detachment (when it may still adhere to the epithelial sheet with one side). These refinements do alter the detachment condition [Eq. (5.43)] quantitatively, but the rationale behind it remains unchanged.

5.5 MESENCHYMAL STRUCTURES

Mesenchymal cells are surrounded by the ground substance matrix, which is a gel-like fluid. They can move through this matrix by extending the filopodia, thereby exerting a traction force which sets the matrix around the cell in motion. At the same time, mesenchymal cells can be transported by the matrix flow just like leaves in the wind, they may disperse due to diffusion, and their density can locally increase due to ingression or cell division. These and other effects involved in the motion of internalized cells can conceivably give rise to two qualitatively different steady-state patterns of the density of mesenchymal cells: uniform and spatially modulated. The former is trivial and cannot be associated with morphogenesis whereas the latter gives rise to an increased density of cells in some parts of the embryo, which can be viewed as a precursor of a body part.

The formation of patterned distribution of the mesenchymal cells can be interpreted in terms of a hydrodynamic theory describing the spatio-temporal evolution of cell and matrix density [175–177] in a manner somewhat analogous to the Turing theory of morphogenesis [178]. Here the cell number density $n = n(\mathbf{r}, t)$ obeys the continuity equation

$$\frac{\partial n}{\partial t} = -\nabla \cdot \mathbf{j} + M, \tag{5.44}$$

where $\mathbf{j}$ is the total flux of cells and M is the number of cells generated at a given point per

---

Turing model

In his seminal 1952 paper [178], Turing proposed a mathematical framework where two or more spatially distributed species (say chemicals or cells) interact with each other, are being degraded, and diffuse about. A modern variant of the rate equations describing the concentration of the two species [A] and [B] reads

$$\frac{\partial[A]}{\partial t} = f_A([A], [B]) - g_A[A] + D_A \nabla^2[A] \tag{5.45}$$

and

$$\frac{\partial[B]}{\partial t} = f_B([A], [B]) - g_B[B] + D_B \nabla^2[B]. \tag{5.46}$$

Here f_A describes the production of species A, which depends on both [A] and [B], and g_A and D_A are the degradation and the diffusion coefficient, respectively. The meaning of f_B, g_B, and D_B is analogous. These equations are an example of a reaction-diffusion system.

Turing showed that under certain conditions, the solution of these equations is characterized by a spatio-temporal pattern such as the stripe or the multidomain hexagonal pattern seen in the quasi-2D chemical reaction of chlorite iodide and malonic acid (below; reproduced from Ref. [179]); also shown is the grain boundary between two hexagonal domains. These patterns are marked by a complexity which cannot be anticipated from the form of the equations alone, and thus they are an example of an emergent phenomenon. The Turing model is a paradigm for pattern formation in chemistry, biology, physics, etc., and one of its best-known manifestations is the Belousov–Zhabotinsky reaction.

---

unit volume in unit time. Cell flux specifies the number of cells passing through a frame of unit area in unit time, and it depends on three processes:

- **Diffusion**: Much like a solute in a solvent, mesenchymal cells dispersed in the matrix should tend toward a uniform spatial distribution. The lowest-order model of the diffusive flux known as Fick's law contains a term proportional to ∇n and points from large to small densities. However, mesenchymal cells are equipped with filopodia which stretch beyond adjacent cells, and each of them does not sense only the gradient of density but also variations across longer distances. This effect can be described by a flow term proportional to $\nabla(\nabla^2 n)$. To understand this *ansatz*, imagine a sinusoidal cell density profile. The second derivative of n is positive in the troughs and negative in the crests so that $\mathbf{j} = D_2 \nabla(\nabla^2 n)$ with a positive coefficient D_2 points from the crests to the troughs, evening out the differences between them. In all, the diffusive flux reads

$$\mathbf{j}_d = -D_1 \nabla n + D_2 \nabla(\nabla^2 n). \tag{5.47}$$

- **Convection**: Here we consider the flux of cells is due to the hydrodynamic flow of the matrix. The convective part of the flux is given by the product of cell density and the local velocity of the matrix:

$$\mathbf{j}_c = n\frac{\partial \mathbf{u}}{\partial t}, \tag{5.48}$$

where $\mathbf{u}(\mathbf{r}, t)$ is the displacement vector of the matrix.

- **Haptotaxis**: If the filopodia of a mesenchymal cell extending in opposite directions do not adhere equally strongly to the surrounding matrix, the cell will move towards the region where adhesion is stronger. This is referred to as haptotaxis. It is plausible that the efficiency of traction generated by a filopodium increases with the density of the matrix ρ at the point of attachment, and thus the associated flux should be proportional to $\nabla\rho$. Naturally, it should also scale with the cell density itself and thus

$$\mathbf{j}_h = an\nabla\rho, \tag{5.49}$$

where $a > 0$.

The source term in Eq. (5.44) is due to cell division and can be described by

$$M = rn(N - n), \tag{5.50}$$

where N is the saturated cell density and r is a rate coefficient such that rN is the cell division rate at small densities.

Like the mesenchyme, the matrix too obeys the principle of mass conservation. The matrix is secreted by the mesenchymal cells so that $\partial\rho/\partial t$ should depend both in fluxes and on sources just like $\partial n/\partial t$ in Eq. (5.44). The secretion rate is however rather slow compared to the other processes involved, and thus the source term analogous to M in Eq. (5.44) can be omitted so that

$$\frac{\partial\rho}{\partial t} = -\nabla \cdot \mathbf{j}_\rho, \tag{5.51}$$

where the mass flux of the matrix is $\mathbf{j}_\rho = \rho\partial\mathbf{u}/\partial t$.

The dynamics of the mesenchyme is described by the two equations of motion, Eqs. (5.44) and (5.51), together with Newton's law which reduces to the equation of mechanical equilibrium [Eq. (2.29)] if we assume that the motion of the matrix is slow and thus quasistatic. The stress appearing in Eq. (2.29) consists of three terms:

- **Elastic stress** which is caused by the deformation of the matrix-cell composite. This composite is an isotropic material and for small deformations, the stress tensor reads

$$\sigma_{ij}^e = \frac{Y}{1+\nu}\left[\frac{1}{2}\left(\frac{\partial u_i}{\partial x_j} + \frac{\partial u_j}{\partial x_i}\right) + \frac{\nu}{1-2\nu}\sum_k \frac{\partial u_k}{\partial x_k}\delta_{ij}\right], \tag{5.52}$$

where Y is Young's modulus and ν is Poisson's ratio.

- **Viscous stress** which depends on the rate of deformation and is given by

$$\sigma_{ij}^v = \eta\left(\frac{\partial}{\partial x_j}\frac{\partial u_i}{\partial t} + \frac{\partial}{\partial x_i}\frac{\partial u_j}{\partial t} - \frac{2}{3}\sum_k \frac{\partial}{\partial x_k}\frac{\partial u_k}{\partial t}\delta_{ij}\right) + \xi\frac{\partial}{\partial x_k}\frac{\partial u_k}{\partial t}\delta_{ij} \tag{5.53}$$

Here η and ξ are the kinematic and the second viscosity, respectively.

By combining σ_{ij}^e with σ_{ij}^v, we model the matrix-cell composite as a viscoelastic solid, i.e., as a material which undergoes an elastic deformation in response to applied forces but approaches the final deformed state in a relaxatory fashion. The characteristic relaxation time is given by η/Y.

- **Traction stress** due to the cells' filopodia stretching out to their neighbors. This stress can be viewed as an isotropic inward pressure causing contraction of the matrix-cell composite proportional to the matrix density ρ and to the cell density n provided that n is small; at large n, traction stress should saturate. This leads to $\sigma_{ij}^t = \tau \rho n/(1+n/n_0)$, where τ measures the traction strength and λ controls stress saturation such that $\sigma_{ij}^t = \tau \rho n$ for $n \ll n_0$ and $\sigma_{ij}^t = \tau \rho n_0$ for $n \gg n_0$. This zero-order approximation can be refined by a non-local term associated with filopodia extending beyond nearest neighbors. The non-local term should be proportional to $\nabla^2 n$ so as to account for the fact that cells at a local maximum of cell density have fewer next-nearest neighbors to hold on to than those at a local minimum. With this in mind, we arrive at

$$\sigma_{ij}^t = \frac{\tau \rho}{1+n/n_0} \left(n + \beta \nabla^2 n \right) \delta_{ij}. \tag{5.54}$$

Here the range of the non-local traction-stress part is given by $1/\beta^{1/2}$.

The equations of motion [Eqs. (5.44) and 5.51)] and mechanical equilibrium [Eq. (2.29)] determine the evolution of n, ρ, and $\mathbf{u}$ with time. The last ingredient to be specified is the external force, typically exerted by a substrate such as the blastula surrounding the internalized mesenchymal lump in Figure 5.38. In the simple case, this effect is represented by a body force proportional to the displacement of the matrix and described by $\mathbf{F} = -s\mathbf{u}$ with $s > 0$.

The key question addressed by this theory is whether a quiescent state with a uniform distribution of cells and a uniform matrix density is stable despite the many processes that can potentially cause it to transform into a segregated state where the concentration of cells in some regions of space is larger than in others. The theory includes several processes that can drive cell aggregation. The most obvious one is cell traction which drags cells together in a self-amplifying fashion [175]. To appreciate this, recall that the traction stress increases with cell density as long as the density is smaller than n_0 and then it levels off as mentioned above. The negative hydrostatic pressure generated by traction is eventually stabilized by the deformation of the ground substance matrix as well as by diffusion. This process can but need not be assisted by cell division, which is faster in regions where cell density is increased, and by haptotactic motion which drags cells toward regions of increased matrix density.

The instabilities are sought by analyzing the time dependence of a sine-like fluctuation of n, ρ, and $\mathbf{u}$ about a uniform value. For example, cell density can be written as

$$n(\mathbf{r}, t) = \bar{n} + \delta n \exp(i\mathbf{k} \cdot \mathbf{r}) \exp(-t/\tau), \tag{5.55}$$

where $\bar{n}$ is the uniform starting density, δn is the magnitude of the fluctuation, and $\mathbf{k}$ and τ are its wavevector and relaxation time, respectively. Analogous *ansätze* are used for the matrix density and for the displacement, the latter being 0 in the reference uniform state. The analysis typically proceeds by expanding the three equations of motion in terms of δn, $\delta \rho$, and $\delta \mathbf{u}$ and retaining only the linear terms in these quantities. This leads to the so-called dispersion relation between τ and $\mathbf{k}$. At a given $\mathbf{k}$, instability is signaled by a negative relaxation time which means that the fluctuation increases with time rather than decays toward the reference state. If there exist several unstable modes, the system is dominated by the fastest growing mode. This approach is common in many related theories such as the Cahn–Hilliard analysis of phase separation in binary mixtures or the Kelvin–Helmholtz hydrodynamic instability.

The details of the analysis are somewhat involved but the main predictions are quite clear [175], one of them being that the long-range diffusion coefficient D_2 [Eq. (5.47)] should not be too large for the instability to take place. Secondly, in a mesenchyme where cells divide, the spacing of cell aggregates is given by

$$L = 2\pi \left(\frac{D_2}{rN}\right)^{1/4}, \tag{5.56}$$

showing that in the more rapidly dividing cells the neighboring aggregates should be close to each other and thus smaller. The third aspect of interest is the effect of confinement. Imagine placing the mesenchyme in a container just barely larger than the spacing predicted by Eq. (5.56). If so, it is evident that there cannot be more than a single cell aggregate. The full analysis of this effect should be done in two or more dimensions where the aspect ratio of the domain in question can be varied. By comparing the growth rates of the Fourier modes, one then finds that, e.g., by increasing just one dimension of the domain beyond the threshold, the number of aggregates jumps from one to two—naturally, for a suitable choice of the other parameters of the system.

An important instance of mesenchymal condensation is the formation of feather placodes arising from the accumulation of mesenchymal cells just below the epithelium. The placodes are thickened areas of the epithelium that later give rise to feathers, and Figure 5.39 shows the placodes in the dorsal part of an about 9 days old chick embryo. The process of placode formation is somewhat more complex than mesenchymal condensation alone because it also involves the interaction between mesenchyme and epithelium as well as chemotaxis [180]. Nonetheless, the main underlying principles are very similar to those outlined above and the regularity of the emerging hexagonal pattern is very pronounced.

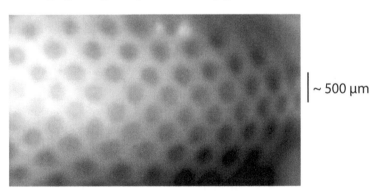

~ 500 µm

Figure 5.39 Hexagonal array of feather placodes in stage 35 chick embryo. (Adapted from Ref. [180].)

This theory has been applied to processes characterized by regularly spaced body structures such as skin scales or feather germs, to limb morphogenesis, to wound healing, etc. In limb morphogenesis, the most interesting aspects of morphogenesis reproduced by the model are condensation of cartilage which relies primarily on cell aggregation rather than on cell division, then branching and segmentation of the cartilage [177]. These processes take place at a more advanced stage of embryonic development than those discussed earlier but the approach employed and the underlying ideas are quite generic. As such, they can be adapted and generalized to interpret a range of processes including patterning of fish skin and sea shells such as *Oliva porphyria* and *Conus aulicus* shown in Figure 5.40 [181]. Many captivating examples of the latter are collected in Meinhardt's book *The Algorithmic Beauty of Sea Shells* [182].

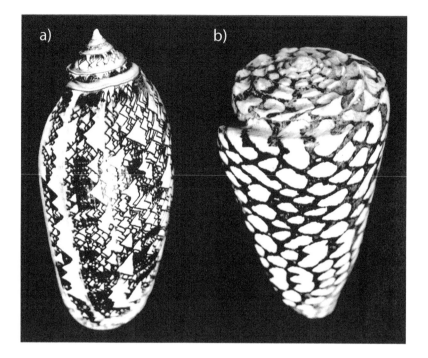

Figure 5.40 Sea shells *Oliva porphyria* (a) and *Conus aulicus* (b) with patterns which can be reproduced using algorithms based on the Turing model. (Samples belong to the Hohenwart collection from the Slovenian Museum of Natural History.)

5.6 EPIBOLY

Epiboly refers to a process where a tissue spreads across a substrate. Given that the total volume of cells in the tissue is constant during this deformation, spreading necessarily implies tissue thinning either at the cell level (in single-cell-thick tissues) or in terms of the number of layers in the tissue (in multilayer tissues); in the latter case, epiboly involves radial intercalation. Epiboly has been extensively studied in the embryonic development of zebrafish where it is one of the key elements of gastrulation [183].

Figure 5.41 shows a zebrafish embryo during the first 2 hours 30 minutes of the gastrula stage. Initially the embryo consists of a multilayer patch of tissue of uniform thickness reminiscent of an inverted cup. The patch stretches across the top hemisphere of the yolk, which is referred to as 50% epiboly (Figure 5.41a). This structure is known as the dome. During epiboly, both EVL and the deep-cell layer stretch so as to encompass an ever larger solid angle, the former by thinning and the latter by radial intercalation. At the beginning of gastrulation about 5 hours after fertilization the patch is axisymmetric but about 1 hour later the symmetry is lost with the establishment of the dorsoventral axis as seen in Figure 5.41b and c. At about 9 hours after fertilization, the embryo reaches 90% epiboly (Figure 5.41d).

The loss of axisymmetry is a clear indication that the deformation of tissue during gastrulation includes other movements in addition to epiboly. Accumulation of cells on the dorsal side (right in Figure 5.41) takes place by way of convergent extension such that the converging direction is ventrodorsal (left to right in Figure 5.41) and thus perpendicular to the epiboly movement, which spreads from the animal pole toward the vegetal pole (top to bottom in Figure 5.41). Yet the most evident process at work is epiboly because at the end of gastrulation at 10 hours after fertilization, the yolk is completely covered by the embryo.

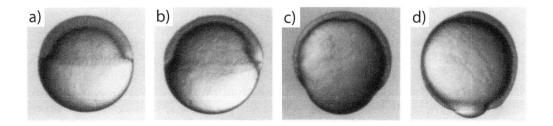

Figure 5.41 Zebrafish embryo during gastrulation at 5.25, 6, 7.7, and 9 hours after fertilization (a, b, c, and d, respectively) with yolk coverages of 50, 50, 70, and 90%, respectively. Panel a shows the dome stage; arrows in panels c and d point to the prechordal plate and the tail bud, respectively. (Reproduced from Ref. [183].)

The most straightforward theoretical explanation of tissue spreading that one can think of is based on the differential adhesion hypothesis. As mentioned in Section 2.4, one of the key cell rearrangement phenomena that can be described by this hypothesis is envelopment of a tissue by another tissue [36] and zebrafish epiboly has some attributes of envelopment, at least in the beginning. In the blastula stage, the mass of deep cells appears to be malleable enough so that the whole embryo can be viewed as a liquid drop sitting on the yolk which too can be regarded as a drop. Thus we arrive at the two-drop model which is formally identical to the description of the cell doublet in Section 2.3.1.

This model is characterized by five parameters: the volumes of the blastodisc and yolk and the tensions of the yolk/blastodisc, yolk/embryonic fluid, and blastodisc/embryonic fluid interfaces denoted by Γ_{yb}, Γ_{yf}, and Γ_{bf}, respectively. At a given blastodisc/yolk volume ratio which is about 0.3, we can vary the tension ratios Γ_{yb}/Γ_{yf} and Γ_{bf}/Γ_{yf}, minimize the total surface energy, and compare the obtained shapes with observations (Problem 5.10).

Gastrulation in zebrafish

The zebrafish zygote sits on top of the yolk cell surrounded by the embryonic fluid. During cleavage the zygote forms a cap of cells. After about 10 divisions, three cell populations can be identified: the single-cell-thick epithelial enveloping layer (EVL) covering the embryo, the yolk syncytial[6] layer (YSL) at the embryo-yolk interface, and the deep cells between EVL and YSL. EVL and YSL are extraembryonic membranes whereas the deep cells eventually form the body. The cap-like blastula gradually transforms into a dome which then spreads toward the vegetal pole, eventually engulfing the yolk (figure reproduced from Ref. [183]).

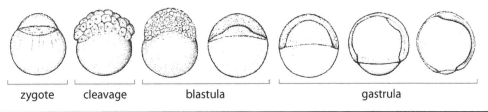

| zygote | cleavage | blastula | gastrula |

---

[6]Syncytium is a mass of cytoplasm that contains several nuclei, which may form either by division of nuclei that is not followed by cytokinesis or by cell fusion.

Some of the results of such an analysis where we tried to reproduce the large features of the embryo such as the radii of curvatures and the thickness of the blastodisc are shown in Figure 5.42 together with the zebrafish embryo at three different times during the blastula stage. While the two-drop model does come close to the observed shapes (Figure 5.42a and b), the details disagree. For example, the model curvature of the interface between the blastodisc and embryonic fluid is different from the true curvature and the precise shape of the yolk/blastodisc interface close to the perimeter also disagrees with observations as witnessed, e.g., by a larger angle between the yolk/blastodisc interface and the blastodisc/embryonic fluid interface. These differences suggest that the two-drop model is incomplete. In part, the failure can be associated with the absence of EVL and YSL which are formed at about 2 and 3 hours post fertilization, respectively.

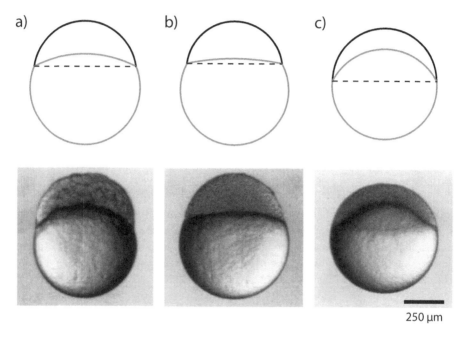

Figure 5.42 Comparison of zebrafish embryo at 2.5, 3.5, and 4.7 hours post fertilization (a, b, and c, respectively) with the two-drop model for blastodisc/yolk volume ratio of 0.3. In panels a, b, and c, the ratio of yolk/blastodisc and the yolk/embryonic fluid tension Γ_{yb}/Γ_{yf} is 0.3, 0.4, and 0.22, respectively; the ratio of blastodisc/embryonic fluid and yolk/embryonic fluid tension Γ_{bf}/Γ_{yf} is 0.8 in all panels. (Micrographs reproduced from Ref. [183].)

It is reasonable to conclude that EVL and YSL modify the shape of the blastoderm, and this is even more evident in the dome stage in Figure 5.42c where the tensions can be tuned so as to stabilize a partly engulfed two-drop structure but the quantitative agreement of the model shape with observations is poor. The cross-section of the real blastodisc reveals a rather uniform tissue thickness except right at the blastodisc edge, and this is clearly not reproduced by the model. The overall profile of the blastodisc/yolk interface including the contact angle close to the blastodisc edge is quite different too. At this stage, EVL and YSL and possibly other elements of the yolk and blastodisc mechanics not included in the two-drop model are evidently essential, suggesting that this simple description is inadequate.

The two-drop model disregards several experimental facts. Most importantly, it does not include the YSL acto-myosin ring running circumferentially around the blastodisc edge, which has been postulated to drive epiboly in a purse-string manner. It is plausible that this mechanism alone could drive epiboly after the sheet has spread beyond the equator, that is, after 50% epiboly. It is also known that the spreading of EVL over the mass of deep cells is distinct from the spreading of the deep cells themselves, which too is absent in the two-drop model. In turn, this simple model does not really account for the active processes in the tissue [184] which were also explored in cell-level models [185]; these processes are only included in a rather indirect way by describing the kinematics of tissue spreading in terms of the time-dependent tensions. However, with the advances of the experimental techniques that can provide quantitative insight into the forces at work (e.g., hydrodynamic regression [186]) the possible mechanisms involved in epiboly can be more readily compared and validated, eventually unraveling the intertwined deformations of the different structures in the embryo.

5.7   INVOLUTION

This morphogenetic process is somewhat similar to the invagination in that the epithelium bends inward. However, during involution the epithelium forms a fold rather than a channel, and this fold is rolled under the epithelium itself in a tank-treading fashion as shown schematically in Figure 5.3. As the epithelial sheet moves toward the edge formed by involution, which is referred to as the lip, it rolls past the lip and advances into the subsurface fold. In many cases, the infolded cells become migratory so that involution cannot only be seen as a mechanism of tissue internalization but also as an instance of epithelial-mesenchymal transition.

A neat example of involution is involved in gastrulation in *Xenopus*. During the blastula stage, this embryo is roughly spherical. The vegetal hemisphere is of a mass of large yolky cells whereas the animal hemisphere consists of a dome-like multilayer cap known as the blastocoel roof enclosing the fluid-filled blastocoel. In the early gastrula stage, the vegetal mass is moved toward the animal pole. This active movement is referred to as the vegetal rotation. At this stage, the rim of the blastocoel roof undergoes involution at the so-called blastopore lip, the involution evidently involving epiboly of the roof, and the involuted tissue forms the mesoderm layer.

This process is shown in Figure 5.43, which zooms in on the fractured *Xenopus* gastrulae. At stage 10.5 (that is, 11 hours post fertilization), the margin of the blastocoel roof has already involuted and the notch in the tissue at the point of involution known as Brachet's cleft is clearly visible. 1 hour 30 minutes later at stage 11.5 the blastopore lip has moved toward the vegetal pole (downward in Figure 5.43), and involution of the blastocoel roof has advanced considerably. Here the involuted tissue and the blastocoel roof can be easily distinguished from each other as the boundary between them is clear. In the late-gastrula embryo at stage 12.5 (14 hours 15 minutes post fertilization), the groove formed by involution referred to as the blastopore is rather deep and its invaginated tip already constitutes the archenteron.

In this case, involution is a rather sophisticated interplay of a few morphogenetic transformations. In part, it appears to be driven by the vegetal rotation of the yolk, and the tank-treading deformation of the blastocoel roof multilayer includes active tissue movements at the blastopore lip that contribute to involution. In addition, involution itself is combined with epithelial-mesenchymal transition and active radial intercalation in the mesoderm [187]. Interestingly enough, involution in *Xenopus* also proceeds if the blastocoel roof is removed, which suggests that epiboly of the roof and some other movements also involved in the process do not contribute to the driving forces [188]. Given the complexity

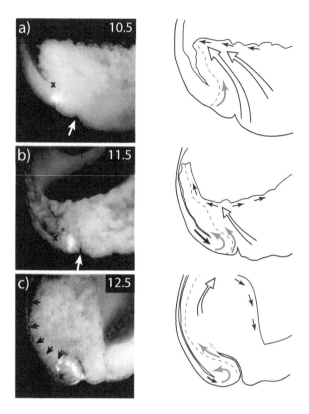

Figure 5.43 Fractured *Xenopus* embryo at the beginning of gastrulation at stages 10.5, 11.5, and 12.5 (a, b, and c, respectively) showing the involution on the ventral side of the embryo. In panel b, the apposing blastocoel roof and involuted mesoderm are readily visible. Black arrowheads in the micrograph in panel c show the location of the boundary between the mesoderm and the endoderm, and white arrowheads in panels a and b point to the blastopore. Tissue displacements are schematized in the right column where the dashed lines indicate the boundary between the mesoderm and the endoderm. (Adapted from Ref. [187].)

of involution, it should not be too surprising that quantitative studies of its kinematic and mechanistic aspects are scarce. At this point, there exist no theories of the process at a level comparable to that in the rest of this chapter.

The new advances in some problems in the mechanics of morphogenesis emphasize the patently uneven theoretical insight into the different movements and processes even more than the above narrative. For example, the detailed analysis of in-plane epithelial deformations in the *Drosophila* dorsal thorax identified the individual contributions of cell divisions, rearrangements, size and shape changes, and cell death, providing an exquisite insight into the global development of the tissue [189]. Although very appealing from the methodological and theoretical perspective, this framework is beyond our scope.

Also left out are some of the more mature topics, not just the recent developments. Among the processes not discussed we mention neurulation, a process in vertebrates where the so-called neural plate infolds and closes so as to form the neural tube, which later develops into brain and spinal cord [190], and branching morphogenesis, that is, the formation of hierarchical structures such as lung or gland ducts [191] as well as organoids [140, 192]. We touched upon the specific features of dividing tissues but not upon the subtleties of the mechanical feedback loop controlling cell size and shape so as to ensure stable and uniform growth [193]. Nonetheless, the diverse range of approaches presented here do illustrate the possible ways of viewing tissue deformations in a comparative fashion, inviting readers to approach their own projects from more than a single angle.

PROBLEMS

5.1 The so-called limiting shapes of lipid vesicles [194] consist of spherical surfaces. Consider two such shapes combining spheres of different radii connected by an infinitesimally small neck, an invaginated one known as the stomatocyte and an evaginated one known as the budded shape:

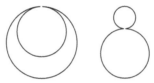

At a given reduced volume $v < 1$ [Eq. (5.8)], calculate $\overline{C}$ for these two shapes [Eq. (5.7)] and evaluate the total energy [Eq. (5.5)] for $C_0 = 0$ and a few fixed values of k_r/k_c, say 1, 3, and 5. Plot the energies at functions of $\overline{C}_0$ and determine the ranges of stability of either shape.

5.2 Consider Lewis' brass-bar-and-rubber-band model (Chapter 4) of *Drosophila* gastrulation where the left side of the first segment coincides with the right side of the N-th segment so that the segments form a ring. In a variant of the model with $N = 27$, $l_{i,0}^a = l_{i,0}^b = 0$, and the ratio of apical to basal spring constant equal to 4/5, the gastrulation-like inward movement of the segments is induced by changing one apical spring constant from 4 to 42 as indicated in the figure below. The minimal energy shape can be characterized by a set of angles as shown in the figure. Discuss the shapes of the segments near the invaginated part in the context of the excitable epithelium model [Eq. (4.29)] and explain how apical constriction may propagate from a given cell to its neighbors.

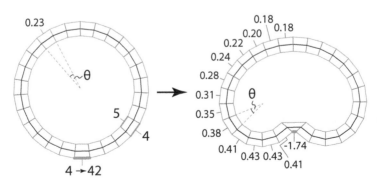

5.3 Consider a flat monolayer tissue formed by three-dimensional cuboidal cells carrying surface energy (Section 4.1.2), and assume that the apical and the basal tensions are identical. Calculate the energy of the sheet in terms of the lateral and the apical/basal tension as well as the equilibrium aspect ratio of cells. Calculate Young's modulus and the bending rigidity of the sheet, and discuss how they depend on its thickness.

5.4 Strips made of paper or plastic foil, with the ends glued so to form a ring of radius R_r, are inserted in a cylindrical confinement of radius $R_c < R_r$, say a bottle cap. Due to confinement, the strips buckle as shown in the photograph (panel e); the traced strip contours at various ratios R_r/R_c are shown in panels a–d.

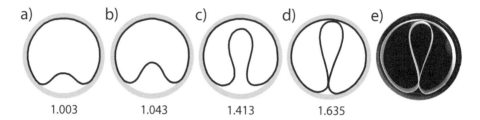

a) 1.003 b) 1.043 c) 1.413 d) 1.635 e)

As soon as the ratio R_r/R_c exceeds 1, the radial symmetry of the problem is broken and a trough is formed, its depth increasing with the ratio. This simple experiment mimics the growth of an elastic cylinder in confinement. The sequence of shapes obtained is reminiscent of gastrulation—the increased length of the paper strip simulates the uniform apico-basal contraction, i.e., the overall thinning of the strip with a simultaneous increase of its length. Estimate how the bending energy of the strip changes with R_r/R_c. Repeat the experiments in order to explore the minimal energy shapes for $R_r/R_c > 1.635$.

5.5 Within the tension-based model of global active deformation from Section 5.1.5, analyze the stresses along lateral, apical, and basal sides of the 80 cells seen in the transverse cross-section of the *Drosophila* embryo; focus on large $\alpha + \beta$ where the cross-section is circular. At $\alpha - \beta = 0$, calculate $\alpha + \beta$ for which the stress along the lateral sides vanishes by assuming that the vitelline membrane is perfectly rigid. The ratio of radii of the vitelline membrane and the yolk is 1.570; both cell and yolk area are constant.

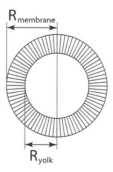

R_{membrane}

R_{yolk}

5.6 Estimate the depth of the ventral furrow in the tension-based model by comparing the perimeter of the embryo cross-section at $\alpha + \beta = 3.8$ and $\alpha - \beta = 0$; all assumptions and parameters are as in Problem 5.5. *Hint:* Calculate the perimeter of the cross-section by treating it as a straight chain of cells.

5.7 The Q tensors examined in Section 5.2.3 were constructed so as to preserve the area of cells in the tissue flow. However, they can also be extended to study cells which shrink and expand in the flow. Construct such tensors and divide the general Q tensor into an area-preserving/shape-changing and a shape-preserving/area-changing part.

5.8 Sketch cell displacement fields similar to those shown in Figure 5.31 for tissues undergoing mediolateral intercalation, say that in the figure where the initial state is overlaid over the final state on the right.

5.9 Calculate the binding energy of an epithelial cell [Eq. (5.41)].

5.10 The two-drop model zebrafish embryo shapes in the top row of Figure 5.42 consist of three spherical caps. Recalculate their radii R_1, R_2, and R_3 as well as the base radius a by minimizing the total surface energy at constant blastula and yolk volumes, with their ratio equal to 0.5. Also calculate the pressures in the blastula and in the yolk according to the Young–Laplace law [Eq. (2.3)]; use the same ratios of surface tensions as in Figure 5.42.

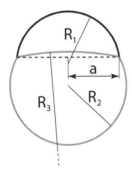

CHAPTER 6

Bulk tissues

THE PREVIOUS CHAPTERS mostly discussed the structure, shape, and shape transformations of epithelia. As already mentioned, these sheet-like tissues may consist of one or more layers of cells. A multilayer epithelium is typical for body surfaces that are exposed to wear and friction; in vertebrates, it is generally found in parts of digestive, urinary, and reproductive tracts and, of course, in skin. Cells in stratified epithelia, say in the epidermis, are packed just as tightly as those in single-cell-thick epithelia, filling out most of the tissue volume so that the intercellular spaces are very small. In some cases, their packing is fairly regular and their shape is evidently polyhedral, especially in the topmost cornified layer. The cornified layers in hamster and mouse epidermises shown in Figure 6.1 illustrate the columnar arrangement of cells stacked one on top of the other in 10 or so layers, with their beveled edges neatly interdigitated. Also clearly visible is the markedly hexagonal in-plane

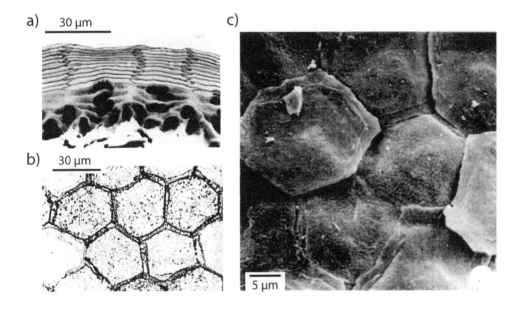

Figure 6.1 Stratified epithelia: cross-section of hamster epidermis expanded in alkaline environment (a) and top view of mouse epidermis impregnated with silver (b; these two panels are reproduced from Ref. [195]). Panel c shows the oblique electron-micrograph view of mouse epidermis (reproduced from Ref. [196]).

structure—in stark contrast with the much more disordered layer below the cornified layer. Naturally, not all epidermises, let alone all stratified epithelia, are as ordered as these two.

The epidermis is a constantly renewing tissue, and this is why it is organized into specialized strata. Cells in the basal layer proliferate, then they detach from the basement membrane and differentiate while migrating toward the surface of the epidermis, and eventually they reach the cornified layer. This layer consists of dead cells which are devoid of nuclei and organelles; instead, they are filled with a protein called keratin, which makes the skin virtually waterproof. In the end, these cells are shed from the topmost surface. The renewal takes about 48 days in humans.

With their closely packed cells, the epidermises in Figure 6.1 illustrate what we mean by a bulk tissue as far as cell shape is concerned: a three-dimensional arrangement of cells, often with small or no intercellular spaces. Yet the thickness of stratified epithelia is limited because there is no direct blood supply to the epithelium, and nutrients from the dermis just underneath the epidermis can only be transported across a certain distance. As a result, the epidermis is, just like skin itself, a sheet of cells thinner than its typical lateral dimension. In a true bulk tissue, the roles of the three dimensions should be roughly equivalent, which implies that the tissue must be self-sufficient in terms of blood supply—and must thus be criss-crossed with a network of blood vessels. This, in itself, means that while cells in such a tissue may constitute a well-connected mass, this mass is only homogeneous across a certain finite length scale given by the typical distance between the vessels. In this respect bulk tissues differ from, say, a crystal with atoms stacked in all three directions; a better comparison is a thick bush with a bulky appearance due to leaves which, in turn, are watered through the branches.

An nice example of such a tissue is liver. The human liver is the largest gland in the body, its many functions including the production of bile. The main parenchymal[1] cell type in liver are hepatocytes, which represent as much as 70 to 85% of the mass. At the same time, they hardly form a uniform, completely contiguous structure. The main building block of the liver is the hepatic lobule, a prismatic body of a diameter of about 1 mm and about 2 mm long. The cross-section of the lobule is roughly hexagonal, with so-called triads consisting of a bile duct, portal venule, and portal arteriole at each corner (Figure 6.2a). In the center of the lobule is the central vein.

The porous structure formed by the hepatocytes is connected in all three directions but it is finely perforated by the vascular network referred to as sinusoids radiating from the central vein seen in Figure 6.2c—to the extent that when viewed in cross-section, the closely-packed hepatocytes seem to form string-like formations like those in Figure 6.2b. These formations are known as hepatic cords although their true three-dimensional structure consists of interconnected sheets. The sheets are only one or a few cells thick, evidently depending on whether one cuts across it through the sinusoid or through the junction of neighboring sheets. In any event, the shape of hepatocytes is polyhedral and cells are roughly isometric. Most hepatocytes have two basal surfaces, one on either side of the cord facing the sinusoids, and their apical surface is midway between the basal surfaces. Bile is secreted at the apical surface and collected by tiny bile canaliculi seen in Figure 6.2b.

Close-packed cells in both plant and animal tissues have been observed by the early microscopists such as Hooke and Grew, both of whom likened them to soap froth. This comparison did occur rather naturally simply because soap froth is the most commonplace cellular partition of space known from the macroscopic domain. The first detailed quantitative study of cell shape was carried out by Lewis [197] a few years before his cucumber epidermis work mentioned in Section 3.2.1. Lewis examined the pith of the American black

---

[1]Parenchymal cells of an organ are those essential for the specific function of the organ, as opposed to the structural role of the supporting and connective tissue.

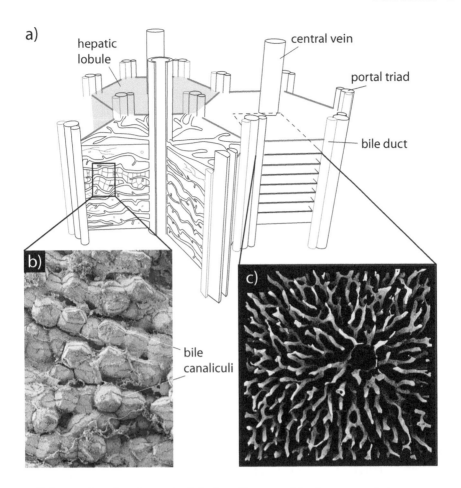

Figure 6.2 Schematic of a hepatic lobule of a roughly hexagonal transverse cross-section, showing the central vein and the portal triad (a). Panel b zooms in on the polyhedral hepatocytes with visible bile canaliculi (mammalian liver; image courtesy of R. C. Wagner), and panel c shows a reconstruction of the sinusoidal blood vessels (adapted from Ref. [135]).

elderberry *Sambucus canadensis* chosen for large cell size and regularity. He sliced this tissue into 10-micrometer-thick sections so that each cell stretched across many sections; 16 on average, to be more precise. Then he projected the sections using a lantern so as to magnify them 250 times, which allowed him to draw them. Finally the drawings were used to mold wax models which were in turn portrayed by hand. An example of such a drawing is shown in Figure 6.3a.

Lewis then counted the number of faces of cells in a sample containing 100 cells; much like what he did a little later when analyzing the cucumber epithelium. He found that the average number of faces is 13.96. The distribution of cells is quite broad, including polyhedra with as few as 6 and as many as 20 faces, but we note that polyhedra with 11 to 16 faces amount to 82% of all shapes (Figure 6.3b). Lewis went on and measured the volume of the cells by immersing wax models in water, finding that the variability of cell volume is considerable and noting that the number of faces correlates with volume; in Section 3.2.1 we have already seen the two-dimensional version of this statement.

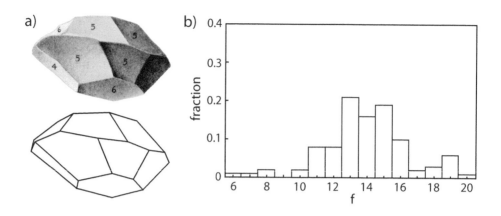

Figure 6.3 Drawing of a 15-sided cell in elder pith from F. T. Lewis' study of polyhedral cell shape, with labels indicating the number of sides of each facet (reproduced from Ref. [197]), and its edge-only reinterpretation (a). Distribution of the number of faces in a sample of 100 cells (b).

In 1944, J. W. Marvin provided a quantitative confirmation of the statement in case of three-dimensional tissues [198]. He studied the shapes and volumes of cells in the pith of *Eupatorium perfoliatum* in a painstaking manner similar to Lewis'. Marvin's method again involved the preparation of wax casts of the cells and the production of scaled-up models but this time by using *camera lucida* on bristol board,[2] excising faces from the board and then constructing the model. In these experiments, cells were magnified 210 times. The sample of cells chosen is not statistically representative of the whole tissue—the emphasis was on the study of "small" cells with less than 14 faces and on "large cells" with more than 14 faces, where "small" and "large" were obviously chosen relative to a mean number of faces of 14, which is typical for tissues as witnessed, e.g., by Lewis' data in Figure 6.3b. Nevertheless, the volumes of all 100 cells in the sample were measured, which allowed Marvin to analyze the correlation between the volume of a cell and the number of its faces f. In Figure 6.4, we reinterpret the data to show that the average volume of cells with f faces $\bar{v}_f$ scales almost linearly with f: in particular, by fitting the data by

$$\bar{v}_f = \mathcal{V}(f - f_0)^p, \tag{6.1}$$

we find that $f_0 \approx 8$ and $p \approx 1$; here $\mathcal{V}$ is a parameter defining the volume of cells.[3] This observation may be regarded as a variant of Lewis' law in three dimensions. Yet the data should be interpreted with some care, as the pith is far from a random or an isotropic tissue—it consists of easily discernible columns of cells, which follow the direction of the plant stem.

The cells with the same number of faces need not have the same distribution of face types, i.e., the same numbers of triangular, quadrilateral, pentagonal... faces. This is immediately seen by comparing, e.g., a cube and a pentagonal pyramid, which both have six faces, yet the cube has six quadrilateral faces whereas the pyramid has one pentagonal and five triangular faces. The average distribution of faces with n edges in a class of cells with f faces can be computed for each f, and four such distributions obtained from the data in Ref. [198] are

---

[2]This ancient method was implemented using an apparatus which combines the microscope with lenses and mirrors, enabling projection of a magnified image of a sample on a drawing surface. The apparatus is sometimes called the drawing tube or the tracing device.

[3]In the line fitted to the data in Figure 6.4, $\mathcal{V} = 190300\ \mu m^3$, $f_0 = 7.79$, and $p = 1$.

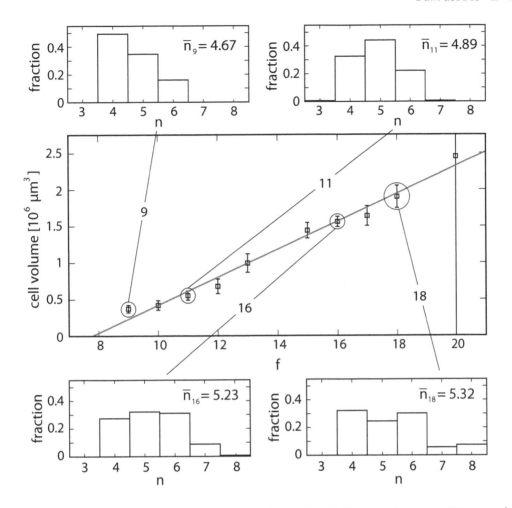

Figure 6.4 Average volume of cells in the pith of *Eupatorium perfoliatum* [198] vs. number of faces of a cell f, f ranging from 9 to 20 (symbols). The errorbars are the standard deviations and the thick line is a linear fit to the data. The histograms show the distribution of faces in four classes of polyhedra ($f = 9, 11, 16$, and 18).

presented in Figure 6.4 for $f = 9, 11, 16$, and 18. Cells with a smaller f have more faces with a smaller number of edges, so that the fraction of quadrilaterals ($n = 4$) decreases from 49.4% in the $f = 9$ class to 27.4% in the $f = 16$ class. At the same time, the fraction of heptagonal ($n = 7$) faces increases from 0.6% in the $f = 11$ class to 8.8% in the $f = 16$ class. The mean number of edges of a face in a polyhedron class $\bar{n}_f$ also increases with f, which is a topological necessity of the Euler formula for polyhedra. The agreement between the experimentally determined $\bar{n}_f$ and Eq. (6.3) is practically perfect, which implies that the vertices with valence of four or more were either extremely rare or absent altogether.

In his studies, Lewis searched for "the typical cell." Such an endeavor is poorly defined, except perhaps at the level that the 14-sided cells seem to have a similar role in three dimensions as the hexagonal cells in two dimensions, appearing as a highly probable mean of the distribution. His search for the typical cell was, at least in part, motivated by Kelvin's work on minimal-area partitions of space, which is discussed in Section 6.1.1. In Lewis' time, this problem was tentatively solved by a 14-sided body best described as a truncated octa-

How many edges per face?

In three dimensions, the Euler formula [Eq. (3.5)] reads

$$-C + F - E + V = 1, \tag{6.2}$$

where C, F, E, and V are the numbers of cells, faces, edges, and vertices of a cellular partition, respectively. Consider a single cell with f faces with an average number of sides given by $\overline{n}$ and assume that in all vertices of the cell, edges meet in threes. Then $V = 2E/3$ because each edge connects two vertices. In turn, $E = \overline{n}f/2$ as each edge is shared by two faces. In a single cell, $C = 1$ so that Eq. (6.2) gives

$$\overline{n} = 6\left(1 - \frac{2}{f}\right). \tag{6.3}$$

This formula does not only describe cells in a random cellular partitions, foams and tissues, but also the more regular polyhedra such as the Platonic solids with three-valent vertices (tetrahedron where $f = 4$ and $n = 3$, cube where $f = 6$ and $n = 4$, and dodecahedron where $f = 12$ and $n = 5$); in those with vertices of a different valence the formula can be modified (Problem 6.6).

hedron but also referred to as the orthic tetrakaidecahedron (Figure 6.5). Its relevance for tissues relied on the superficial phenomenological similarity of tissues and soap froths known to be controlled by surface tension where the problem of finding the minimal-energy partition of space reduces to the minimal-area problem.

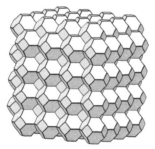 

Figure 6.5 Froth of Kelvin's minimal-area truncated octahedra, with the magnified image of a single bubble showing the fine curvature of the hexagonal faces.

This intriguing and mathematically appealing view was popular among biologists interested in such topics. Thompson was no exception. In *On Growth and Form*, he discusses Kelvin's truncated octahedron in some detail but notes that its absence in Lewis' observations may be due to "restraint of degrees of mobility or fluidity," that is, by the inability of the cells to reach the minimal-energy configuration [3]. When we return to the Kelvin problem in Section 6.1.1, we will see that these expectations were unjustified. Before doing so, we cannot avoid mentioning one of the most impressive experimental feats carried out so as to experimentally test the connection of the minimal-area problem to biological tissues.

In 1945, E. B. Matzke reported on an extraordinary experiment [199]. He used a syringe to blow soap bubbles of identical volume one by one, placing them into dishes one next to the other so as to construct a froth containing a thousand or more bubbles. He then observed the bubbles through a microscope and drew the most interesting shapes, producing a simple yet elegant representation of what he saw. He counted the number of faces in these bubbles, much like what Lewis did in elder pith. In Figure 6.6b we replot his results in the central part of the 1000-bubble sample, excluding the 400 peripheral bubbles at the walls and at the free surface. Compared to Lewis' results in Figure 6.3b, the soap froth is more ordered in the sense that the distribution is narrower but it clearly consists of more than a single type of polyhedron. In Matzke's words, none of them "may be singled out as *the type*" [199].

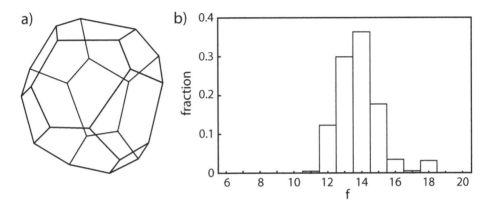

Figure 6.6   Matzke's drawing of a 14-sided soap bubble (a; adapted from Ref. [199]) and the fractions of polyhedral shapes in the bulk of his monodisperse soap froth (b).

Matzke's main quantitative conclusions may be summarized as follows: the average number of faces in the froth was 13.70 and the most frequent polyhedra were a 13-sided bubble with 1 quadrilateral, 10 pentagonal, and 2 hexagonal faces, a 14-sided bubble with 1 quadrilateral, 10 pentagonal, and 3 hexagonal faces, and another one with 2 quadrilateral, 8 pentagonal, and 4 hexagonal faces. Together these three shapes represented 42.5% of all bubbles in his experiment. Matzke also counted the number of edges in the faces and he found that 10.5% were quadrilateral, 66.9% were pentagonal, 22.1% were hexagonal, and 0.4% were octagonal. He repeated the experiment with bidisperse froth with bubbles of volume ratio of 8 and he varied the number ratio of small and large bubbles, invariably finding that the small ones have a smaller number of faces. This again resonates with Lewis' law (Section 3.2.1) and its three-dimensional variant in Eq. (6.1).

Matzke's soap froths were inconsistent with Kelvin's prediction, suggesting that if certain tissues can be likened to a froth then they are unable to reach the minimal-energy ordered state consisting of Kelvin's truncated octahedra. Yet at the same time, his results did accurately reflect the structure of real random froths, except that they were monodisperse by construction. Matzke was a botanist and he continued to pursue a similar line of research when studying biological samples, say the dividing apical meristem of the large-flowered waterweed *Egeria densa* [200].

At around the same time, alternative mechanical interpretations of tissue structure were considered too. One of them—the compressed-lead-shot model—was elaborated by Marvin in 1939 [201], five years before his studies of cells in the pith of *Eupatorium perfoliatum* [198]. Inspired by the 18th-century pea-compression experiments of S. Hales [202], this experimental model consisted of identical spherical lead shots either 1.27 or 2.54 mm in diameter, which were put in a steel cylinder and pressed with a plunger. As the load was increased, the shots were deformed to increasingly more faceted polyhedral shapes so as to fill space, the packing fraction approaching unity. After that, the sample was disassembled and Marvin then examined the shots one by one, counting the number of faces and other quantities of interest.

Marvin's sample contained a few 100 shots in a cylinder measuring 25 mm in diameter so that the finite size of the cylinder did matter. Indeed he found that the shots in the peripheral region close to the wall were considerably different from those in the center but those a little farther from the wall were already more similar to shots in the central part of the sample; the latter were then considered independent of the wall and thus representative. Marvin compared shots that were shaken when poured into the cylinder with those that were not shaken. The distribution of shot shapes in the two samples were not identical but they were not very different either (Figure 6.7b), both peaking at 14-sided polyhedra and a little broader than the distribution of shapes in Matzke's soap froth (Figure 6.6b) but not as broad as in Lewis' elder pith (Figure 6.3b).

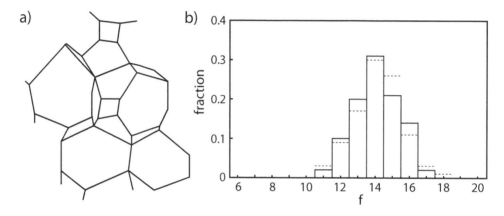

Figure 6.7 Edge-only drawing of a small part of Marvin's 2.54 mm lead-shot sample at the largest load examined (a) and the fractions of polyhedra shapes in the bulk in the 1.27 mm-shot shaken and unshaken samples (b; solid lines correspond to shaken samples and the dashed bars to the samples that were not shaken) [201].

Marvin performed his work under the direction of Matzke, and thus the common logic pursued in the soap-froth and the lead-shot models is no surprise. On the other hand, the mechanical features of the models are quite different. The shape of froth is governed by surface tension and the interior of the bubbles is completely unaffected by shape changes as long as the volume is unchanged. The shots however undergo a plastic deformation when compressed, and here the bulk of the shots is dominant whereas the surface effects are negligible. Despite these essential differences in the underlying mechanics, the final distributions of polyhedra are quite similar, which suggests that the structure of both froth and compressed lead shots is controlled by more fundamental considerations analogous to the no-gap, no-overlap, and edge-sharing constraints in the case of tilings (Section 3.2).

In the above discussion, we tacitly focused on tissues where the volume fraction of parenchymal cells is close to unity. From the physical perspective, these tissues are more interesting because cells need to deform in some way so as to fill space. In tissues with a smaller cell volume fraction, cells are dispersed within the matrix and their direct physical interaction is weak. As a result, the packing issues do not arise in the same sense as in the close-packed tissues.

6.1 SPACE-FILLING PATTERNS

In every possible mechanical model of the structure of close-packed bulk tissues, be that they postulate minimization of a suitable energy or some other morphogenetic process such as cell division or some other type of rearrangement, one seeks candidate solutions among space-filling partitions. Just like tilings of the plane, these partitions may be ordered, based either on a single polyhedron or on two or more polyhedra, or they may be random. The current insight into the solutions of the different proposed models is very unbalanced, the surface-tension/minimal-area model being the most studied one by far.

6.1.1 The Kelvin problem

One of the key problems of physics at the end of the 19th century was the question of luminiferous ether, and among the many interested in this hypothetical substance was Kelvin. In his search of a medium with mechanical properties consistent with the propagation of light, he thought of a soap froth, eventually asking himself what could be the structure of an ideal froth consisting of bubbles of identical volume. Soon he became more interested in the froth itself. Given that the mechanical energy of a froth consists of the surface term $E = \Gamma A$ alone so that minimization of energy is the same as minimization of area, he formulated the question that is presently known as the Kelvin problem:

> What partition of space into cells of equal volume has the smallest area?

Kelvin's work on the topic is summarized in his 1887 paper [203] where he proposed a tentative solution, building on the insight into the mechanics of soap froths published by Plateau in 1873 [47]. This insight can be succinctly stated as follows: in equilibrium, soap-bubble walls meet in threes at an edge called the Plateau border, forming angles of $120°$, and Plateau borders meet at vertices in fours such that the angle between them is $\arccos(-1/3) = 109.47°$.

Space-filling polyhedra

Just like a plane cannot be tiled with arbitrary tiles, three-dimensional space cannot be partitioned into cells of arbitrary shape. The simplest example of a space-filling polyhedron is the cube, but cuboids are also admissible and so is any quadrilateral prism. Other easily visualizable examples include the triangular prism, hexagonal prism, rhombic dodecahedron (left; all images courtesy of T. Hutton), truncated octahedron (middle), and a few more shapes; a little less intuitive is the triakis truncated tetrahedron (right). Also possible are partitions into two or more polyhedra, say into octahedra and tetrahedra in the ratio of 1:2. A stack of space-filling polyhedra is referred to as a honeycomb.

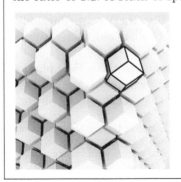  

In his paper, Kelvin first considered the rhombic dodecahedron as a possible solution of the problem (Problem 6.4) but he soon realized that a partition based on this space-filling polyhedron is unstable because its vertices are eight-valent instead of four-valent.[4] He then went on to examine the instability of these vertices and found that they must disintegrate into 4 four-valent vertices at the corners of the quadrilateral replacing each eight-valent vertex; the four-valent vertices satisfy the Plateau rules whereas the eight-valent ones do not. After some more work he arrived at the truncated octahedron in Figure 6.5. This body consists of 14 faces, that is of 6 flat quadrilateral faces with slightly bent edges and of 8 slightly non-planar hexagonal faces. The curvature of the hexagonal faces is required by the Plateau rules, and Kelvin noted that while it cannot satisfactorily be depicted by any shading, it can be shown "beautifully, and illustrated in great perfection, by making a skeleton model (...) and dipping it in soap solution" [203].

Just how good a solution of the problem is the truncated octahedron? A suitable figure of merit for the minimal-area partitions is

$$\Omega = \frac{A}{V^{2/3}}, \tag{6.4}$$

which may be regarded as the dimensionless surface area of the body.[5] To appreciate Kelvin's work, calculate Ω for the sphere and the cube as two simple bodies. In a sphere, $A = 4\pi R^2$ and $V = 4\pi R^3/3$ so that $\Omega_{\text{sphere}} = 6^{2/3}\pi^{1/3} \approx 4.836$. As the sphere is the body of smallest area at given volume, this figure represents the lowest bound of Ω. Naturally, since spheres of equal size do not fill space, the value of Ω corresponding to any partition of space should be larger than this, most likely not just by a small fraction. A partition into cubes is easy to visualize and one quickly finds that in this case $\Omega = 6$. This partition does not satisfy the Plateau rules and hence does not minimize the energy. Yet it still helps us to see that the candidate solutions of the Kelvin problem will probably not outcompete one another by a large margin.

Indeed, by calculating Ω for the rhombic dodecahedron (which, as mentioned above, is unstable as a soap bubble) we find that $\Omega = 2^{5/6}3 \approx 5.345$, exceeding Ω_{sphere} by about 10% (Problem 6.4). In Kelvin's truncated octahedron, $\Omega \approx 5.306$, which is a 0.7% improvement compared to the rhombic dodecahedron.

For over 100 years the truncated octahedron was considered the solution of the Kelvin problem. But then in 1994 D. Weaire and R. Phelan discovered that a froth consisting of dodecagonal (twelve-faced) and decatetrahedral (fourteen-faced)[6] bubbles with $\Omega \approx 5.288$ is better than the truncated octahedron by about 0.3% [204]. This froth is more complicated than the Kelvin froth: the bubbles form a cubic lattice with the dodecagonal ones at the vertices of the unit cell and in the center whereas the decatetrahedral ones are arranged in pairs along the bisectors of the faces, forming three mutually perpendicular columns of bubbles (Figure 6.8). Just like the truncated octahedron can be constructed by inflating bubbles sitting at the vertices of the body-centered cubic lattice, the Weaire–Phelan froth is based on the A15 lattice. This complex structure can only be analyzed numerically, and this was done using the Surface Evolver package [26].

These results reignited the interest in the Kelvin problem and in theoretical physics of soaps [205], and they in part explained why the truncated octahedron was not really seen in experiments. At the same time, the numerical approaches that became available during

---

[4]Here the valence of a vertex is meant in the space-filling sense. For example, the vertices of an isolated cube are three-valent but the vertices of a tessellation of space into cubes are six-valent.

[5]Apart from the numerical prefactor, the thus defined Ω is analogous to elongation of polygons defined by $P/\sqrt{4\pi A} \propto P/A^{1/2}$, where P and A are polygon perimeter and area, respectively (Box on p. 56).

[6]The designations "decatetrahedron" and "tetrakaidecahedron" are synonymous, the latter being based in ancient Greek. Both terms are used in the literature.

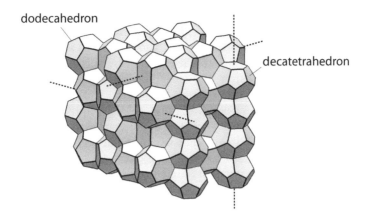

dodecahedron

decatetrahedron

Figure 6.8 Weaire–Phelan minimal-area froth, with the dashed lines indicating columns formed by decatetrahedral bubbles. On the right face of the cluster, one can spot four partial cradles for the dodecahedral bubbles; just search for pentagonal faces next to each other.

the past two or three decades, especially the Surface Evolver package, allowed one to more systematically study not only periodic but also random foams. In view of the observations and experimental results mentioned in the beginning of this chapter, the disordered cellular structures are generally prevalent in most tissues, and considering random foams as their model is a reasonable starting point. In fact, the basic fixed-volume area-minimizing version of the soap-froth problem may not apply so well to real foams because they coarsen with time as everyone knows. In this process, large bubbles grow at the expense of the small ones and gas diffuses across the film due to pressure difference. These effects can be included in the models but as far as tissues are concerned, they need not be because such coarsening does not take place in them.

Far more important is a deeper understanding of the topological disorder seen in Matzke's work [199], and this understanding can only be obtained numerically.

6.1.2 Random foam

Matzke's method of froth preparation by adding one bubble at a time produced a disordered foam but of course each bubble remained in the spot where it was introduced into the existing froth, adapting to its neighbors as well as to bubbles added later. In a system as delicate as the soap froth, any perturbation done so as to help the froth reach the minimal-energy state would inevitably make some or even most bubbles collapse. But we learned from Marvin's lead-shot experiment that the method of preparation does matter. In particular, the fractions of polyhedron classes in the shaken and non-shaken samples were not exactly the same as shown in Figure 6.7b.

On a computer, one can mimic Marvin's method even in a soap froth because the froth is most conveniently generated starting from a set of points in space put there using a suitable algorithm, and then partitioning the space based on this set. From an idealized perspective, Marvin's non-shaken samples were prepared by a protocol known as random sequential adsorption (RSA) where shots are added to an existing batch one at a time such that they do not overlap with those that are already there. This generally produces a looser, small-density packing because it does not allow a given local configuration to rearrange so as to make space for an additional shot even if such a rearrangement required an allowed displacement of a single shot to a nearby unoccupied position. On the other hand, if the shots are first poured into the container and then shaken, they can move about so as to fill space more efficiently. This leads to a denser, more compact ensemble including the random close packing (RCP); we will refer to this method by this acronym. The two methods are easily implemented on a computer, the second one involving a suitable dynamics driven by, say, a thermostat.

The reference computational study of random foams employed both protocols [206]. The foam was constructed in four steps including i) generation of a random packing of monodisperse spheres in a cube-shaped container with periodic boundary conditions; ii) construction of a three-dimensional Voronoi tessellation (Box on p. 83) based on the sphere packing, and iii) enforcement of the constant-volume constraint for each bubble and minimization of the energy of the tessellation so as to satisfy the Plateau rules and ensure mechanical equilibrium. The last step was further complemented by iv) mechanical annealing where topological rearrangements were promoted by a few cycles involving large-strain uniaxial dilation-compression sequences. The structure of the foam starting from an RSA hard-sphere system after steps ii), iii), and iv) is shown in panels a, b, and c in Figure 6.9, respectively; panels d, e, and f are the RCP versions of the foam after the three steps.

From these snapshots, it is hard to tell the non-annealed foam from the annealed one but the initial RSA Voronoi tessellation, which is not a foam at all, is quite different; the distribution of face sizes is clearly considerably broader as witnessed by many large as well as small faces. At the same time, the initial Voronoi tessellation based on the RCP

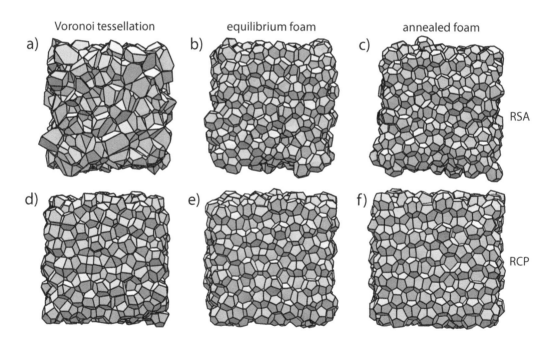

Figure 6.9 Initial Voronoi tessellation based on an RSA ensemble of 512 hard spheres (a), equilibrated foam obtained from panel a by energy minimization and local T1 topological transformation needed to satisfy the Plateau rules (b), and its annealed variant (c). The bottom row shows the three stages of the foam prepared using the RCP protocol. (Panels a–d are adapted from Ref. [206], and panels e and f are courtesy of A. M. Kraynik.)

protocol produces a much more foam-like tessellation. This is evident from the snapshots themselves, where neither equilibration nor annealing seem to alter the appearance of the partition very much. This conclusion is consistent with the apparently small differences between the two-dimensional Voronoi tessellation of hard disks and the in-plane structure of disordered epithelia (Figure 3.30).

The qualitative similarity of the two equilibrium and annealed foams in Figure 6.9 is also reflected in the fractions of polyhedron classes (Figure 6.10). While the initial RSA Voronoi tessellation is characterized by a broad distribution of polyhedra centered at $f = 15$, its RCP counterpart is considerably narrower and peaks at $f = 14$ much like in Matzke's froth (Figure 6.6b). The equilibrated and annealed random foams, each with a slightly different distribution of fractions, all approach Matzke's experimental results rather well, with the annealed RSA foam being closest to them.

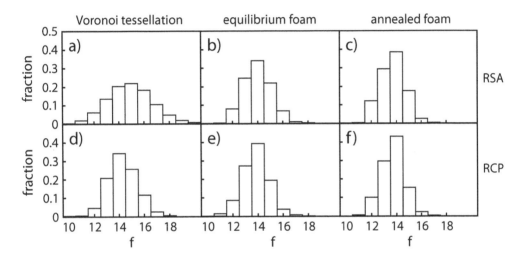

Figure 6.10 Polyhedron fractions in the initial Voronoi tessellation (RSA and RCP in panels a and d, respectively), equilibrated foam (b and e), and annealed foam (c and f). (Adapted from Ref. [206].)

From these distributions, it is evident that the average number of faces in polyhedra is around 14. The precise value depends on the preparation of the initial hard-sphere ensemble, system size, annealing, etc., and ranges from 13.69 to 14.00 in the equilibrated and annealed foams [206], which is again consistent with Matzke's value of 13.70. More interesting is the dimensionless area per bubble Ω shown in Figure 6.11, which too is determined by the preparation protocol and is between 5.323 and 5.380 in most samples. The lower boundary of Ω in random foams is less than 0.7% and 0.3% above the dimensionless area of the Weaire–Phelan and the Kelvin froth, respectively. These differences are very small and they may explain why random foams are seen so often whereas the regular minimal-energy foams, especially the Weaire–Phelan froth, are difficult to observe.

Equally interesting as the fractions of polyhedron classes, which are the basic structural feature of the foam, are the fractions of polygon classes, partly because they may be easily determined in freeze-fracture samples such as the hepatocytes in Figure 6.2b. Figure 6.12 shows the fractions in annealed RCP foam dominated by pentagons, which amount to about 65% of all faces. Apart from pentagons, the only polygon classes present in the foam are quadrilaterals, hexagons, and heptagons.

The soap-froth model is very instructive for several reasons, firstly because we can com-

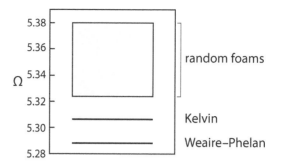

Figure 6.11 Dimensionless area per bubble Ω in numerical random-foam samples lie within the box, the exact value depending on the preparation protocol. Plotted for comparison are the values for Kelvin's truncated octahedron and the Weaire–Phelan froth. (Adapted from Ref. [206].)

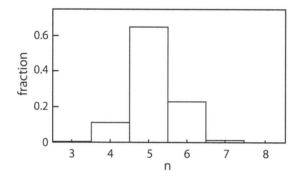

Figure 6.12 Fractions of polygon classes in annealed RCP random foam (adapted from Ref. [206]).

pare the tentative minimal-energy Weaire–Phelan froth with the disordered foams and conclude that the former is only slightly better than the latter. The second lesson learned from the numerical studies of random foams is that just by judging from the structure represented by the fractions of polyhedron classes, a Voronoi tessellation devoid of any energy-related feature may well compare very favorably to the experimental data, provided that the preparation protocol is suitable—this is demonstrated by the similarity of panels d, e, and f in Figure 6.10. In fact, such a Voronoi tessellation may come quite close to the true equilibrium froth obtained numerically by ensuring balance of forces across all of the froth.

These conclusions teach us that it will be difficult to decide whether a given tissue is indeed shaped by surface tension by relying solely on structural data such as the fractions of polyhedra classes; even more so because the experimental data may carry a considerable error related to the tricky identification of the small faces of cells. In addition, it may happen that alternative mechanical theories (two of which are discussed below) provide a description that is essentially just as good as that of the foam-based model, simply because tessellations of space are so constrained and their features such as the distribution of polyhedron classes are to a large extent a geometric necessity. In a proliferating tissue, the impact of cell mechanics on the structure may well be additionally obscured by cell division.

6.1.3 Melzak's problem

The relevance of minimal-area partitions for tissues relies on the assumption that cell membranes or better cell-cell contacts behave as constant-tension interfaces, which is a simplification but still a well-established working hypothesis from the perspective of physics. We can, however, think of other possibilities. The acto-myosin ring of epithelial cells is an example of a linear structural element distributed around the perimeter, and if we visualize it as a contractile filament of a fixed line tension then the minimal-energy configuration corresponds to the minimal perimeter. Now we make a bold generalization, if only for the sake of argument, and consider such an energy in cells in bulk-like tissues and in all of their edges. Another source of an edge energy may be the bending energy of the cell membrane in a tissue consisting of closely adhering polyhedral cells, which is dominated by a term proportional to the total edge length. This type of edge energy is discussed in Section 6.1.4 but it is somewhat more delicate because the energy also depends on the number of polyhedra meeting at an edge.

In the simplest variant of the model, we assume that the energy carried by a cell depends only on the total length of its edges and we look for the tessellation of minimal edge length at fixed cell volume. This question is related to Melzak's problem also referred to as the waste storage problem. Here one seeks the shape of a polyhedral vessel such that the total edge length is minimal at given volume, say so as to construct a container from flat plates by using as little glue or sealant as possible [207]. This problem is formulated for an isolated container but we can just as well consider it within the context of tessellations, simply comparing the edge lengths of space-filling rather than arbitrary polyhedra. The dimensionless edge length as the figure of merit analogous to Ω defined by Eq. (6.4) can be introduced by

$$\zeta = \frac{L}{V^{1/3}}, \tag{6.5}$$

where L is the total edge length per polyhedron.

For a few evident candidates ζ is straightforward. In the cube, $\zeta = 12$ whereas in the equilateral triangular prism $\zeta = 2^{2/3}3^{11/6} \approx 11.896$; it is easy to verify that any right regular length-minimizing prism is equilateral (Problem 6.7) but showing that it is right and regular is more demanding. The dimensionless edge lengths for a few polyhedra are summarized in Table 6.1, and so far the triangular prism appears to be the best.

Table 6.1 shows that the minimal-length tessellations have a small number of faces—the minimal-length equilateral triangular prism has just 5—whereas in the Weaire–Phelan minimal-area structure $f = 13.5$. Indeed, the tentative solutions of the two problems are

Table 6.1 Dimensionless edge length for some polyhedra; all prisms in the table are equilateral. The tentative solution of Melzak's problem is the triangular prism [208].

polyhedron	ζ	space-filling
triangular prism	11.896	yes
cube	12.000	yes
tetrahedron	12.238	no
pentagonal prism	12.518	no
square pyramid	12.814	no
hexagonal prism	13.093	yes
triangular dipyramid	14.342	no
dodecahedron	15.217	no
octahedron	15.419	no

quite different: the best minimal-length space-filling shapes are all prisms, which are consistent with a layered structure, whereas the Kelvin froth and the Weaire–Phelan froth are both of cubic symmetry. At this juncture, we allow ourselves a small divertissement. Imagine that the shape of cells in a tissue is indeed determined by the minimal-energy condition and that the energy includes both a surface and an edge term:

$$E(A, L) = \Gamma A + \lambda L, \tag{6.6}$$

where Γ and λ are the interfacial and the line tension, respectively.[7] In general, the minimizer is neither the Weaire–Phelan froth favored by the surface term nor the triangular prism favored by edge term but some in-between shape depending on the ratio

$$\frac{\Gamma V^{1/3}}{\lambda}, \tag{6.7}$$

where $V^{1/3}$ is the characteristic length scale.

In Figure 6.13 we plot the dimensionless area and the dimensionless length of five space-filling polyhedra. This diagram shows that the differences between their areas are generally smaller than the differences between their lengths, and it emphasizes the distinction between the length- and area-minimizers. It also suggests that in a system where the typical magnitudes of the length and the area terms are similar, the triangular prism will most likely not be the minimizer because its advantage over the cube as the second-best minimal-length shape is very small compared to the area difference.

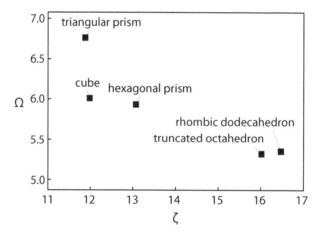

Figure 6.13 Area-length diagram of five space-filling polyhedra; for consistency, we plot the data for flat-face truncated octahedron rather than for the Kelvin minimal-area shape shown in Figure 6.5.

The length-minimizing tessellations discussed here are all based on a single cell type and ordered. One could, of course, also consider the more complex ordered tessellations as well as the disordered minimal-length analogs of the random foam. As far as we know, these structures have not been explored yet but they can be constructed, again starting from the Voronoi tessellation based on packings of hard spheres.

---

[7]This reasoning can be generalized to other forms of interfacial and edge energies as long as they are increasing functions of area and length, respectively.

6.1.4   Vesicle-aggregate model

With the exception of the triangular prism as the tentative solution of Melzak's problem, all energy-minimizing tessellations mentioned so far involve fairly isometric shapes. In order to generalize these models to anisometric cells such as those in the cornified layer of epidermis in Figure 6.1 where the height-to-diameter ratio is only about 0.1, we need to introduce differential interfacial tensions so as to distinguish between the apical, basal, and lateral faces like we did in Section 4.1.2. In the case of epidermis, it is plausible that these tensions are dissimilar and if the apical and basal tensions are much smaller than the lateral tension, then cells should be flattened like those in Figure 6.1.

Lipid vesicles

In water solution, amphiphilic molecules of cylindrical shape self-organize into a bilayer membrane so that their hydrophobic tails are buried within the membrane whereas their hydrophilic heads face water on either side. Closed membranes referred to as vesicles occur naturally in cells where they are involved in transport of matter between organelles but also in endo- and exocytosis, that is absorption and secretion of matter. Artificial vesicles can be fabricated in the laboratory, say using the film rehydration technique, and they may be as large as cells themselves.

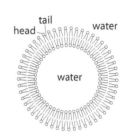

Depending on their reduced volume defined by Eq. (5.8) and on the difference in the number of molecules in the inner and the outer monolayer, vesicles may assume many different shapes. Shown below are a prolate, an oblate biconcave, and an invaginated shape referred to as the cigar, the discocyte, and the stomatocyte, respectively (images courtesy of M. Imai).

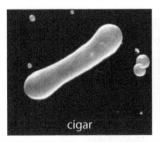
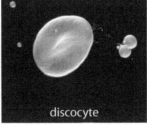
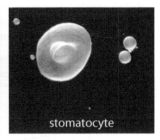

cigar   discocyte   stomatocyte

Yet another possibility leading to anisometric shapes is to impose additional constraints, say by assuming that cells are characterized by a certain finite preferred cell area A_0 and by setting it to a value larger than $\sim V^{2/3}$. Like in the area- and perimeter-elasticity model (Section 3.3.1), the penalty for the deviation of actual area from the preferred value can be represented by a surface energy of the form

$$\frac{\kappa_A}{2}\frac{(A - A_0)^2}{A_0},\qquad(6.8)$$

where κ_A is the area expansivity modulus. If the preferred cell area A_0 is smaller than the value corresponding to a sphere of volume V, which is $A_s = \pi^{1/3}(6V)^{2/3}$, then the actual area A is necessarily larger than A_0. For small deviations from some reference value, say A_s, Eq. (6.8) can be linearized and reads $\kappa_A(A_s/A_0 - 1)A + const$. Here $\kappa_A(A_s/A_0 - 1)$ is the effective surface tension and the constant term has been disposed of so as to better emphasize the similarity with the area term in Eq. (6.6). Thus if A_0 is smaller than A_s, surface energy is minimized by shapes with as small an area as possible consistent with the imposed constraints. In this case, the preferred-area model is not too different from the surface-tension model, especially in the context of tessellations with cells of identical volumes.

However, if the preferred cell area is larger than A_s then Eq. (6.8) is minimized at $A = A_0$. In cases where the magnitudes of all other mechanical energies involved are much smaller than the typical value of the surface energy [Eq. (6.8)], we may treat the cell area as a constraint rather than as a variable just like we treat cell volume. This approximation is conventionally employed in lipid vesicles characterized by the membrane bending energy described by the area-difference-elasticity (ADE) theory [194], which consists of the local Helfrich term and of the non-local term associated with the relative stretching of the monolayers [Eq. (5.5)].[8] When minimized at constant vesicle volume and area, these two terms combined accurately predict the equilibrium shapes of vesicles as well as the transitions between them.

To see how the edge energy arises in an aggregate of vesicles, consider identical lipid vesicles adhering strongly to each other so that their shape is faceted to the extent that it can be approximated by a polyhedron with rounded edges and vertices of radius R; in construction, this roundedness is referred to as the bullnose trim. Assume for simplicity that the polyhedron is a cuboid so that all edges are symmetric and four-valent (Figure 6.14), and disregard the non-local bending energy term such that the vesicle energy reads

$$E = \frac{\kappa}{2} \oint (C_1 + C_2)^2 \mathrm{d}A - \frac{\Gamma A_c}{2}. \tag{6.9}$$

Here κ is the local bending constant, C_1 and C_2 are the principal curvatures, Γ is the adhesion strength, and A_c is the area of the contact zone; the adhesion energy is weighted by $1/2$ so as to divide it equally between the two vesicles adhering to each other at any of the contact zones.

In equilibrium, balance of forces at the rim of the contact zone requires that the curvature be discontinuous, the discontinuity being given by $\Delta C = \sqrt{\Gamma/\kappa}$ [209]. This condition deter-

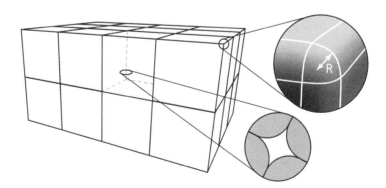

Figure 6.14 Model vesicle aggregate consisting of rounded cuboids with identical radii of curvature in all vertices and edges. In this structure, all edges are four-valent and symmetric so that the angles between adjacent contact zones are 90°.

---

[8]Formally, the non-local term in Eq. (5.5) describes the energy penalty associated with a deviation of the integrated curvature from its preferred value but this is equivalent to a deviation of the area difference of the monolayers $\Delta A = h \oint (C_1 + C_2)\mathrm{d}A$ from a preferred value; here h is the bilayer thickness.

mines the radius of the rounded edges and vertices in the aggregate, which reads

$$R = \frac{1}{\Delta C} = \sqrt{\frac{\kappa}{\Gamma}}. \tag{6.10}$$

The limit where all edges of the cuboidal vesicles are seen as quarter-cylinder stripes is consistent with the strong-adhesion limit where R is much smaller than the vesicle size conventionally defined as the radius of the sphere that has the same area as the vesicle, which reads $R_s = \sqrt{A/(4\pi)}$.

The bending energy of each vesicle is localized at the cylindrical surfaces running along the edges and at the spherical surfaces at vertices, which are here briefly referred to as edges and vertices, respectively. The total solid angle of the vertices is 4π irrespective of their number so that their bending energy amounts to $8\pi\kappa$ independent of R; to see this, note that the bending energy per unit area is proportional to curvature squared, that is, to $1/R^2$, whereas the area of a spherical surface is proportional to R^2. The energy of the edges, on the other hand, is proportional to the total edge length L and is $\pi\kappa L/(2R) = \pi\sqrt{\kappa\Gamma}L/2$ [210]. The total bending energy of a vesicle

$$E_b = 8\pi\kappa \left(1 + \frac{1}{16} \sqrt{\frac{\Gamma}{\kappa}} L \right) \tag{6.11}$$

is thus minimized by minimal-length shapes.

The adhesion energy too depends on the edge length. The total area of each vesicle can be divided into the area of the contact zone and the non-contact area, the latter consisting of a vertex term and an edge term. The combined area of vertices is $4\pi R^2 = 4\pi\kappa/\Gamma$ whereas the area of the edges is $\pi RL/2 = \pi\sqrt{\kappa/\Gamma}L/2$. Thus the area of the contact zone $A_c = A - 4\pi R^2 - \pi RL/2$ decreases with increasing edge length, and the adhesion energy $-\Gamma A_c/2$ contains two terms that are independent of L and another one proportional to L:

$$E_a = -\frac{\Gamma A}{2} + 2\pi\sqrt{\kappa\Gamma} + \frac{\pi}{4}\sqrt{\kappa\Gamma}L. \tag{6.12}$$

We see that vesicle-vesicle adhesion also favors as short an edge length as possible.

A more complete discussion where non-cuboidal vesicle shapes are considered too includes the three- and six-valent edges as well as the four-valent edges, all of which are present in single-cell-type aggregates [210]. As shown in Figure 6.15, these edges carry different line energies because the angles between the adjacent contact zones are not the same.

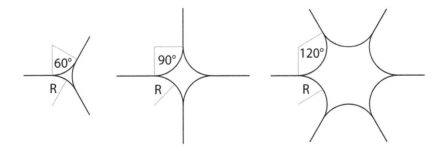

Figure 6.15 Transverse cross-sections of regular three-, four-, and six-valent edges with the angle between the adjacent contact zones being equal to $60°, 90°$, and $120°$, respectively. This illustrates that the edge energy increases with valence.

After these differences are taken into account, we find that the total energy depends on the weighted sum of edge length conveniently defined in dimensionless form as

$$\ell = \frac{1}{R_s}\left(\frac{L_3}{6} + \frac{L_4}{4} + \frac{L_6}{3}\right), \tag{6.13}$$

where $L_3, L_4,$ and L_6 are the total lengths of three-, four-, and six-valent edges in a vesicle, and $R_s = \sqrt{A/(4\pi)}$.

Now our task is to calculate the length of a suitable set of space-filling polyhedra characterized by a given area A and volume V, or rather by the reduced volume defined as the ratio of vesicle volume and the volume of the sphere of identical area:

$$v = \frac{V}{4\pi R_s^3/3} = \frac{6\sqrt{\pi}V}{A^{3/2}}. \tag{6.14}$$

In the single-parameter space-filling shapes such as the triangular, quadrilateral, and hexagonal prism, the reduced volume is determined by the ratio of height and base edge length. The zero-parameter shapes like the rhombic dodecahedron or the truncated octahedron can be modified in various ways so as to decrease the reduced volume. Figure 6.16 illustrates how three different shapes—two prolate and one oblate—can be constructed from the rhombic dodecahedron by cutting it in two and then either filling out the gap between the halves or trimming and reassembling them.

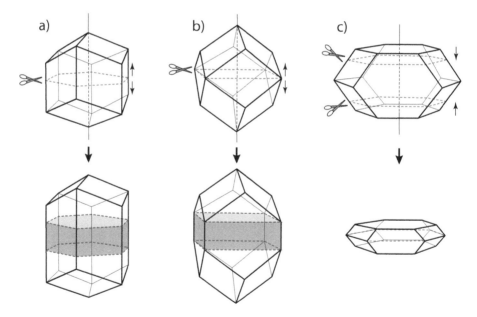

Figure 6.16 Two prolate shapes obtained from the rhombic dodecahedron by cutting it perpendicular to the three- and four-fold axis (a and b, respectively), and an oblate shape obtained from the second prolate shape (c). (Adapted from Ref. [210].)

The results are presented in Figure 6.17. Evidently, the shapes can be classified into the gentle oblate and the steep prolate branch, the former being generally more favorable. At large reduced volumes, say above 0.6 or so, the differences in total edge length do not exceed about 20% whereas at small v below 0.4 they are much more dramatic. Within the oblate branch, the three shapes consisting of two large hexagonal faces and a belt of small

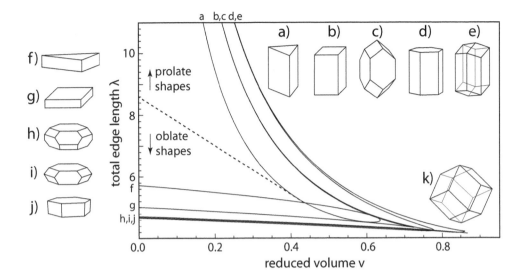

Figure 6.17 Total edge length vs. reduced volume for a few space-filling trial shapes including triangular, square, and hexagonal prism (a and f, b and g, and d and j, respectively), flattened truncated octahedron (h), and three polyhedra derived from the rhombic dodecahedron (c, e, and i; the rhombic dodecahedron itself is shown in inset k). (Adapted from Ref. [210].)

lateral faces (insets h, i, and j in Figure 6.17) are very close in energy, the right hexagonal prism being marginally lower than the other two.

The ordered minimal-length vesicle aggregates are qualitatively similar to the solution of Melzak's problem in that they too consist of flattened shapes as long as the reduced volume is small enough. On the other hand, the optimal shapes (h, i, and j in Figure 6.17) have an approximate six-fold rather than three-fold rotational symmetry like the triangular prism of Melzak's problem. In this respect, they agree better with the shape of epidermis cells seen in Figure 6.1b and c—although, as we mentioned earlier, these cells may well be flattened primarily by a stronger adhesion at the apical and basal sides compared to that at the lateral sides.

6.1.5 Sequential cell sheet formation

In the cornified layer, positional order of cells may also arise from a mechanism that is unrelated to the precise energy of cells and cell-cell contacts, and relies instead on packing considerations combined with differentiation and migration of cells through the layer [211]. As cells in the proliferating basal layer become keratinized, they flatten and pack against the existing cornified layer. Figure 6.1a shows that while cells in the basal layer are rather malleable as witnessed by their deformed contours, their nuclei do not appear to be. Representing cells in the cornified layer and the prospective members of the layer by their nuclei alone is thus a reasonable approximation, which allows us to examine how cells arrange as they form a new bottommost sheet of the layer [211].

To this end, we consider a simple rule requiring that each incoming cell fits in the gap between its top neighbors and does not overlap with its in-plane neighbors; it suffices to treat the nuclei as hard disks and assume that they lie in equidistant planes within the cornified layer. A simple two-dimensional simulation of this process is shown in Figure 6.18,

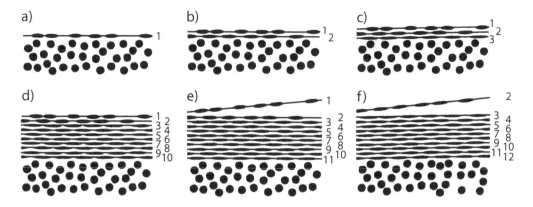

Figure 6.18 Early (a–c) and late stages (d–f) of a two-dimensional simulation of the emergence of positional order in the epidermis' cornified layer where in each step a new sheet of cells is added to the layer. The first sheet is disordered (a) but in the seventh or eighth sheet (d) the cells are essentially equidistant. (Adapted from Ref. [211].)

where cells in the cornified layer are represented by flattened ellipses such that they can be distinguished from those in the basal layer which are drawn as circles. The simulation starts with randomly distributed cells in the top sheet (Figure 6.18a). In the next step, a new sheet of the cornified layer is formed according to the above rule (Figure 6.18b). It is easy to see that the center-to-center distances between cells in the new plane are more uniform than those in the top sheet. The third sheet is even more ordered and after a large enough number of sheets have been formed—ten in this simulation—the top sheets are shed one after another. As a result, a highly ordered cornified layer is formed in just a dozen generations (Figure 6.18f) without relying on any specific mechanical cell-level features.

This model can also be implemented in three dimensions. Here the position of each new member of the cornified layer is given by the centroid of the triangle formed by its top neighbors, provided that the in-plane distance from centroids of adjacent triangles is larger than the hard-core diameter of the cell. The cells and their in-plane order can be better visualized by the Voronoi tessellation based on cell centers than by centers themselves. This tessellation mimics cell contours and is thus more readily comparable with experiments. Figure 6.19 shows the tessellation in sheets 2, 19, 79, 109, and 159, demonstrating how quickly the initially random-tiling pattern becomes ordered. Also evident from this simulation is that cell density in the first couple of sheets is larger than in the more ordered late-stage sheets where the cells arrange relative to both top and in-plane neighbors.

The mechanism of sequential cell sheet addition is much more robust than the minimal-length model from Section 6.1.4. In addition, it is conceptually consistent with experiments showing that cell division within the basal layer is stochastic so that the proliferating cells end up in multiple columns seen in Figure 6.1a rather than in a single one immediately above the spot where a given mother cell has divided [212]. If cell shape and order relied on energy minimization alone, the random behavior of cells within the basal layer would probably lead to a more disordered cornified layer, especially in view of the small differences of the energies of the oblate shapes in the vesicle-aggregate model. This would be at odds with observations (Figure 6.1). At the same time, the sequential cell sheet addition mechanism alone does not explain the flattened cell shape; here this is taken for granted.

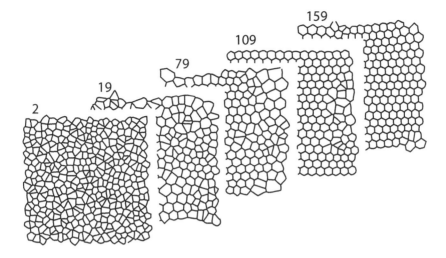

Figure 6.19 Five sheets in the three-dimensional simulation of the cornified layer represented by the two-dimensional Voronoi tessellation of cell centroids; the sheets are labeled by the sequential number, the late ones being almost perfectly ordered. (Adapted from Ref. [211].)

6.2 TISSUE ELASTICITY

Most animal tissues are packed far less tightly than the liver and the epidermis. An example of a tissue that is rather homogeneous in terms of composition and contains incompletely faceted cells is white adipose tissue, which stores body fat. A type of connective tissue, it consists of white adipose cells also known as adipocytes as well as of fibroblasts, leukocytes, endothelial cells, and nerves. Adipocytes are the most numerous of all types of cells in this tissue and account for up to 95% of the tissue. An adipocyte has a nucleus and a cytoplasm just like other cells, but its main constituent is a drop of lipids such as triglycerides which are in a semiliquid state at body temperature (Figure 6.20). The typical adipocyte size is comparable to the thickness of human hair, that is about 0.1 mm, and the lipid content represents about 85% of cell mass.

Given the structure of adipocytes, it should not be too surprising that individual cells assume a preferentially spherical shape as demonstrated by adipocytes grown in culture [213]. Figure 6.21a shows a close-up of human white adipose tissue with a visible collagen mesh which binds the adjacent adipocytes such that they deform so as to conform to each other but not as completely as, e.g., Marvin's lead shots [201]. The degree of cell faceting in the adipose tissue varies and is generally smaller than in compressed lead shots or in a dry foam. As a result, pockets of space are left between neighboring adipocytes and the cell-level structure of the tissue qualitatively resembles a polydisperse wet foam.[9] At the scale of about 1 mm, i.e., at a length scale of about 10 adipocytes, the tissue is divided into lobules enveloped with another structure known as the interlobular septa, which consists primarily of type I collagen fibers (Figure 6.21b).

The hierarchical two-level cellular structure defined by the cells and the lobules is reflected in the response of the tissue to shear stress. While the mechanics of epidermis and

---

[9]In a dry foam, the volume occupied by the liquid films that separate the bubbles is much smaller than the volume of the bubbles; the relative volume of the liquid is known as the liquid fraction φ and is zero in the limiting case. In wet foams, $\varphi \gtrsim 0.15$ but smaller than about 0.26; at large φ where the bubbles no longer touch each other, we speak of a bubbly liquid [214].

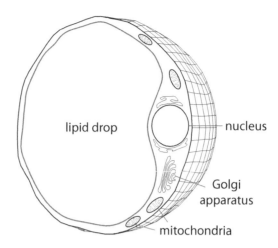

Figure 6.20 Schematic of white adipose cell with the nucleus, organelles, and lipid drop.

liver discussed earlier in this chapter is not very interesting because these tissues do not have a structural role, the mechanics of the adipose tissue is important. The subcutaneous adipose tissue, for example, is located just below the skin and is not used solely as storage of fat but also as padding and as thermal insulation. Although it is not a load-bearing structure, the subcutaneous adipose tissue serves as a substrate for the dermis and the epidermis, and this is why its mechanical nature is involved in the form of these two layers as illustrated by the models of nontrivial epithelial shapes in Section 4.1. In addition, the mechanics of adipose tissues may also figure in liver and heart conditions related to the increased deposition of fat around these organs. The hypothesized chain of events leading to it starts with i) an enhanced deposition of collagen in adipose tissue in obese individuals, which results in ii) an increased rigidity of the tissue and its iii) inability to accommodate

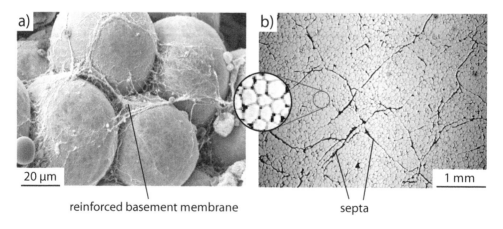

Figure 6.21 Human white adipose tissue showing collagen mesh around adipocytes referred to as the reinforced basement membrane (a; adapted from Ref. [215]) and porcine adipose tissue at a smaller magnification where the collagen septa dividing the lobules are seen as dark streaks (b; adapted from Ref. [216]).

additional dietary lipids. Instead these lipids iv) accumulate ectopically, that is, in areas where they are normally not present, and they interfere with the function or state of the organ in question [217].

The mechanical response of the adipose tissue is one of the elements in this development and this is why understanding it in detail is important. Not all types of adipose tissue are the same in this respect: given its hierarchical structure, we may expect that the deformation may take place either at the cell level or at the lobule level depending on tissue type. Indeed a tensile test performed on two types of human adipose tissue revealed considerable differences [217]. The subcutaneous tissue underwent large adipocyte deformation and displacement, whereas in the omental tissue which covers colon and small intestine, the adipocytes themselves remained spherical and the strain resulted in the stretching of the collagen fibers between the lobules. Despite the different deformation mechanisms, both subcutaneous and omental tissue showed pronounced strain stiffening, with the large-deformation Young's modulus at 30% strain that is about an order of magnitude larger than the small-deformation modulus.

Equally interesting as strain stiffening itself are the plastic deformation and the stress relaxation of the tissue at these large strains. The observed plasticity is quite prominent: after dilation to strains a little over 30%, the omental adipose tissue reaches the zero-stress state at a residual strain of about 20% once the stress is removed. This means that only a part of the deformation is reversible. The plastic response is also seen in the decay of stress following an application of large, 30% strain, with the fast stress-relaxation time of about 100 s. The subcutaneous adipose tissue exhibits a similar behavior with an almost identical residual strain and about twice slower stress-relaxation time while the magnitude of stress at a given strain is about a third of that in the omental tissue [217].

These measurements suggest that at large strains, the adipose tissues undergo considerable rearrangements so that they cannot be viewed as elastic materials. Naturally, the stroma considered in Section 4.1.1 does not consist solely of the adipose tissue and it is possible that in the other components the yield strain is larger than in the adipose tissue. At the same time, in most folds and villi the strain is evidently considerably larger than a few 10% when measured relative to a flat state as witnessed by the shapes in Figure 4.1. The magnitude of the strain depends on the unknown position of the reference flat state, which may be below the grooves of the folds, above the crests, or at some intermediate height. However, in the well-developed nontrivial shapes such as those in Figure 4.1, one can safely conclude that the strain is very large irrespective of the position of the reference state. Together with the pronounced stress relaxation measured in the adipose tissue [217], this suggests that the stroma is a malleable material and that treating it as an elastic solid is generally inappropriate.

In view of their structure, the simplest cell-level model that one could relate to the plasticity of the adipose tissue is based on liquid foams. In a liquid foam, bubbles can slide past each other if shear stress is large enough so as to overcome the barrier associated with bubble deformation. Adipose cells held together by collagen fibrils of the reinforced basement membrane should not be too different from bubbles provided that the work expended by the external stress suffices to break the bonds between fibrils and cells, which can then be re-established once cells rearrange in a new, stress-free structure. If tissue rigidity is primarily due to the collagen-rich septa like in the omental tissue in Ref. [217], then the foam picture can be applied at the lobule level by viewing the septa as liquid-like films with the collagen-collagen bonds within each septum broken and re-established in response to stress.

In both cases, the analogy between liquid foams and tissues is partial rather than complete but still insightful; another structural model that one can use is the open-cell foam discussed in Section 6.2.2. Before introducing the open-cell foams, let us look into the response of liquid foams to shear stress.

6.2.1 Liquid foams

The basic mechanism of foam plasticity and bubble rearrangement due to shear stress can be illustrated with a two-dimensional argument proposed by H. M. Princen, one of the pioneers of the physics of foams [218]. We consider a planar pattern of identical hexagonal cells of area A shown in Figure 6.22a, and we assume for simplicity that the foam is perfectly dry so that the area contained in the borders between the bubbles vanishes. Now examine a simple-shear deformation of a unit cell containing a single hexagon (shaded area in Figure 6.22a). The unit cell is a 60°–120° rhombus of edge length $\sqrt{3}a$, with a being the length of the edge of the undeformed hexagon.

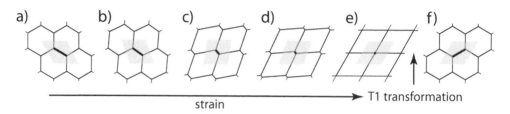

Figure 6.22 Simple-shear deformation of a two-dimensional hexagonal foam illustrated by a sequence of states at increasing strain (a–e); in panel a strain is zero whereas in panel e it reaches the yield strain $\epsilon^c_{xy} = 2/\sqrt{3}$. The unit cell is shaded and the shrinking edge is highlighted. Panel f shows the foam after the four-valent vertex in panel e has been resolved into a pair of three-valent vertices, restoring the shape of cells in panel a.

As the unit cell is sheared, four of the hexagon edges elongate and two shrink such that the angles between then remain at 120° so as to satisfy the Plateau rules. This deformation is depicted by the sequence of states in Figure 6.22a–e, with panel e showing the limiting shear deformation where the short edge has shrunk to a point so that all vertices are four-valent and thus unstable. At this stage, the vertices are resolved into two three-valent ones (Figure 6.22f), thus re-establishing a zero-stress hexagonal foam of identical energy as the original foam in Figure 6.22a but consisting of rearranged bubbles because those in the top row have jumped past those in the bottom row. The initial and the final states in panels a and f, respectively, are evidently related by the T1 topological transformation (Box on p. 60).

The strain regime up to the yield strain (Figure 6.22a–e) is elastic as the topology of the tiling remains unchanged, but the decay of the four-valent vertex into two three-valent vertices signals a plastic deformation. From these sketches, we can deduce the yield strain. To this end, note that the initial rhombic unit cell in Figure 6.22a and that in Figure 6.22e are identical except for the orientation; they must be identical because the angles between their edges are the same. Thus the horizontal displacement of any point on the top edge of the unit cell is equal to the edge length itself: $u_x = \sqrt{3}a$. Since the distance between the top and the bottom edge is $3a/2$, the yield strain for this deformation reads

$$\epsilon^c_{xy} = \frac{2}{\sqrt{3}}. \tag{6.15}$$

The corresponding yield stress is given by

$$\sigma_{xy}^c = \frac{\Gamma}{\sqrt{3}a} \tag{6.16}$$

(Problem 6.8). A persistent strain exceeding ϵ_{xy}^c causes the bubbles to rearrange over and over again, thus giving rise to flow, and it is easy to see that in the idealized case of a hexagonal foam the stress-strain curve would be sawtooth-shaped. As expected, the yield strain and the yield stress depend on the orientation, and a more detailed analysis shows that they have a discontinuity at shear stress applied along a diagonal direction at 45° with respect to the x axis [219].

This simple example illustrates why plastic deformation and flow are so prominent in foams at large strains. The qualitative features of the stress-strain curve in a random three-dimensional foam, where the sawtooth-like dependence mentioned above is averaged out, are sketched in Figure 6.23. For a monotonically increasing strain, the linear small-strain elastic regime is followed by a peak at yield stress and then by a saturation at a slightly smaller limit stress where the foam flows. If the temporal dependence of the strain is non-monotonic, the foam behaves as an elastic solid as soon as the stress falls below the yield value [220].

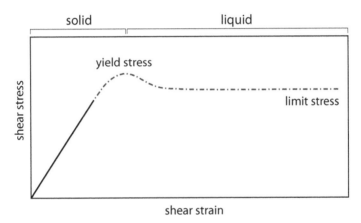

Figure 6.23 Schematic stress-strain curve for a random foam and a monotonically increasing stress, with the elastic regime at small strains (solid line) followed by the liquid-like regime at large strains.

The foam stress-strain curve captures the essence of the elastic and plastic deformation regimes including, e.g., atomic crystals where atoms arranged in neighboring lattice planes slip past each other too if the shear stress is large enough. As such, the foam model is quite instructive. Yet its applicability to adipose tissue is restricted to samples where the collagen fibers form more or less uniform sheets wrapped around either cells or lobules, depending on the length scale of interest. However, these sheets usually consist of bundled collagen fibrils as shown in Figure 6.21a so that they are more similar to a network of struts than to a tessellation where adjacent cells are separated by a well-defined wall. Such networks are referred to as open foams. The interlobular septa in the white adipose tissue are an even better example of an open foam. Materials characterized by such a structure are commonplace in both animals and plants, an often cited example being trabecular bone [221].

6.2.2 Cellular solids

The mechanics of solids with a compartmentalized structure is a rich subject, pertaining to many biological and man-made materials [222]. Cellular solids may be divided into open- and closed-cell foams, and their response to stress translates into the deformation at the cell level. In the case of an open-cell foam consisting of struts, Young's modulus, Poisson's ratio, yield strain, and other material coefficients all depend on the distortion of the struts, which brings us back to the topics discussed in Chapter 2. An important parameter of interest is the relative density of the foam, that is the ratio

$$\frac{\rho^*}{\rho_s}, \tag{6.17}$$

where ρ^* and ρ_s are the densities of the foam and of the solid material that the struts are made of. To calculate the effective Young's modulus of the foam, it is sufficient to consider a cubic unit cell with struts of thickness t and length L arranged at the edges, the neighboring cells being connected by hinges (Figure 6.24a). We assume that the struts are thin so that their thickness is much smaller than their length ($t \ll L$). In this case,

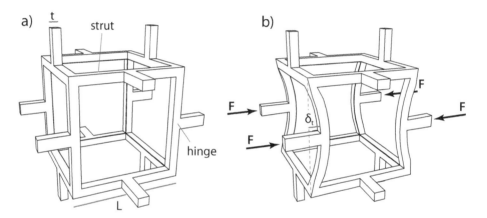

Figure 6.24 Cubic unit cell of an open-cell foam showing struts and hinges (a) and a deformed cell under uniaxial compression (b) [221].

a uniaxial load applied along, e.g., the horizontal axis translates into a lengthwise force F on each horizontal strut and a transverse force on each vertical strut (Figure 6.24b). The former leads to a compression or dilation of the strut given by

$$\delta_l = \frac{F}{Y_s t^2} L, \tag{6.18}$$

whereas the latter results in strut bending such that the displacement of the hinge relative to the corner is proportional to the cube of the length of the section in question, that is, to $(L/2)^3$. In particular,

$$\delta_t \sim \frac{F}{Y_s t^4} L^3, \tag{6.19}$$

which follows from Eq. (2.67); we took into account that the area moment of inertia of a strut of square cross-section equals $t^4/12$ and we dropped the numerical prefactor. In turn, the overall numerical prefactor in Eq. (6.19) depends on the boundary conditions at the ends—but they need not really be specified in order to see that in slender struts, the ratio of the two displacements is much smaller than 1:

$$\frac{\delta_l}{\delta_t} \sim \frac{L/t^2}{L^3/t^4} = \frac{t^2}{L^2} \ll 1. \tag{6.20}$$

Thus we find that the deformation of an open foam is dominated by strut bending, which is hardly a surprise.

Now it is easy to consider the compression or dilation of the cell as a whole, again in a scaling-argument style. The strain is given by δ_t/L and the stress reads F/L^2 so that the effective Young's modulus is

$$Y^* \sim \frac{F/L^2}{\delta_t/L} = Y_s \left(\frac{t}{L}\right)^4. \tag{6.21}$$

Since the mass of struts scales as $\rho_s t^2 L$ and the cell volume is L^3, the density of the foam reads $\rho^* \sim \rho_s(t/L)^2$. We see that the ratio of Young's moduli of the foam and of the raw material is

$$\frac{Y^*}{Y_s} = C \left(\frac{\rho^*}{\rho_s}\right)^2, \tag{6.22}$$

where C is a numerical constant which depends on geometry but is not too different from 1 [221].

Equation (6.22) agrees well with the measurements, and the line of reasoning employed to derive it can be used to obtain the compressive yield stress, which is more accurately referred to as the elastic collapse stress. To this end, we start from the critical force for the Euler instability (Figure 4.5) which can be estimated as suggested in Problem 4.3 and reads

$$F_c = \frac{\pi^2 Y I}{L^2} \tag{6.23}$$

for a strut with one end fixed; here I is the area moment of inertia. The critical stress is given by $F_c/L^2 \propto Y(t/L)^4$ so that $\sigma_c/Y_s \propto (\rho^*/\rho_s)^2$. This result is interesting because the critical stress for two other modes of mechanical failure—the brittle crushing and the so-called plastic collapse where the hinges of the network are deformed—both scale as $(\rho^*/\rho_s)^{3/2}$. By analyzing the dependence of the yield stress on density, one can thus decide between the different failure modes in, e.g., trabecular bone where data point to elastic collapse mediated by the Euler instability of individual struts rather than to brittle crushing as one would expect based on layman's experience [221].

These elementary yet insightful results can now be employed to interpret the experimentally observed elastic response of the porcine adipose tissue, which can be regarded as a combination of a closed-cell foam and an open-cell foam [216]. The reinforced basement membrane around adipocytes can be approximated by a solid closed-cell foam. At small deformations where the bonds between the collagen fibrils in the membrane are still intact, the membrane behaves as a solid rather than as a liquid sheet so that the mechanics of this foam is different from that of the liquid foams from Section 6.2.1. In turn, the collagen fibers in the septa separating the adjacent lobules can be treated either as an interconnected network or as an open-cell foam.

As the closed-cell reinforced-basement-membrane foam and the septa are interwoven, they constitute two mechanical elements connected in parallel like in the Kelvin–Voigt model (Box on p. 30), and if the viscosity of the lipid drops is disregarded then these elements reduce to springs (Figure 6.25). Here the strains in the basement-membrane foam and in the septa network must be the same, say $\epsilon_{xx}^m = \epsilon_{xx}^s$ in an uniaxial deformation, and

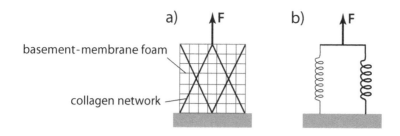

Figure 6.25 Adipose tissue consisting of a closed-cell solid foam interwoven with a septa network regarded as two elastic materials (a) can be represented by two springs connected in parallel (b).

each of them is proportional to the respective part of the stress such that $\epsilon_{xx}^m = \sigma_{xx}^m/Y_m$ and $\epsilon_{xx}^s = \sigma_{xx}^s/Y_s$, where Y_m and Y_s are Young's moduli of the membrane foam and the septa network, respectively. Given that the total stress is a sum of the partial stresses σ_{xx}^m and σ_{xx}^s, we see that the effective modulus of the tissue is given by

$$Y = Y_m + Y_s. \tag{6.24}$$

The interesting part of the analysis is in the actual values of these two moduli. Experiments suggest that at small strains, the effective Young's modulus of the porcine subcutaneous adipose tissue from the jowl is about 10^3 N/m^2 [216]. The effective modulus of the septa considered as an open-cell foam is described by Eq. (6.22) and if we use 10^9 N/m^2 as the modulus of solid collagen and a density ratio $\rho^*/\rho_s = 3 \times 10^{-4}$, we find that $Y_s = 90$ N/m^2, which is an order of magnitude less than the measured value. Alternatively, the septa can be described as a random network of wavy fibers but this gives an even smaller value of the part of the modulus that is associated with the septa [216]. These estimates suggest that the elastic response is dominated by the reinforced basement membrane—just like in human subcutaneous adipose tissue where too the adipocytes themselves undergo considerable deformation when the tissue is stressed [217] as mentioned at the beginning of Section 6.2.

In a solid foam where the faces of the cells are filled, a dilation or compression of the cell gives rise to face stretching [222]. The strain scales as δ_t/L and if the wall thickness is the same as the strut thickness t, the elastic energy per face is proportional to $Y(\delta_t/L)^2 \times L^2 t = Y\delta_t^2 t$. Then the force reads $\sim Y\delta_t t$ and the effective stress per cell is $F/L^2 \sim Y\delta_t t/L^2$, and after dividing by the strain we find that the ratio of the effective Young's modulus and of the raw material is

$$\frac{Y^*}{Y_m} \sim \frac{t}{L} \sim \frac{\rho^*}{\rho_m}. \tag{6.25}$$

The scaling of Y^*/Y_m with relative density follows from the mass of the cell which is proportional to $\rho_m t L^2$ so that the foam density is $\rho_m t/L$. Note that this scaling law is different from Eq. (6.22); in a real closed foam, the two terms should be combined as the foam generally consists of both cell walls and struts.

The ratio ρ^*/ρ_m of the reinforced basement membrane is much larger than that of the septa and is around 0.1. If we now assume that the effective modulus of the tissue is entirely due to the membrane then we can conclude that Young's modulus of the membrane must be 10^4 N/m^2. This appears to be a reasonable estimate [216], which allows us to conclude that microscopic models of tissue structure can indeed be used to rationalize its mechanical behavior.

A conceptually antipodal starting point is to regard tissues as continuous media and search for a suitable theory of elasticity consistent with their symmetry. This approach is well known from the classical elasticity of materials and has been traditionally explored mainly within the context of the harder tissues in animal bodies such as bones, cartilage, ligaments, and tendons [223]. Building from the laws of mass, momentum, and energy conservation as well as suitable constitutive equations relating stress and strain, one can construct very accurate models of these tissues, accounting for their porosity and inhomogeneity. These theories are invaluable from the perspective of biomechanical engineering and the insight that they offer is complementary to that of the models pursued in this chapter, which all start out from the cell as the building block of the tissue.

At the same time, the phenomenological continuum-mechanics theories are also interesting because they may provide guidance for the development of microscopic models. In the past, many of the key theoretical developments in the mechanics of large deformations were motivated by the scientific and industrial importance of rubber [224], perhaps the most widespread material that can sustain strains exceeding several 100%. To understand these data, a generalization of the small-deformation Hookean elasticity (Box on p. 29) is needed and many classes of models were proposed to this end [225], some of them with a neat molecular-scale interpretation [226]. All of them involve the so-called invariants of the right Cauchy–Green deformation tensor $F^T F$ mentioned in Section 5.1.5, which can be expressed in terms of the principal stretches λ_1, λ_2, and λ_3. These quantities can be introduced as follows: locally, any deformation of a material can be decomposed into a rotation and a stretch or vice versa, and the principal stretches are the eigenvalues of the tensor describing stretch. The invariants are functions of the principal stretches that remain unchanged under any transformation of the coordinate system that maintains the symmetry of the material. A simple example of such an elastic theory for an isotropic material is the neo-Hookean model with the energy density given by

$$u_V = \frac{\mu}{2} \left(\lambda_1^2 + \lambda_2^2 + \lambda_3^2 - 3 \right), \tag{6.26}$$

where μ is the shear modulus.

The measurements of the shear modulus of murine brain and fat tissue undergoing gradual stepwise compressions and dilations reaching 40% revealed a considerable difference between the two regimes. Under compression, the effective modulus increases by about a factor of 5 in brain and 10 in fat tissue as the strain reaches 40% [227] whereas under tension the variation of the modulus is much smaller. This behavior cannot be reproduced by any of the simpler models with one or two elastic constants such as the neo-Hookean, Mooney–Rivlin, Fung, and Gent models; in the neo-Hookean model, for example, the shear modulus is independent of strain, which is evidently inconsistent with these experimental observations. On the other hand, the various variants of the Ogden model where the elastic energy per unit volume reads

$$u_V = \sum_{p=1}^{N} \frac{C_p}{2m_p} \left(\lambda_1^{2m_p} + \lambda_2^{2m_p} + \lambda_3^{2m_p} - 3 \right) \tag{6.27}$$

$(N, C_p,$ and m_p being constants[10]) provide a rather good description of the measurements. The agreement is illustrated in Figure 6.26 showing the obese adipose tissue data and the $N = 4$ Ogden model fit, which has four free parameters because the four exponents m_p were fixed to $-2, -1, 1$, and 2. The $N = 6$ and $N = 8$ variants of the model do not further improve the agreement very much.

---

[10]The reader will notice that for $N = 1$ and $m_1 = 1$ the Ogden model reduces to the neo-Hookean model [Eq. (6.26)]; the $N = 2, m_1 = 1$, and $m_2 = -1$ variant is the incompressible Mooney–Rivlin model. The coefficients m_p need not be integers but they must satisfy the relation $\sum_{p=1}^{N} C_p m_p = \mu$; in incompressible materials where $\lambda_1 \lambda_2 \lambda_3 = 1$ they need not be positive [225].

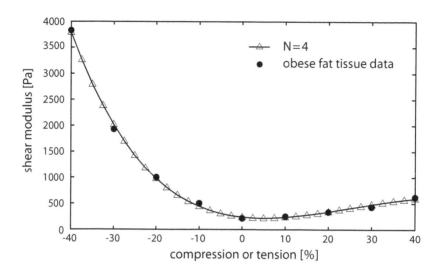

Figure 6.26 Measured shear modulus of uniaxially deformed obese murine adipose tissue vs. strain (circles) compared to the $N = 4$ variants of the Ogden model with the material coefficients determined by the least-square fit. (Adapted from Ref. [227].)

At this point, we need to stress that the number of coefficients of the $N = 4, 6$, and 8 Ogden models is larger than in the simpler models and thus a better agreement should not be too surprising. In addition, the measurements themselves should be interpreted with care as the tissues studied are viscoelastic rather than elastic and every increase in strain is followed by stress relaxation. Stress relaxation was especially prominent at large compressions and the data in Figure 6.26 were all recorded 100 s after application of strain, the delay being reasonable yet arbitrary. Finally, like in other adipose tissues, the murine samples too show some hysteresis in the shear modulus, which may well be a sign of a plastic deformation. Despite these caveats, studies revolving around continuum-mechanics theories are very helpful, partly because they may prompt and steer the search of improved or entirely new cell-level models.

The more elaborate theories of tissue structure and elasticity go quite far from the early investigations of models of tissues, say the foam-based cellular partitions of space first systematically studied by Matzke. Yet liquid foam remains one of the main themes of this chapter. Although it does not capture all of the mechanics of tissues—given their diversity, expecting this would be an impossibly tall order—it still nicely combines the particulate character of cells, elastic deformation at small strains, and plasticity at large strains. In view of its clear albeit not simple physical basis consisting of the surface energy and fixed-cell-volume constraints, this is quite an achievement and when generalized to wet foams, it offers an even richer spectrum of behaviors. In many respects, it ranks along with some of the eminent models mentioned in Section 1.3.

PROBLEMS

6.1 The bubble polyhedra in the annealed RCP random foam (Figure 6.12) and the cell polyhedra in *Eupatorium perfoliatum* (Figure 6.4) consist primarily of quadrilateral, pentagonal, and hexagonal faces. The fractions of other polygons are small, especially in polyhedra with a small number of faces (less than about 14 in case of pith cells studied by Marvin [198]).

Consider now a set of polyhedra such that their faces are either quadrilateral, pentagonal, or hexagonal and their vertices three-valent.

Show that for each particular polyhedron in the set with F_4 quadrilateral, F_5 pentagonal, and F_6 hexagonal faces, the following relation has to hold:

$$2F_4 + F_5 = 12. \tag{6.28}$$

If $F_4 = 0$ and the total number of faces is larger than 11, a polyhedron from this set consists of 12 pentagons and an arbitrary number of hexagons. This includes the dodecahedron[11] with $F_6 = 0$, but also fullerene-like and other complicated polyhedra.[12]

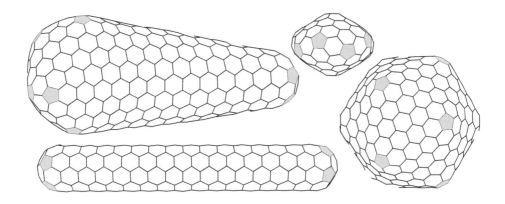

If $F_5 = 0$, a polyhedron consists of 6 quadrilaterals and an arbitrary number of hexagons. A truncated octahedron, a flat-faced version of Kelvin's minimal-area shape, belongs to this category of polyhedra, and so does the polyhedron in Figure 6.17h.

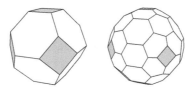

---

[11]Does a polyhedron with 12 pentagonal and 1 hexagonal face exist?
[12]This type of geometry is typical for viruses, fullerenes, and carbon nanotubes [228].

Discuss how the ratio of hexagonal faces in a polyhedron with $F = F_4 + F_5 + F_6$ faces changes as F is increased and relate the answer to Eq. (6.3). Prove that the number of polyhedron subsets, defined by pairs (F_4, F_5) for a given F_6, is smaller than 8. Show that the generalization of Eq. (6.28) where we include all possible face types in a polyhedron with three-valent vertices yields [3]

$$\sum_i (6 - i)F_i = 12. \tag{6.29}$$

6.2 Derive the formula for the average number of sides in a polyhedron with four-valent vertices such as an octahedron.

6.3 Consider the two-dimensional version of the Kelvin problem and calculate the dimensionless perimeter $\omega = P/A^{1/2}$ for the regular hexagon and for the square. Also calculate the dimensionless perimeter for rectangles as a function of aspect ratio b/a.

6.4 Calculate the area A and the volume V of the rhombic dodecahedron, and then show that for this shape $\Omega = A/V^{2/3} = 2^{5/6}3$.

6.5 Construct the Voronoi tessellation (Box on p. 83) of the low- and the high-density configurations of hard disks in the figure and describe the differences between them.

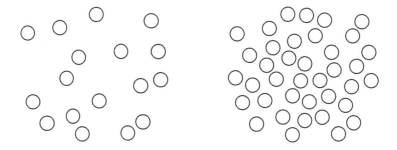

6.6 Count the number of triangular, quadrilateral, pentagonal... faces visible in the random foams in panels b, c, e, and f in Figure 6.9, compute the fractions of polygons, and compare them with the histogram in Figure 6.12.

6.7 Calculate the volume of an n-sided right regular prism and show that the total edge length is minimized if the prism is equilateral.

6.8 Calculate the yield stress for the two-dimensional hexagonal foam sheared as shown in Figure 6.23. *Hint:* Parametrize the deformation in terms of the acute angle in the unit cell, calculate the lengths of all edges and then the energy of the deformed hexagon.

6.9 Use Eq. (6.21) to discuss the elastic properties of a network of filaments (e.g., in the cytoskeleton) interlinked strongly with proteins, so that the filaments cannot slide over each other in response to stress. Can this equation be used to explain the fact that Young's modulus of the cortex is orders of magnitude smaller than Young's moduli of individual filaments as mentioned on p. 28?

CHAPTER 7

Afterthoughts

I F THE MOST PLEASURABLE MOMENT in research is discovery, the second best is when one realizes that a given phenomenon is a reincarnation of something that is already known. By establishing these analogies, which are sometimes very evident and sometimes less so, we gradually build a system that helps us better understand the awe-inspiring diversity of nature. This approach is highly praised in physics, and it is probably one of the reasons why physics is quite successful. Another reason for success of physics is the delineation of its territory, i.e., the part of the universe that it strives to explain. Life and biology have mostly been outside these borders yet never completely so—partly because it is impossible to define impermeable borders and partly because physics still expands, but mostly because living matter too is physical. However, the universe beyond the traditional domain of physics appears to be more difficult to talk about. This may be due to the language that we still have to formulate and to the difficulty that we face even in defining life as the concept of interest.

Our book is by no means original in addressing certain aspects of living matter from the safe position of physics. This view has been usually termed "mechanistic"—often in a derogatory manner, implying that it leaves out many biologically essential processes and phenomena. Here is a citation from a reviewer's report on a manuscript that eventually emerged as one of the pioneering theoretical studies of the mechanics of embryogenesis (Ref. [124]):[1] "The authors are attempting to apply Newton's laws to embryos, but as all biologists know, biological systems don't obey the laws of physics" [229]. This was in 1980 and it is fair to say that since then, the attitude behind the but-clause has changed as we witness a closer and closer cooperation between life sciences, physical sciences, and mathematics which occasionally amalgamate into science without adjectives. Just science.

We try to promote this spirit but as all biologists know, physicists usually take their business quite seriously. Evidently we are no exception as this book too is full of physicists' business, especially Hooke's and Newton's laws in all sorts of disguises. The application of these laws to cells and their mutual interactions, to tissues and their shapes, and to growth and formation of structures certainly clarifies many issues. It demystifies parts of these topics by relating them to analogous phenomena from the non-living world, and it explains how exactly organisms fit within the realm of the possible which is demarcated by the laws of physics. The aim of this book was to contribute to this endeavor, and we know that we reached it only partly. One of the reasons for this is the sheer diversity and richness of the field, which is not easily covered comprehensively. Another reason is that life is extremely ingenious in filling the evolutionary niches: The breadth of the physical phenomena that it exploits in the struggle for existence is truly overwhelming. Sonar and

---

[1]This paper was co-authored by G. Oster, whose ideas are included in Sections 5.1, 5.5, and 5.6.

echo-location were discovered by the ancestors of bats and whales long before the theory of sound was developed. Flight was invented by insects, dinosaurs, gliding squirrels, and rotating maple seeds well before Leonardo and the Wright brothers thought about it. (In fact, it still happens that the features of a certain flying mode appear to defy what we know from theories.)

It all becomes even weirder and weirder at micrometer and nanometer scales. After all, life started from these length scales, spending three billion years there before exploring the multicellular ways of existence. It did have some time, therefore, to scrutinize the domain of the small. In some cases, this domain is also manifest at the macroscopic scale, e.g., in a gecko which uses nano-structured hair at its feet to exploit the van der Waals interaction so as to stick to vertical walls. Sight too is an evolved molecular phenomenon built on chemistry and on a cascade induced by the absorption of a single photon. This could of course go on and on, filling many pages of some other book. We know that life is rooted in physics, yet how it steers it to its own advantage by using genes and other mechanisms is still poorly understood, especially when it comes to the evolved macroscopic organisms equipped with a bagful of tricks, which are selected and adapted so precisely that they indeed look like a miracle. The formation of a fully developed, immensely complex organism such as the *Drosophila* from a single cell is perhaps the greatest miracle of all, and we know that it must ultimately be described by the laws of physics.

Yet a cell is not like anything that we know from physics. Is it like bubble gum? Is it like a drop of honey? Is it like a soap bubble? Well... Yes and no. It is not like a soap bubble, but it is still useful to resort to surface tension when describing certain aspects of cell mechanics. Tissue is not like a soap froth, but it is still a good idea to invoke an image of a froth when thinking of tissues. There exists no comprehensive physical model of cells or tissues at this time; all that we have are approximate representations which are of use in certain circumstances. Many of these representations reach beyond cells and tissues, posing new questions and opening new perspectives. For example, F. T. Lewis' investigations of the in-plane structure of epithelia (Section 3.2.1) or Marvin's analysis of the shape of compressed lead shots (Chapter 6) address rather fundamental issues important not only for biology and physics but also from the mathematical standpoint, as they point to the role of geometrical necessity in packings of cells and other deformable bodies.

Despite all this, the physical modeling of cells and tissues has come a long way, and we tried to emphasize this in the book. We included many of the early ideas and observations, which fit well with our goal of providing a broad, often pedagogical account of the mechanics of tissues instead of giving an overview of the state-of-the-art technologies and theories. We attempted to describe a few legs of this meandering historical path. We tried to re-enact F. T. Lewis' and Marvin's reconstruction of elderberry pith cells, with a scenography of lanterns, wax molds, and precise hand drawings, to tell the reader about W. H. Lewis' rubber bands, tubes, and brass bars that he used to simulate the mechanics behind apical constriction in gastrulation, about Kelvin's wire frames and minimal-volume soap bubbles, about Matzke's delicate froth... Some of these milestones are included in the timeline in Figure 1.7 which shows the continuity of the mechanistic view of tissues.

Then there are the more recent advances. Peeking into them is truly enlightening, if only so as to grasp the level of complexity that can be handled with modern techniques. Take, for example, the 2015 quantitative analysis of the morphogenesis of the *Drosophila* dorsal thorax and pupal wing [189]. Some figures: in the dorsal thorax epithelium, the contours of about 18400 cells were tracked for 20 hours with the temporal resolution of 5 minutes, and a total of about 3000000 contours were used to compute the local rate of strain tensor in the tissue. This tensor describes tissue growth and morphogenesis, and the strain rates associated with each of the four processes involved—cell division, rearrangement, changes of size and shape, as well as apoptosis and delamination—was separately identified frame by frame. (Here one cannot but think again of Marvin and his painstakingly molded 100 wax cells, precisely scaled so as to enable mechanical manipulation and measurement [198].)

Figure 7.1 shows the dorsal thorax epithelium with the superimposed pattern of shear rate due to topological rearrangements [mentioned in Section 5.2.3, say in Eq. (5.29)] averaged over 20 hours of development, which varies considerably from point to point across the thorax. Locally, this tensor is represented by a bar such that its length encodes the magnitude of the shear rate whereas its orientation shows the average direction of a new junction between cells established by T1 transformations. An in-depth analysis of the patterns corresponding to each of the four processes suggests that in some parts of the dorsal thorax epithelium, cell division takes place such that it contributes to tissue elongation (that is, the average center-to-center line between daughter cells points along the direction of tissue elongation) whereas in other parts it is almost exactly the opposite. With this level of detail, a comprehensive understanding of the mechanisms of tissue development including its theoretical basis is well within reach.

In addition to the increasingly more detailed and less invasive experimental techniques, the field often requires fresh and innovative theoretical approaches. For example, one of the

Figure 7.1 *Drosophila* dorsal thorax epithelium during pupal morphogenesis, showing cell contours and an overlaid pattern of the shear rate due to cell rearrangement R averaged over a few 10 cells. Bar length represents the magnitude of the rate and its orientation points along the average direction of the junctions formed between the new neighbors after topological transformations. Some bars at the perimeter are in lighter gray because they were averaged over fewer cells, as some of them drifted out of view. (Image courtesy of B. Guirao.)

key quantities used to analyze the above *Drosophila* dorsal thorax data is the recently introduced notion of texture as a suitable descriptor of the response to stress in materials where plastic deformation is prominent [230], say in foams discussed in Section 6.2.1. Constructed from the network of links between sites in a given cellular pattern, such as a biological tissue or a foam, this quantity does not depend on the reference undeformed state like the usual strain tensor but it still conveys the intuitive idea of the deformation, nicely illustrating how thinking of the salient mechanical properties of foams, tissues, and related materials led to a novel theoretical concept.

We opened this book with the starfish and the bay leaf but we could just as well have done it with the sponge. Sponges are among the simplest animals, and most people will think twice before even regarding them as animals; Haeckel erroneously classified them in the kingdom of Protista which contains unicellular organisms and colonies of single cells [231]. Nonetheless, sponges would make an excellent starting point because they embody the physics of cell adhesion and give us clues to the evolution of multicellularity and thus tissues. Their existence relies on cell adhesion molecules as the "glue" that holds the cells together. The presence of a glue was first observed in 1907 by H. V. Wilson who showed that when sponges are mechanically broken into individual cells dispersed in sea water, the randomly mixed cells of different types eventually re-aggregate and reform new sponges [232]. This insight paved the way to further studies of cell aggregation, sorting, and self-assembly, and eventually led to the present molecular-level understanding of these processes which are, in turn, included in the cell surface and interfacial tension as the overarching coarse-grained concepts in this book.

Yet mentioning sponges at the end is appropriate too, as it allows us to return one more time to Hooke's *Micrographia*. Hooke did not only introduce and name cells for the first time, he also described a sponge (Figure 7.2a). His drawing shows a complex network, which probably represents the spongin fibers. With a clear perspective of depth, it helps the reader to appreciate the random open-cell-foam structure of the sponge, visualizing its strut-and-joint architecture. A scanning electron micrograph of a sponge seen in Figure 7.2b is so much more advanced than Hooke's drawing yet strikingly similar at the same time, especially if we scale them and place them side by side; as if it came full circle.

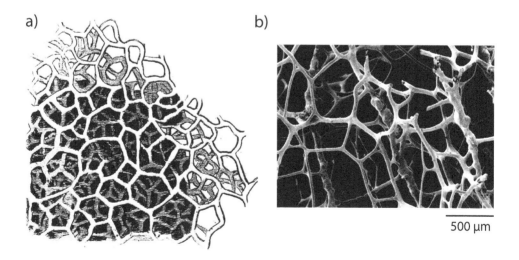

Figure 7.2   Hooke's drawing of a sponge from *Micrographia* [4] (a) and a scanning electron micrograph of the *Hyatella cribriformis* sponge (b; reproduced from Ref. [233]).

Equally crisp as the drawing itself is Hooke's account of sponge structure: "(...) it consists of an infinite number of small short fibres (...) curiously jointed or contex'd together in the form of a Net (...) The joints are, for the most part, where three fibres only meet, for I have very seldom met with any that had four. (...) The meshes likewise, and holes of this reticulated body, are not less various and irregular: some bilateral, others trilateral, and quadrilateral (...) I have observ'd some meshes to have 5, 6, 7, 8, or 9. sides (...)" [4]. After having learned about the various cellular partitions in two and three dimensions, polygon and polyhedron classes, vertex valence, and the different topological intricacies discussed in Chapters 3 and 6, the reader will perhaps more easily appreciate the meaning of Hooke's words and where they fit into what is now known about the structure of tissues.

The line of thought exposed in this book is admittedly rather theoretical and focused on basic science. Yet we would like to think that it may also be of use in practice or even in applications, especially within the context of tissue engineering including organoids [234] as well as embryoids and gastruloids [235], that is, bodies of differentiating embryonic stem cells akin to organoids but resembling embryos rather than organs. For example, a more precise control of tissue deformation would clearly be welcome so as to program the form of the growing organoids such as that in Figure 7.3. It seems that many of the ideas discussed here—apico-basal cell polarization, coupling between tissue curvature and cell division, fixed-volume constraints, etc.,—can be exploited so as to establish such a control. How these ideas can be materialized may well again be a topic of some other book, although mechanics is just a part of the physics involved in the organoid technology [236]. We just

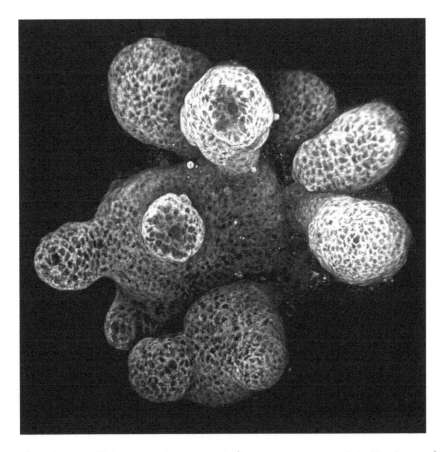

Figure 7.3 Murine small intestinal organoid (image courtesy of J. Heuberger).

reiterate that any displacement or deformation of cells, organoids, or tissues as well as soccer balls or planets is caused by a physical force: in this respect, mechanics is indispensable. It seems appropriate to close with Thompson's words, which resonate perhaps even more powerfully today than 100 years ago when they were first written: "In our own day the philosopher neither minimises nor unduly magnifies the mechanical aspect of the Cosmos; nor need the naturalist either exaggerate or belittle the mechanical phenomena which are profoundly associated with Life, and inseparable from our understanding of Growth and Form" [3].

References

[1] Y. Klein, E. Efrati, and E. Sharon, *Shaping of Elastic Sheets by Prescription of Non-Euclidean Metrics*, Science **315**, 1116 (2007).

[2] W. His, *Unsere Körperform und das Physiologische Problem Ihrer Entstehung* (F. C. W. Vogel, Leipzig, 1874).

[3] D. W. Thompson, *On Growth and Form* (Cambridge University Press, Cambridge, 1942).

[4] R. Hooke, *Micrographia: Or Some Physiological Descriptions of Minute Bodies Made by Magnifying Glasses. With Observations and Inquiries Thereupon* (Royal Society, London, 1665).

[5] J. Stegmaier, F. Amat, W. C. Lemon, K. McDole, Y. Wan, G. Teodoro, R. Mikut, and P. J. Keller, *Real-Time Three-Dimensional Cell Segmentation in Large-Scale Microscopy Data of Developing Embryos*, Dev. Cell **25**, 225 (2016).

[6] M. B. Hallett, C. J. von Ruhland, and S. Dewitt, *Chemotaxis and the Cell Surface-Area Problem*, Nat. Rev. Mol. Cell Biol. **9**, 662 (2008).

[7] V. Dugina, I. Alieva, N. Khromova, I. Kireev, P. W. Gunning, and P. Kopnin, *Interaction of Microtubules with the Actin Cytoskeleton via Cross-Talk of EB1-containing +TIPs and γ-actin in Epithelial Cells*, Oncotarget **7**, 72699 (2016).

[8] E. Fuchs and D. W. Cleveland, *A Structural Scaffolding of Intermediate Filaments in Health and Disease*, Science **279**, 514 (1998).

[9] K. E. Kasza, A. C. Rowat, J. Liu, T. E. Angelini, C. P. Brangwynne, G. H. Koenderink, and D. A. Weitz, *The Cell as a Material*, Curr. Opin. Cell Biol. **19**, 101 (2007).

[10] K. Luby-Phelps, *Cytoarchitecture and Physical Properties of Cytoplasm: Volume, Viscosity, Diffusion, Intracellular Surface Area*, Int. Rev. Cytol. **192**, 189 (2000).

[11] E. Evans and A. Yeung, *Apparent Viscosity and Cortical Tension of Blood Granulocytes Determined by Micropipet Aspiration*, Biophys. J. **56**, 151 (1989).

[12] T. Soderlund, J.-M. I. Alakoskela, A. L. Pakkanen, and P. K. J. Kinnunen, *Comparison of the Effects of Surface Tension and Osmotic Pressure on the Interfacial Hydration of a Fluid Phospholipid Bilayer*, Biophys. J. **85**, 2333 (2003).

[13] R. M. Hochmuth, *Micropipette Aspiration of Living Cells*, J. Biomech. **33**, 15 (2000).

[14] T. Kalwarczyk, N. Ziębacz, A. Bielejewska, E. Zaboklicka, K. Koynov, J. Szymański, A. Wilk, A. Patkowski, J. Gapiński, H.-J. Butt, and R. Hołyst, *Comparative Analysis of Viscosity of Complex Liquids and Cytoplasm of Mammalian Cells at the Nanoscale*, Nano Lett. **11**, 2157 (2011).

[15] D. Arcizet, B. Meier, E. Sackmann, J. O. Rädler, and D. Heinrich, *Temporal Analysis of Active and Passive Transport in Living Cells*, Phys. Rev. Lett. **101**, 248103 (2008).

[16] M. Herant, V. Heinrich, and M. Dembo, *Mechanics of Neutrophil Phagocytosis: Behavior of the Cortical Tension*, J. Cell Sci. **118**, 1789 (2005).

[17] J. Lam, M. Herant, M. Dembo, and V. Heinrich, *Baseline Mechanical Characterization of J774 Macrophages*, Biophys. J. **96**, 248 (2009).

[18] J. Guck, R. Ananthakrishnan, H. Mahmood, T. J. Moon, and C. C. Cunningham, *The Optical Stretcher: A Novel Laser Tool to Micromanipulate Cells*, Biophys. J. **81**, 767 (2001).

[19] S. Suresh, *Biomechanics and Biophysics of Cancer Cells*, Acta Biomater. **3**, 413 (2007).

[20] A. G. Clark, K. Dierkes, and E. K. Paluch, *Monitoring Actin Cortex Thickness in Live Cells*, Biophys. J. **105**, 570 (2013).

[21] B. Audoly and Y. Pomeau, *Elasticity and Geometry: From Hair Curls to Non-Linear Response of Shells* (Oxford University Press, Oxford, 2010).

[22] M. L. Gardel, K. E. Kasza, C. P. Brangwynne, A. J. Liu, and D. A. Weitz, *Mechanical Response of Cytoskeletal Networks*, Meth. Cell Biol. **89**, 487 (2008).

[23] L. D. Landau and E. M. Lifshitz, *Theory of Elasticity* (Butterworth–Heinemann, Oxford, 1986).

[24] C. Klinger, A. V. Cherian, J. Fels, P. M. Diesinger, R. Aufschnaiter, N. Maghelli, T. Keil, G. Beck, I. M. Tolić-Norrelykke, M. Bathe, and R. Wedlich-Soldner, *Isotropic Actomyosin Dynamics Promote Organization of the Apical Cell Cortex in Epithelial Cells*, J. Cell Biol. **207**, 107 (2014).

[25] A. Saha, M. Nishikawa, M. Behrndt, C.-P. Heisenberg, F. Jülicher, and S. W. Grill, *Determining Physical Properties of the Cell Cortex*, Biophys. J. **110**, 1421 (2016).

[26] K. A. Brakke, *The Surface Evolver*, Exp. Math. **1**, 141 (1992).

[27] L. Shapiro and W. I. Weis, *Structure and Biochemistry of Cadherins and Catenins*, Cold Spring Harbor Perspect. Biol. **1**, a003053 (2009).

[28] G. M. Cooper, *The Cell: A Molecular Approach* (ASM Press, Washington, 1997).

[29] U. S. Schwarz and S. A. Safran, *Physics of Adherent Cells*, Rev. Mod. Phys. **85**, 1327 (2013).

[30] O. du Roure, A. Saez, A. Buguin, R. H. Austin, P. Chavrier, P. Silberzan, and B. Ladoux, *Force Mapping in Epithelial Cell Migration*, Proc. Natl. Acad. Sci. USA **102**, 2390 (2004).

[31] J. Fu, Y.-K. Wang, M. T. Yang, R. A. Desai, X. Ye, Z. Liu, and C. S. Chen, *Mechanical Regulation of Cell Function with Geometrically Modulated Elastomeric Substrates*, Nat. Meth. **79**, 733 (2010).

[32] O. du Roure, A. Saez, A. Buguin, R. H. Austin, P. Chavrier, P. Silberzan, and B. Ladoux, *Force Mapping in Epithelial Cell Migration*, Proc. Natl. Acad. Sci. USA **102**, 2390 (2005).

[33] C. M. Edwards and U. S. Schwarz, *Force Localization in Contracting Cell Layers*, Phys. Rev. Lett. **107**, 128101 (2011).

[34] W. N. de Vries, A. V. Evsikov, B. E. Haac, K. S. Fancher, A. E. Holbrook, R. Kemler, D. Solter, and B. B. Knowles, *Maternal β-Catenin and E-Cadherin in Mouse Development*, Development **131**, 4435 (2004).

[35] P. L. Townes and J. Holtfreter, *Directed Movements and Selective Adhesion of Embryonic Amphibian Cells*, J. Exp. Zool. **128**, 53 (1955).

[36] M. S. Steinberg, *Reconstruction of Tissues by Dissociated Cells*, Science **141**, 401 (1963).

[37] R. A. Foty, C. M. Pfleger, G. Forgacs, and M. S. Steinberg, *Surface Tensions of Embryonic Tissues Predict Their Mutual Envelopment Behavior*, Development **122**, 1611 (1996).

[38] D. Duguay, R. A. Foty, and M. S. Steinberg, *Cadherin-Mediated Cell Adhesion and Tissue Segregation: Qualitative and Quantitative Determinants*, Dev. Biol. **253**, 309 (2003).

[39] H. M. Phillips and M. S. Steinberg, *Embryonic Tissues as Elastoviscous Liquids: I. Rapid and Slow Shape Changes in Centrifuged Cell Aggregates*, J. Cell Sci. **30**, 1 (1978).

[40] R. A. Foty and M. S. Steinberg, *The Differential Adhesion Hypothesis: A Direct Evaluation*, Dev. Biol. **278**, 255 (2005).

[41] P. Katsamba, K. Carroll, G. Ahlsen, F. Bahna, J. Vendome, S. Posy, M. Rajebhosale, S. Price, T. M. Jessell, A. Ben-Shaul, L. Shapiro, and B. H. Honig, *Linking Molecular Affinity and Cellular Specificity in Cadherin-Mediated Adhesion*, Proc. Natl. Acad. Sci. USA **106**, 11594 (2009).

[42] J. D. Amack and M. L. Manning, *Knowing the Boundaries: Extending the Differential Adhesion Hypothesis in Embryonic Cell Sorting*, Science **338**, 212 (2012).

[43] V. Maruthamuthu, B. Sabass, U. S. Schwarz, and M. L. Gardel, *Cell-ECM Traction Force Modulates Endogenous Tension at Cell-Cell Contacts*, Proc. Natl. Acad. Sci. USA **108**, 4708 (2011).

[44] M. L. Manning, R. A. Foty, M. S. Steinberg, and E.-M. Schoetz, *Coaction of Intercellular Adhesion and Cortical Tension Specifies Tissue Surface Tension*, Proc. Natl. Acad. Sci. USA **107**, 12517 (2010).

[45] T. Lecuit and P.-F. Lenne, *Cell Surface Mechanics and the Control of Cell Shape, Tissue Patterns and Morphogenesis*, Nat. Rev. Mol. Cell Biol. **8**, 633 (2007).

[46] M. Yoneda, *Tension at the Surface of Sea-Urchin Egg: A Critical Examination of Cole's Experiment*, J. Exp. Biol. **41**, 893 (1964).

[47] J. A. F. Plateau, *Statique expérimentale et théorique des liquides soumis aux seules forces moléculaires* (Gauthier–Villars, Paris, 1873).

[48] P. C. F. Moller and L. B. Oddershede, *Quantification of Droplet Deformation by Electromagnetic Trapping*, EPL **88**, 48005 (2009).

[49] L. A. Segel and G. H. Handelman, *Mathematics Applied to Continuum Mechanics* (Dover Publications, New York, 1987).

[50] F. T. Lewis, *The Effect of Cell Division on the Shape and Size of Hexagonal Cells*, Anat. Rec. **33**, 331 (1926).

[51] F. T. Lewis, *The Correlation between Cell Division and the Shapes and Sizes of Prismatic Cells in the Epidermis of Cucumis*, Anat. Rec. **38**, 341 (1928).

[52] D. Boal, *Mechanics of the Cell* (Cambridge University Press, Cambridge, 2001).

[53] P. H. Raven, R. F. Evert, and H. Curtis, *Biology of Plants* (Worth Publishers, New York, 1981).

[54] M. P. Miklius and S. Hilgenfeldt, *Epithelial Tissue Statistics: Eliminating Bias Reveals Morphological and Morphogenetic Features*, Eur. Phys. J. E **34**, 1 (2011).

[55] D. Weaire and N. Rivier, *Soap, Cells and Statistics—Random Patterns in Two Dimensions*, Contemp. Phys. **25**, 59 (1984).

[56] N. Rivier, R. Ocelli, J. Pantaloni, and A. Lissowski, *Structure of Bénard Convection Cells, Phyllotaxis and Crystallography in Cylindrical Symmetry*, J. Phys. (Paris) **45**, 49 (1984).

[57] K. M. Cadigan, *Wnt Signaling—20 Years and Counting*, Trends Genet. **18**, 340 (2002).

[58] J. A. Zallen and R. Zallen, *Cell-Pattern Disordering during Convergent Extension in Drosophila*, J. Phys.: Condens. Matter **16**, S5073 (2004).

[59] A.-K. Classen, K. I. Anderson, E. Marois, and S. Eaton, *Hexagonal Packing of Drosophila Wing Epithelial Cells by the Planar Cell Polarity Pathway*, Dev. Cell **9**, 805 (2005).

[60] F. T. Lewis, *A Comparison between the Mosaic of Polygons in a Film of Artificial Emulsion and the Pattern of Simple Epithelium in Surface View (Cucumber Epidermis and Human Amnion)*, Anat. Rec. **50**, 235 (1931).

[61] P. Pina and M. A. Fortes, *Characterization of Cells in Cork*, J. Phys. D: Appl. Phys. **29**, 2507 (1996).

[62] J. C. M. Mombach, M. A. Z. Vasconcellos, and R. M. C. de Almeida, *Arrangement of Cells in Vegetable Tissues*, J. Phys. D: Appl. Phys. **23**, 600 (1990).

[63] R. Farhadifar, J.-C. Röper, B. Aigouy, S. Eaton, and F. Jülicher, *The Influence of Cell Mechanics, Cell-Cell Interactions, and Proliferation on Epithelial Packing*, Curr. Biol. **17**, 2095 (2007).

[64] A. B. Patel, W. T. Gibson, M. C. Gibson, and R. Nagpal, *Modeling and Inferring Cleavage Patterns in Proliferating Epithelia*, PLoS Comput. Biol. **5**, e1000412 (2009).

[65] S. Kim, M. Cai, and S. Hilgenfeldt, *Lewis' Law Revisited: The Role of Anisotropy in Size-Topology Correlations*, New J. Phys. **16**, 015024 (2014).

[66] K. A. Newhall, L. L. Pontani, I. Jorjadze, S. Hilgenfeldt, and J. Brujic, *Size-Topology Relations in Packings of Grains, Emulsions, Foams, and Biological Cells*, Phys. Rev. Lett. **108**, 268001 (2012).

[67] S. J. Cox, D. Weaire, and M. F. Vaz, *The Transition from Two-Dimensional to Three-Dimensional Foam Structures*, Eur. Phys. J. E **7**, 311 (2002).

[68] A. E. Roth, B. G. Chen, and D. J. Durian, *Structure and Coarsening at the Surface of a Dry Three-Dimensional Aqueous Foam*, Phys. Rev. E **88**, 062302 (2013).

[69] C. R. Desch, *The Solidification of Metals from the Liquid State.*, J. Inst. Met. **22**, 241 (1919).

[70] P. Feltham, *Grain Growth in Metals*, Acta Metall. **5**, 97 (1957).

[71] K. Y. Szeto and W. Y. Tam, *Lewis' Law Versus Feltham's Law in Soap Froth*, Physica A **221**, 256 (1995).

[72] M. Durand, J. Käfer, C. Quilliet, S. Cox, S. Ataei Talebi, and F. Graner, *Statistical Mechanics of Two-Dimensional Shuffled Foams: Prediction of the Correlation between Geometry and Topology*, Phys. Rev. Lett. **107**, 168304 (2011).

[73] M. Rauzi, P. Verant, T. Lecuit, and P.-F. Lenne, *Nature and Anisotropy of Cortical Forces Orienting Drosophila Tissue Morphogenesis*, Nat. Cell Biol. **10**, 1401 (2008).

[74] D. D. O'Keefe, E. Gonzalez-Niño, M. Burnett, L. Dylla, S. M. Lambeth, E. Licon, C. Amesoli, B. A. Edgar, and J. Curtiss, *Rap1 Maintains Adhesion between Cells to Affect Egfr Signaling and Planar Cell Polarity in Drosophila*, Dev. Biol. **333**, 143 (2009).

[75] D. A. Aboav, *The Arrangement of Grains in a Polycrystal*, Metallography **3**, 383 (1970).

[76] D. Weaire, *Some Remarks on the Arrangement of Grains in a Polycrystal*, Metallography **7**, 157 (1974).

[77] C. J. Lambert and D. Weaire, *Order and Disorder in Two-Dimensional Random Networks*, Philos. Mag. B **47**, 445 (1983).

[78] N. Rivier and A. Lissowski, *On the Correlation Between Sizes and Shapes of Cells in Epithelial Mosaics*, J. Phys. A: Math. Gen. **15**, L143 (1982).

[79] M. A. Peshkin, K. J. Strandburg, and N. Rivier, *Entropic Predictions for Cellular Networks*, Phys. Rev. Lett. **67**, 1803 (1991).

[80] R. M. C. de Almeida and J. R. Iglesias, *Towards Statistical Mechanics of a 2D Random Cellular Structure*, J. Phys. A: Math. Gen. **21**, 3365 (1988).

[81] S. F. Edwards and R. B. S. Oakeshott, *Theory of Powders*, Physica A **19**, 1080 (1989).

[82] R. Blumenfeld and S. F. Edwards, *Geometric Partition Functions of Cellular Systems: Explicit Calculation of the Entropy in Two and Three Dimensions*, Eur. Phys. J. E **19**, 23 (2006).

[83] M. C. Gibson, A. B. Patel, R. Nagpal, and N. Perrimon, *The Emergence of Geometric Order in Proliferating Metazoan Epithelia*, Nature **442**, 1038 (2006).

[84] T. Aegerter Wilmsen, A. C. Smith, A. J. Christen, C. M. Aegerter, E. Hafen, and K. Basler, *Exploring the Effects of Mechanical Feedback on Epithelial Topology*, Development **137**, 499 (2010).

[85] W. T. Gibson, J. H. Veldhuis, B. Rubinstein, H. N. Cartwright, N. Perrimon, G. W. Brodland, R. Nagpal, and M. C. Gibson, *Control of the Mitotic Cleavage Plane by Local Epithelial Topology*, Cell **144**, 427 (2011).

[86] B. Strauss, R. J. Adams, and N. Papalopulu, *A Default Mechanism of Spindle Orientation Based on Cell Shape Is Sufficient to Generate Cell Fate Diversity in Polarised Xenopus Blastomeres*, Development **133**, 3883 (2006).

[87] B. Dubertret and N. Rivier, *The Renewal of the Epidermis: A Topological Mechanism*, Biophys. J. **73**, 38 (1997).

[88] R. Cowan and V. B. Morris, *Division Rules for Polygonal Cells*, J. Theor. Biol. **131**, 33 (1988).

[89] A. Hočevar and P. Ziherl, *Degenerate Polygonal Tilings in Simple Animal Tissues*, Phys. Rev. E **80**, 011904 (2009).

[90] A. Hočevar, S. El Shawish, and P. Ziherl, *Morphometry and Structure of Natural Random Tilings*, Eur. Phys. J. E **33**, 369 (2010).

[91] H. J. Hilhorst, *Asymptotic Statistics of the n-Sided Planar Poisson–Voronoi Cell: II. Heuristics*, J. Stat. Mech.: Theory Exp. P05007 (2009).

[92] H. Honda, *Geometrical Models for Cells in Tissues*, Int. Rev. Cytol. **81**, 191 (1983).

[93] K. Binder, S. Sengupta, and P. Nielaba, *The Liquid-Solid Transition of Hard Discs: First-Order Transition or Kosterlitz–Thouless–Halperin–Nelson–Young Scenario?*, J. Phys.: Condens. Matter **14**, 2323 (2000).

[94] M. Engel, J. A. Anderson, S. C. Glotzer, M. Isobe, E. P. Bernard, and W. Krauth, *Hard-Disk Equation of State: First-Order Liquid-Hexatic Transition in Two Dimensions with Three Simulation Methods*, Phys. Rev. E **87**, 042134 (2013).

[95] F. Moučka and I. Nezbeda, *Detection and Characterization of Structural Changes in the Hard-Disk Fluid under Freezing and Melting Conditions*, Phys. Rev. Lett. **94**, 040601 (2005).

[96] T. Hayashi and R. W. Carthew, *Surface Mechanics Mediate Pattern Formation in the Developing Retina*, Nature **431**, 647 (2004).

[97] H. Honda, H. Yamanaka, and G. Eguchi, *Transformation of a Polygonal Cellular Pattern During Sexual Maturation of the Avian Oviduct Epithelium: Computer Simulation*, J. Embryol. Exp. Morphol. **98**, 1 (1986).

[98] F. Graner and J. A. Glazier, *Simulation of Biological Cell Sorting Using a Two-Dimensional Extended Potts Model*, Phys. Rev. Lett. **69**, 2013 (1992).

[99] J. A. Glazier and F. Graner, *Simulation of the Differential Adhesion Driven Rearrangement of Biological Cells*, Phys. Rev. E **47**, 2128 (1993).

[100] J. Käfer, T. Hayashi, A. F. M. Marée, R. W. Carthew, and F. Graner, *Cell Adhesion and Cortex Contractility Determine Cell Patterning in the Drosophila Retina*, Proc. Natl. Acad. Sci. USA **104**, 18549 (2007).

[101] S. Hilgenfeldt, S. Erisken, and R. W. Carthew, *Physical Modeling of Cell Geometric Order in an Epithelial Tissue*, Proc. Nat. Acad. Sci. USA **105**, 907 (2008).

[102] G. Salbreux, L. K. Barthel, P. A. Raymond, and D. K. Lubensky, *Coupling Mechanical Deformations and Planar Cell Polarity to Create Regular Patterns in the Zebrafish Retina*, PLoS Comput. Biol. **8**, e1002618 (2012).

[103] J. K. Mason, R. Ehrenborg, and E. A. Lazar, *A Geometric Formulation of the Law of Aboav-Weaire in Two and Three Dimensions*, J. Phys. A: Math. Theor. **45**, 065001 (2012).

[104] P. Ziherl, *Aggregates of Two-Dimensional Vesicles: Rouleaux, Sheets, and Convergent Extension*, Phys. Rev. Lett. **99**, 128102 (2007).

[105] A. E. Shyer, T. Tallinen, N. L. Nerurkar, Z. Wei, E. S. Gil, D. L. Kaplan, C. J. Tabin, and L. Mahadevan, *Villification: How the Gut Gets its Villi*, Science **342**, 212 (2013).

[106] J. S. Babel, *Reproduction, Life History, and Ecology of the Round Stingray, Urolophus halleri Cooper*, Fish Bull. (California Department of Fish and Game) **137**, 102 (1967).

[107] T. H. Ermak, *Cell Proliferation in the Digestive Tract of Styela clava (Urochordata: Ascidiacea) as Revealed by Autoradiography with Tritiated Thymidine*, J. Exp. Zool. **194**, 449 (1976).

[108] T. H. Ermak, *The Renewing Cell Populations of Ascidians*, Am. Zool. **22**, 795 (1982).

[109] B. Li, Y.-P. Cao, X.-Q. Feng, and H. Gao, *Mechanics of Morphological Instabilities and Surface Wrinkling in Soft Materials: A Review*, Soft Matter **8**, 5728 (2012).

[110] Q. Wang and X. Zhao, *A Three-Dimensional Phase Diagram of Growth-Induced Surface Instabilities*, Sci. Rep. **5**, 8887 (2015).

[111] W. H. Lewis, *Mechanics of Invagination*, Anat. Rec. **97**, 139 (1947).

[112] S. P. Timoshenko and J. M. Gere, *Theory of Elastic Stability* (McGraw–Hill, Singapore, 1963).

[113] N. Bowden, S. Brittain, A. G. Evans, J. W. Hutchinson, and G. M. Whitesides, *Spontaneous Formation of Ordered Structures in Thin Films of Metals Supported on an Elastomeric Polymer*, Nature **393**, 146 (1998).

[114] E. Hannezo, J. Prost, and J. F. Joanny, *Instabilities of Monolayered Epithelia: Shape and Structure of Villi and Crypts*, Phys. Rev. Lett. **107**, 078104 (2011).

[115] P. Ciarletta, V. Balbi, and E. Kuhl, *Pattern Selection in Growing Tubular Tissues*, Phys. Rev. Lett. **113**, 248101 (2014).

[116] R. Sbarbati, *Morphogenesis of the Intestinal Villi of the Mouse Embryo: Chance and Spatial Necessity*, J. Anat. **135**, 477 (1982).

[117] A. E. Shyer, T. R. Huycke, C. Lee, L. Mahadevan, and C. J. Tabin, *Bending Gradients: How the Intestinal Stem Cell Gets Its Home*, Cell **161**, 569 (2015).

[118] M. Krajnc, N. Štorgel, A. Hočevar Brezavšček, and P. Ziherl, *A Tension-Based Model of Flat and Corrugated Simple Epithelia*, Soft Matter **9**, 8368 (2013).

[119] E. Hannezo, J. Prost, and J.-F. Joanny, *Theory of Epithelial Sheet Morphology in Three Dimensions*, Proc. Natl. Acad. Sci. USA **110**, 27 (2014).

[120] O. Luu, R. David, H. Ninomiya, and R. Winklbauer, *Large-Scale Mechanical Properties of Xenopus Embryonic Epithelium*, Proc. Natl. Acad. Sci. USA **108**, 4000 (2011).

[121] N. Štorgel, M. Krajnc, P. Mrak, J. Štrus, and P. Ziherl, *Quantitative Morphology of Epithelial Folds*, Biophys. J. **110**, 269 (2016).

[122] O. Saglam, E. Samayoa, S. Somasekar, S. Naccache, A. Iwasaki, and C. Y. Chiu, *No Viral Association Found in a Set of Differentiated Vulvar Intraepithelial Neoplasia Cases by Human Papillomavirus and Pan-Viral Microarray Testing*, PLoS One **10**, e0125292 (2015).

[123] M. Basan, J.-F. Joanny, J. Prost, and T. Risler, *Undulation Instability of Epithelial Tissues*, Phys. Rev. Lett. **106**, 158101 (2011).

[124] G. M. Odell, G. Oster, P. Alberch, and B. Burnside, *The Mechanical Basis of Morphogenesis. I. Epithelial Folding and Invagination*, Dev. Biol. **85**, 446 (1981).

[125] J. M. Sawyer, J. R. Harrell, G. Shemer, J. Sullivan-Brown, M. Roh-Johnson, and B. Goldstein, *Apical Constriction: A Cell Shape Change That Can Drive Morphogenesis*, Dev. Biol. **341**, 5 (2010).

[126] L. C. Prosser, *Smooth Muscle*, Annu. Rev. Physiol. **36**, 503 (1974).

[127] W. F. Jackson, *Ion Channels and Vascular Tone*, Hypertension **35**, 173 (1999).

[128] J. E. Magney, S. L. Erlandsen, M. L. Bjerknes, and H. Cheng, *Scanning Electron Microscopy of Isolated Epithelium of the Murine Gastrointestinal Tract: Morphology of the Basal Surface and Evidence for Paracrinelike Cells*, Am. J. Anat. **177**, 43 (1986).

[129] S.-J. Dunn, P. L. Appleton, S. A. Nelson, I. S. Näthke, D. J. Gavaghan, and J. M. Osborne, *A Two-Dimensional Model of the Colonic Crypt Accounting for the Role of the Basement Membrane and Pericryptal Fibroblast Sheath*, PLoS Comput. Biol. **8**, e1002515 (2012).

[130] B. I. Shraiman, *Mechanical Feedback as a Possible Regulator of Tissue Growth*, Proc. Natl. Acad. Sci. USA **102**, 3318 (2005).

[131] D. Drasdo, *Buckling Instabilities of One-Layered Growing Tissues*, Phys. Rev. Lett. **84**, 4244 (2000).

[132] S. Wolfram, *A New Kind of Science* (Wolfram Media, Champaign, 2002).

[133] D. C. Walker, J. Southgate, G. Hill, M. Holcombe, D. R. Hose, S. M. Wood, S. Mac Neil, and R. H. Smallwood, *The Epitheliome: Agent-Based Modelling of the Social Behaviour of Cells*, BioSystems **76**, 89 (2004).

[134] J. Galle, M. Loeffler, and D. Drasdo, *Modeling the Effect of Deregulated Proliferation and Apoptosis on the Growth Dynamics of Epithelial Cell Populations in vitro*, Biophys. J. **88**, 62 (2005).

[135] S. Hoehme, M. Brulport, A. Bauer, E. Bedawy, W. Schormann, M. Hermes, V. Puppe, R. Gebhardt, S. Zellmer, M. Schwarz, E. Bockamp, T. Timmel, J. G. Hengstler, and D. Drasdo, *Prediction and Validation of Cell Alignment along Microvessels as Order Principle to Restore Tissue Architecture in Liver Regeneration*, Proc. Natl. Acad. Sci. USA **107**, 10371 (2010).

[136] A. Šiber and H. Buljan, *Theoretical and Experimental Analysis of a Thin Elastic Cylindrical Tube Acting as a Non-Hookean Spring*, Phys. Rev. E **83**, 067601 (2011).

[137] A. Šiber, *Shapes of Minimal-Energy DNA Ropes Condensed in Confinement*, Sci. Rep. **6**, 29012 (2016).

[138] A. Kaul-Strehlow and T. Stach, *A Detailed Description of the Development of the Hemichordate Saccoglossus Kowalevskii Using SEM, TEM, Histology and 3D-reconstructions*, Front. Zool. **10**, 1 (2013).

[139] J. F. Colas and G. C. Schoenwolf, *Towards a Cellular and Molecular Understanding of Neurulation*, Dev. Dyn. **221**, 117 (2001).

[140] T. Sato, R. G. Vries, H. J. Snippert, M. van de Wetering, N. Barker, D. E. Stange, J. H. van Es, A. Abo, P. Kujala, P. J. Peters, and H. Clevers, *Single Lgr5 Stem Cells Build Crypt-Villus Structures in vitro Without a Mesenchymal Niche*, Nature **459**, 262 (2009).

[141] J. E. Johndrow, C. R. Magie, and S. M. Parkhurst, *Rho GTPase Function in Flies: Insights from a Developmental and Organismal Perspective*, Biochem. Cell Biol. **82**, 643 (2004).

[142] J. J. Muñoz, K. Barrett, and M. Miodownik, *A Deformation Gradient Decomposition Method for the Analysis of the Mechanics of Morphogenesis*, J. Biomech. **40**, 1372 (2007).

[143] A. Hočevar Brezavšček, M. Rauzi, M. Leptin, and P. Ziherl, *A Model of Epithelial Invagination Driven by Collective Mechanics of Identical Cells*, Biophys. J. **103**, 1069 (2012).

[144] L. Rhumbler, *Zur Mechanik des Gastrulationsvorganges insbesondere der Invagination*, Archiv für Entwicklungsmechanik der Organismen **14**, 401 (1902).

[145] A. R. Moore and A. S. Burt, *On the Locus and Nature of the Forces Causing Gastrulation in the Embryos of Dendraster excentricus*, J. Exp. Zool. **82**, 159 (1939).

[146] L. A. Davidson, M. A. R. Koehl, R. Keller, and G. F. Oster, *How do Sea Urchins Invaginate? Using Biomechanics to Distinguish between Mechanisms of Primary Invagination*, Development **121**, 2005 (1995).

[147] M. C. Lane, M. A. R. Keohl, F. Wilt, and R. Keller, *A Role for Regulated Secretion of Apical Extracellular Matrix during Epithelial Invagination in the Sea Urchin*, Development **14**, 1049 (1993).

[148] L. Pauchard and Y. Couder, *Invagination during the Collapse of an Inhomogeneous Spheroidal Shell*, Europhys. Lett. **66**, 667 (2004).

[149] A. V. Pogorelov, *Bendings of Surfaces and Stability of Shells* (American Mathematical Society, Providence, 1988).

[150] H. Takata and T. Kominami, *Shrinkage and Expansion of Blastocoel Affect the Degree of Invagination in Sea Urchin Embryos*, Zool. Sci. **18**, 1097 (2001).

[151] B. Božič, J. Derganc, and S. Svetina, *Blastula Wall Invagination Examined on the Basis of Shape Behavior of Vesicular Objects with Laminar Envelopes*, Int. J. Dev. Biol. **50**, 143 (2006).

[152] L. A. Davidson, G. F. Oster, R. Keller, and M. A. R. Koehl, *Measurements of Mechanical Properties of the Blastula Wall Reveal Which Hypothesized Mechanisms of Primary Invagination Are Physically Plausible in the Sea Urchin Strongylocentrotus purpuratus*, Dev. Biol. **209**, 221 (1999).

[153] A. Polyakov, B. He, M. Swan, J. W. Shaevitz, M. Kaschube, and E. Wieschaus, *Passive Mechanical Forces Control Cell-Shape Change during Drosophila Ventral Furrow Formation*, Biophys. J. **107**, 998 (2014).

[154] D Sweeton, S. Parks, M. Costa, and E. Wieschaus, *Gastrulation in Drosophila: The Formation of the Ventral Furrow and Posterior Midgut Invaginations*, Development **112**, 775 (1991).

[155] G. W. Brodland, V. Conte, P. G. Cranston, J. Veldhuis, S. Narasimhan, M. S. Hutson, A. Jacinto, F. Ulrich, B. Baum, and M. Miodownik, *Video Force Microscopy Reveals the Mechanics of Ventral Furrow Invagination in Drosophila*, Proc. Natl. Acad. Sci. USA **107**, 22111 (2010).

[156] K. K. Chiou, L. Hufnagel, and B. I. Shraiman, *Mechanical Stress Inference for Two Dimensional Cell Arrays*, PLoS Comput. Biol. **8**, e1002512 (2012).

[157] J. Huisken, J. Swoger, F. Del Bene, J. Wittbrodt, and E. H. K. Stelzer, *Optical Sectioning Deep Inside Live Embryos by Selective Plane Illumination Microscopy*, Science **305**, 1007 (2004).

[158] M. Rauzi, U. Krzic, T. E. Saunders, M. Krajnc, P. Ziherl, L. Hufnagel, and M. Leptin, *Embryo-Scale Tissue Mechanics during Drosophila Gastrulation Movements*, Nat. Commun. **6**, 8677 (2015).

[159] R. J. Tetley, G. B. Blanchard, A. G. Fletcher, R. J. Adams, and B. Sanson, *Unipolar Distributions of Junctional Myosin II Identify Cell Stripe Boundaries That Drive Cell Intercalation Throughout Drosophila Axis Extension*, eLife **5**, e12094 (2016).

[160] R. Keller, L. Davidson, A. Edlund, T. Elul, M. Ezin, D. Shook, and P. Skoglund, *Mechanisms of Convergence and Extension by Cell Intercalation*, Philos. Trans. R. Soc. B **355**, 897 (2000).

[161] M. Zajac, G. L. Jones, and J. A. Glazier, *Model of Convergent Extension in Animal Morphogenesis*, Phys. Rev. Lett. **85**, 2022 (2000).

[162] M. Zajac, G. L. Jones, and J. A. Glazier, *Simulating Convergent Extension by Way of Anisotropic Differential Adhesion*, J. Theor. Biol. **222**, 247 (2003).

[163] J. Shih and R. Keller, *Cell Motility Driving Mediolateral Intercalation in Explants of Xenopus laevis*, Development **116**, 901 (1992).

[164] P. J. Keller, A. D. Schmidt, J. Wittbrodt, and E. H. K. Stelzer, *Reconstruction of Zebrafish Early Embryonic Development by Scanned Light Sheet Microscopy*, Science **322**, 1065 (2008).

[165] L. C. Butler, G. B. Blanchard, A. J. Kabla, N. J. Lawrence, D. P. Welchman, L. Mahadevan, R. J. Adams, and B. Sanson, *Cell Shape Changes Indicate a Role for Extrinsic Tensile Forces in Drosophila Germ-Band Extension*, Nat. Cell Biol. **11**, 859 (2009).

[166] M. Popović, A. Nandi, M. Merkel, R. Etournay, S. Eaton, F. Jülicher, and G. Salbreux, *Active Dynamics of Tissue Shear Flow*, New J. Phys. **19**, 033006 (2017).

[167] M. Merkel, R. Etournay, M. Popović, G. Salbreux, S. Eaton, and F. Jülicher, *Triangles Bridge the Scales: Quantifying Cellular Contributions to Tissue Deformation*, Phys. Rev. E **95**, 032401 (2017).

[168] J. Salençon, *Handbook of Continuum Mechanics* (Springer, Berlin, 2001).

[169] R. Etournay, M. Popović, M. Merkel, A. Nandi, C. Blasse, B. Aigouy, H. Brandl, G. Myers, G. Salbreux, F. Jülicher, and S. Eaton, *Interplay of Cell Dynamics and Epithelial Tension during Morphogenesis of the Drosophila Pupal Wing*, eLife **4**, e07090 (2015).

[170] C. Collinet, M. Rauzi, P.-F. Lenne, and T. Lecuit, *Local and Tissue-Scale Forces Drive Oriented Junction Growth during Tissue Extension*, Nat. Cell Biol. **17**, 1247 (2015).

[171] M. Weliky, S. Minsuk, R. Keller, and G. Oster, *The Mechanical Basis of Cell Rearrangement. I. Epithelial Morphogenesis During Fundulus Epiboly*, Development **113**, 1231 (1991).

[172] E. Walck-Shannon and J. Hardin, *Cell Intercalation from Top to Bottom*, Nat. Rev. Mol. Cell Biol. **15**, 34 (2014).

[173] J. Firmino, D. Rocancourt, M. Saadaoui, C. Moreau, and J. Gros, *Cell Division Drives Epithelial Cell Rearrangements during Gastrulation in Chick*, Dev. Cell **36**, 249 (2016).

[174] J. T. Blankenship, S. T. Backovic, J. S. P. Sanny, O. Weitz, and J. A. Zallen, *Multicellular Rosette Formation Links Planar Cell Polarity to Tissue Morphogenesis*, Dev. Cell **11**, 459 (2006).

[175] G. F. Oster, J. D. Murray, and A. K. Harris, *Mechanical Aspects of Mesenchymal Morphogenesis*, J. Embryol. Exp. Morphol. **78**, 83 (1983).

[176] J. D. Murray, G. F. Oster, and A. K. Harris, *A Mechanical Model for Mesenchymal Morphogenesis*, J. Math. Biol. **17**, 125 (1983).

[177] J. D. Murray, P. K. Maini, and R. T. Tranquillo, *Mechanochemical Models for Generating Biological Pattern and Form in Development*, Phys. Rep. **171**, 59 (1988).

[178] A. M. Turing, *The Chemical Basis of Morphogenesis*, Philos. Trans. R. Soc. B **237**, 37 (1952).

[179] G. H. Gunaratne, Q. Ouyang, and H. L. Swinney, *Pattern Formation in the Presence of Symmetries*, Phys. Rev. E **50**, 2802 (1994).

[180] C.-M. Lin, T. X. Jiang, R. E. Baker, P. K. Maini, R. B. Widelitz, and C.-M. Chuong, *Spots and Stripes: Pleomorphic Patterning of Stem Cells via p-ERK-Dependent Cell Chemotaxis Shown by Feather Morphogenesis and Mathematical Simulation*, Dev. Biol. **334**, 369 (2009).

[181] S. Kondo and T. Miura, *Reaction-Diffusion Model as a Framework for Understanding Biological Pattern Formation*, Science **239**, 1616 (2010).

[182] H. Meinhardt, *The Algorithmic Beauty of Seashells* (Springer, Berlin, 1995).

[183] C. B. Kimmel, W. W. Ballard, S. R. Kimmel, B. Ullmann, and T. F. Schilling, *Stages of Embryonic Development of the Zebrafish*, Dev. Dyn. **203**, 253 (1995).

[184] M. Behrndt, G. Salbreux, P. Campinho, R. Hauschild, F. Oswald, J. Roensch, S. W. Grill, and C.-P. Heisenberg, *Forces Driving Epithelial Spreading in Zebrafish Gastrulation*, Science **338**, 257 (2012).

[185] M. Weliky and G. Oster, *The Mechanical Basis of Cell Rearrangement. I. Epithelial Morphogenesis During Fundulus Epiboly*, Development **109**, 373 (1990).

[186] A. Hernández-Vega, M. Marsal, P.-A. Pouille, S. Tosi, J. Colombelli, T. Luque, D. Navajas, I. Pagonabarraga, and E. Martín-Blanco, *Polarized Cortical Tension Drives Zebrafish Epiboly Movements*, EMBO J. **36**, 25 (2017).

[187] H. Ibrahim and R. Winklbauer, *Mechanisms of Mesendoderm Internalization in the Xenopus Gastrula: Lessons from the Ventral Side*, Dev. Biol. **240**, 108 (2001).

[188] R. Keller and S. Jansa, *Xenopus Gastrulation Without a Blastocoel Roof*, Dev. Dyn. **195**, 162 (1992).

[189] B. Guirao, S. Rigaud, F. Bosveld, A. Bailles, J. López-Gay, S. Ishihara, K. Sugimura, F. Graner, and Y. Bellaïche, *Unified Quantitative Characterization of Epithelial Tissue Development*, eLife **4**, e08519 (2015).

[190] G. C. Schoenwolf and J. L. Smith, *Mechanics of Neurulation: Traditional Viewpoint and Recent Advances*, Development **109**, 243 (1990).

[191] V. D. Varner and C. M. Nelson, *Cellular and Physical Mechanisms of Branching Morphogenesis*, Development **141**, 2750 (2014).

[192] T. Sato and H. Clevers, *Growing Self-Organizing Mini-Guts from a Single Intestinal Stem Cell: Mechanism and Applications*, Science **340**, 1190 (2013).

[193] T. Lecuit and L. Le Goff, *Orchestrating Size and Shape During Morphogenesis*, Nature **450**, 189 (2007).

[194] S. Svetina and B. Žekš, *Membrane Bending Energy and Shape Determination of Phospholipid Vesicles and Red Blood Cells*, Eur. Biophys. J. **17**, 101 (1989).

[195] I. C. Mackenzie, *Ordered Structure of the Epidermis*, J. Invest. Dermatol. **65**, 45 (1975).

[196] T. D. Allen and C. S. Potten, *Significance of Cell Shape in Tissue Architecture*, Nature **264**, 545 (1976).

[197] F. T. Lewis, *The Typical Shape of Polyhedral Cells in Vegetable Parenchyma and the Restoration of That Shape following Cell Division*, Proc. Am. Acad. Arts Sci. **58**, 537 (1923).

[198] J. W. Marvin, *Cell Shape and Cell Volume Relations in the Pith of Eupatorium Perfoliatum L.*, Am. J. Bot. **31**, 208 (1944).

[199] E. B. Matzke, *The Three Dimensional Shape of Bubbles in Foams*, Proc. Natl. Acad. Sci. USA **31**, 281 (1945).

[200] E. B. Matzke and R. M. Duffy, *Progressive Three-Dimensional Shape Changes of Dividing Cells within the Apical Meristem of Anacharis densa*, Am. J. Bot. **43**, 205 (1956).

[201] J. W. Marvin, *The Shape of Compressed Lead Shots and its Relation to Cell Shape*, Am. J. Bot. **43**, 280 (1939).

[202] S. Hales, *Vegetable Staticks: Or, An Account of Some Statistical Experiments on the Sap Vegetables* (W. Innys, R. Manby, T. Woodward, and J. Peele, London, 1727).

[203] W. Thomson, *On the Division of Space with Minimum Partitional Area*, Philos. Mag. **24**, 121 (1887).

[204] D. Weaire and R. Phelan, *A Counter-Example to Kelvin's Conjecture on Minimal Surfaces*, Philos. Mag. Lett. **69**, 107 (1994).

[205] D. Weaire (ed.), *The Kelvin Problem: Foam Structures of Minimal Surface Area* (Taylor & Francis, London, 1996).

[206] A. M. Kraynik, D. A. Reinelt, and F. van Swol, *Structure of Random Monodisperse Foam*, Phys. Rev. E **67**, 031403 (2003).

[207] S. B. Berger, *Edge Length Minimizing Polyhedra* (PhD Thesis, Rice University, Houston, 2001).

[208] F. Morgan, *Surfaces Minimizing Area plus Length of Singular Curves*, Proc. Am. Math. Soc. **122**, 1153 (1994).

[209] U. Seifert, *Adhesion of Vesicles in Two Dimensions*, Phys. Rev. A **43**, 6803 (1991).

[210] A. Hočevar and P. Ziherl, *Periodic Three-Dimensional Assemblies of Polyhedral Lipid Vesicles*, Phys. Rev. E **83**, 041917 (2011).

[211] H. Honda, M. Tanemura, and S. Imayama, *Spontaneous Architectural Organization of Mammalian Epidermis from Random Cell Packing*, J. Invest. Dermatol. **106**, 312 (1996).

[212] D. P. Doupé, A. M. Klein, B. D. Simons, and P. H. Jones, *The Ordered Architecture of Murine Ear Epidermis Is Maintained by Progenitor Cells with Random Fate*, Dev. Cell **18**, 317 (2010).

[213] K. A. Carswell, M.-J. Lee, and S. K. Fried, *Culture of Isolated Human Adipocytes and Isolated Adipose Tissue*, Methods Mol. Biol. **203**, 806 (2012).

[214] W. Drenckhan and S. Hutzler, *Structure and Energy of Liquid Foams*, Adv. Colloid Interface Sci. **224**, 1 (2015).

[215] P. Panettiere, D. Accorsi, L. Marchetti, A. M. Minicozzi, G. Orsini, P. Bernardi, D. Benati, G. Conti, and A. Sbarbati, *The Trochanteric Fat Pad*, Eur. J. Histochem. **55**, e16 (2011).

[216] K. Comley and N. A. Fleck, *A Micromechanical Model for the Young's Modulus of Adipose Tissue*, Int. J. Solids Struct. **47**, 2982 (2010).

[217] N. Alkhouli, J. Mansfield, E. Green, J. Bell, B. Knight, N. Liversedge, J. C. Tham, R. Welbourn, A. C. Shore, K. Kos, and C. P. Winlove, *The Mechanical Properties of Human Adipose Tissues and their Relationships to the Structure and Composition of the Extracellular Matrix*, Am. J. Physiol. Endocrinol. Metab. **305**, E1427 (2013).

[218] H. M. Princen, *Rheology of Foams and Highly Concentrated Emulsions I. Elastic Properties and Yield Stress of a Cylindrical Model System*, J. Colloid Interf. Sci. **91**, 160 (1983).

[219] S. A. Khan and R. C. Armstrong, *Rheology of Foams: I. Theory for Dry Foams*, J. Non-Newton. Fluid **22**, 1 (1986).

[220] D. Weaire, *The Rheology of Foam*, Curr. Opin. Colloid In. **13**, 171 (2008).

[221] L. J. Gibson, *Biomechanics of Cellular Solids*, J. Biomech. **38**, 377 (2005).

[222] L. J. Gibson and M. F. Ashby, *Cellular Solids* (Cambridge University Press, Cambridge, 1997).

[223] S. C. Cowin and S. B. Doty, *Tissue Mechanics* (Springer, New York, 2007).

[224] R. W. Ogden, *Large Deformation Isotropic Elasticity: On the Correlation of Theory and Experiment for Compressible Rubberlike Solids*, Proc. R. Soc. A **328**, 567 (1972).

[225] R. W. Ogden, *Non-Linear Elastic Deformations* (Dover Publications, Mineola, 1997).

[226] M. Rubinstein and R. H. Colby, *Polymer Physics* (Oxford University Press, Oxford, 2003).

[227] L. A. Mihai, L. Chin, P. A. Janmey, and A. Goriely, *A Comparison of Hyperelastic Constitutive Models Applicable to Brain and Fat Tissues*, J. R. Soc. Interface **12**, 20150486 (2015).

[228] A. Šiber, *Energies of sp^2 Carbon Shapes with Pentagonal Disclinations and Elasticity Theory*, Nanotechnology **17**, 3598 (2006).

[229] R. Nuzzo, *Profile of George Oster*, Proc. Natl. Acad. Sci. USA **103**, 1672 (2006).

[230] M. Aubouy, Y. Jiang, J. A. Glazier, and F. Graner, *A Texture Tensor to Quantify Deformations*, Granul. Matter **5**, 67 (2003).

[231] T. Cavalier-Smith, *Origin of Animal Multicellularity: Precursors, Causes, Consequences—The Choanoflagellate/Sponge Transition, Neurogenesis and the Cambrian Explosion*, Philos. Trans. R. Soc. B **372**, 20150476 (2017).

[232] H. V. Wilson, *On Some Phenomena of Coalescence and Regeneration in Sponges*, J. Exp. Zool. **5**, 245 (1907).

[233] R. Pallela and V. R. Janapala, *Comparative Ultrastructural and Biochemical Studies of Four Demosponges from Gulf of Mannar, India*, Int. J. Mar. Sci. **3**, 295 (2013).

[234] M. A. Lancaster and J. Knoblich, *Organogenesis in a Dish: Modeling Development and Disease Using Organoid Technologies*, Science **345**, 1247125 (2014).

[235] M. Simunovic and A. H. Brivanlou, *Embryoids, Organoids and Gastruloids: New Approaches to Understanding Embryogenesis*, Development **144**, 976 (2017).

[236] S. Dahl-Jensen and A. Grapin-Bottton, *The Physics of Organoids: A Biophysical Approach to Understanding Organogenesis*, Development **144**, 946 (2017).

Index

Common words that are used very often (protein, polygon, force, energy, etc.) are not indexed for clarity. However, a few keywords central to the material covered—such as bending, cell, tiling, tissue, and a few others—are indexed only once at the place where they are described in detail or defined as appropriate, with the respective page number appearing in the index in italics.

A

Aboav–Weaire law, 64–66, 75, 95
actin, 11
active
    force, 178
    matter, 8, 11, 31, 36, 37, 141
    motion, 11, 178
    process, 31, 161, 167, 173, 178, 179, 195
    rearrangement, 179–182
    stress, 36, 179–182
acto-myosin
    network, 89–91
    ring, 11, 18, 44, 68, 69, 71, 95, 113, 186, 195, 215
acto-myosin ring, 11
adherens junction, 35
adhesion
    belt, 35
    cell-cell, 16, 33, 34, 38, 44, 51, 90, 95, 96, 111, 115, 147, 238
    differential, 38
    patch, 35
    receptor, 35
    strength, 37–39, 51, 68, 88–91, 96, 97, 187, 218
adipocyte, 223–225, 229, 230
African clawed frog, *see Xenopus*
Agave attenuata, 62
agent-based model, 8, 9, 119, 120, 129
aggregate, 38, 39, 42, 43, 218, 219, 221
airway, 108
alcohol, 46
Aloe arborescens, 62

Ambystoma tigrinum, *see* tiger salamander
American black elderberry, 203, 236
amnion, 61
animal
    pole, 145
animal pole, 147
annealing, 212, 213
cf. Anthurium, 62
apical
    constriction, 7, 144, 151–155, 157–162, 165, 167, 197, 236
    lamina, 147, 148, 151–154
apico-basal
    contraction, 144, 151–153, 158, 159, 161, 162, 167, 198
    elongation, 158, 159, 162
    polarity, 103, 111, 122, 133
    tension, 113–115, 119
Apollonian gasket, 54
apoptosis, 117, 123, 133, 236, *see also* cell death
archaea, 15
archenteron, 145, 151, 156, 167, 169, 195
area elasticity, 69–71, 73, 74, 79, 87
area-difference-elasticity (ADE) theory, 218
aspect ratio, 56, 80, 96, 171, 191, 234
    cell, 170, 171, 198
Astropecten aranciacus, 1
ATP, 8, 11, 35
axis
    anteroposterior, 165, 168, 173, 174, 185, 186